# Benchmark Papers in Genetics

**Series Editor: David L. Jameson**
**University of Houston**

Volume
1 GENETICS AND SOCIAL STRUCTURE: Mathematical Structuralism in Population Genetics and Social Theory / *Paul Ballonoff*
2 GENES AND PROTEINS / *Robert P. Wagner*
3 DEMOGRAPHIC GENETICS / *Kenneth M. Weiss and Paul A. Ballonoff*
4 MUTAGENESIS / *John W. Drake and Robert E. Koch*
5 EUGENICS: Then and Now / *Carl Jay Bajema*
6 CYTOGENETICS / *Ronald L. Phillips and Charles R. Burnham*
7 STOCHASTIC MODELS IN POPULATION GENETICS / *Wen-Hsiung Li*
8 EVOLUTIONARY GENETICS / *D. L. Jameson*
9 GENETICS OF SPECIATION / *D. L. Jameson*
10 HUMAN GENETICS / *William J. Schull and Ranajit Chakraborty*
11 HYBRIDIZATION: An Evolutionary Perspective / *Donald A. Levin*

**RELATED TITLES IN OTHER BENCHMARK SERIES**

MICROBIAL GENETICS / *Morad Abou-Sabé*
CONCEPTS OF SPECIES / *C. N. Slobodchikoff*
MULTIVARIATE STATISTICAL METHODS: Among-Groups Covariation / *William R. Atchley and Edwin H. Bryant*
MULTIVARIATE STATISTICAL METHODS: Within-Groups Covariation / *Edwin H. Bryant and William R. Atchley*
POPULATION REGULATION / *Robert H. Tamarin*

## Benchmark Papers
## in Genetics / 11

**A BENCHMARK ® Books Series**

# HYBRIDIZATION
# An Evolutionary Perspective

Edited by

## DONALD A. LEVIN

**University of Texas at Austin**

Dowden, Hutchinson & Ross, Inc.

STROUDSBURG, PENNSYLVANIA

LIBRARY OF CONGRESS CATALOGING IN PUBLICATIONS DATA

Main entry under title:
Hybridization.
   (Benchmark papers in genetics ; v. 11)
   Includes indexes.
   1.  Hybridization—Addresses, essays, lectures.
2.  Evolution—Addresses, essays, lectures.
I.  Levin, Donald A.
QH421.H9       575.2'8        78-10947
ISBN 0-87933-341-3

Distributed world wide by Academic Press,
a subsidiary of Harcourt Brace Jovanovich,
Publishers.

# SERIES EDITOR'S FOREWORD

The study of any discipline assumes the mastery of the literature of the subject. In many branches of science, even one as new as genetics, the expansion of knowledge has been so rapid that there is little hope of learning of the development of all phases of the subject. The student has difficulty mastering the textbook, the young scholar must tend to the literature near his own research, the young instructor barely finds time to expand his horizons to meet his class-preparation requirements, the monographer copes with a wider literature but usually from a specialized viewpoint, and the textbook author is forced to cover much the same material as previous and competing texts to respond to the user's needs and abilities.

Few publishers have the dedication to scholarship to serve primarily the limited market of advanced studies. The opportunity to assist professionals at all stages of their careers has been recognized by Dowden, Hutchinson & Ross and by a distinguished group of editors knowledgeable in specific portions of the genetic literature. These editors have selected papers and portions of papers that demonstrate both the development of knowledge and the atmosphere in which that knowledge was developed. There is no substitute for reading great papers. Here you can learn how questions are asked, how they are approached, and how difficult and essential it is to obtain definitive answers and clear writing.

This volume contains selected papers on the role of hybridization in evolution, experimental studies of hybridization, stabilization of hybrid derivatives, introgressive hybridization, hybrid zones, and reproductive character displacement. The discussion emphasizes the tenuous nature of our understanding of the processes and significance of hybridization, indicating once again how little we understand of the role of genes in natural populations.

The volume is an integral part of the Benchmark Papers in Genetics series, supplementing and complementing the completed volumes on speciation and evolutionary genetics and providing foundations to future volumes on polyploidy, plant breeding, animal breeding, and experimental population genetics.

DAVID L. JAMESON

# PREFACE

The treatment of hybridization in the literature is rich and multi-faceted, reflecting the maturation of evolutionary theory and the development of new technology during the past two hundred years. My purposes are to provide the reader with insight into a spectrum of prime topics and issues as follows: the role of hybridization in experimental studies on hybridization in relation to evolution; stabilization of hybrid derivatives; introgressive hybridization; hybrid zones; and the reinforcement of barriers to hybridization. By presenting these topics in the aforementioned sequence, I hope to convey a logical thread leading from a conceptual view of hybridization through its documentation and experimentation in local populations, to phenomena involving multiple populations over a considerable range. Unfortunately, space limitations preclude both the full development of each topic and the consideration of other topics of which hybridization is an integral part (e.g., the biological species concept, genome incompatibility in hybrids).

The papers chosen for this volume represent a sampling of conceptual, experimental, and descriptive endeavors of botanists and zoologists. The papers reflect my impression of quality and past or potential impact on evolutionary theory. Choosing papers has been a difficult task since quality defies quantification and impact is difficult to assess in recently published papers. As I still wonder whether I have indeed chosen the best papers according to my criteria, no doubt others will disagree with my choices. Moreover, the space and restrictions inherent in this compilation favors the incorporation of relatively short papers. I have not attempted to limit the contributions of each author to a single paper, as certain figures have been more prominent than others. I had considered balancing the botanical and zoological contributions for each topic, but it was soon evident that symmetry would perforce reduce the volume's overall quality. Polyploidy, with its ties to hybridization will be considered in a forthcoming Benchmark volume by Stebbins. Hybridization, with reference to agriculture, will be considered in a forthcoming Benchmark volume by Heiser.

The author is grateful to Professor C. B. Heiser for his helpful discussions on content and narrative, and for critically reading the manuscript. Professors V. Grant and B. L. Turner also provided valuable suggestions.

DONALD A. LEVIN

# CONTENTS

Series Editor's Foreword                                                                    v
Preface                                                                                   vii
Contents by Author                                                                       xiii

Introduction                                                                               1

## PART I: THE ROLE OF HYBRIDIZATION IN EVOLUTION

Editor's Comments on Papers 1, 2, and 3                                                     8

1    ANDERSON, E.:   Hybridization of the Habitat                                          11
        *Evolution* **2**:1–9 (1948)

2    ANDERSON, E., and G. L. STEBBINS, JR.:   Hybridization as an Evolution-
        ary Stimulus                                                                       20
        *Evolution* **8**:378–388 (1954)

3    RATTENBURY, J. A.:   Cyclic Hybridization as a Survival Mechanism
        in the New Zealand Forest Flora                                                    31
        *Evolution* **16**:348–363 (1962)

## PART II: EXPERIMENTAL HYBRIDIZATION STUDIES

Editor's Comments on Papers 4, 5, and 6                                                    48

4    STEPHENS, S. G.:   The Cytogenetics of Speciation in *Gossypium*. I.
        Selective Elimination of the Donor Parent Genotype in
        Interspecific Backcrosses                                                          51
        *Genetics* **34**:627–637 (1949)

5    GRANT, V.:   The Origin of a New Species of *Gilia* in a Hybridization
        Experiment                                                                         62
        *Genetics* **54**:1189–1199 (1966)

6    RAO, S. V., and P. DeBACH:   Experimental Studies on Hybridization
        and Sexual Isolation Between Some *Aphytis* Species
        (Hymenoptera: Aphelinidae). III. The Significance of
        Reproductive Isolation Between Interspecific Hybrids
        and Parental Species                                                               73
        *Evolution* **23**:525–533 (1969)

*Contents*

## PART III: STABILIZATION OF HYBRID DERIVATIVES

**Editor's Comments on Papers 7 Through 10**                                    84

7   GRANT, V.:   The Role of Hybridization in the Evolution of the Leafy-
    stemmed Gilias                                                             87
    *Evolution* 7:51–64

8   STRAW, R. M.:   Hybridization, Homogamy, and Sympatric Speciation         101
    *Evolution* **9**:441–444 (1955)

9   GRANT, V., and K. A. GRANT:   Dynamics of Clonal Microspecies in
    *Cholla* Cactus                                                           105
    *Evolution* **25**:144–155 (1971)

10  PARKER, E. D., JR., and R. K. SELANDER:   The Organization of Genetic
    Diversity in the Parthenogenetic Lizard *Cnemidophorus*
    *Tesselatus*                                                              117
    *Genetics* **84**:791–805 (1976)


## PART IV: INTROGRESSIVE HYBRIDIZATION

**Editor's Comments on Papers 11 Through 15**                                   134

11  HUBBS, C. L.:   Hybridization Between Fish Species in Nature              139
    *Syst. Zool.* **4**:1–20 (1955)

12  ANDERSON, E., and L. HUBRICHT:   Hybridization in *Tradescantia*.
    III. The Evidence for Introgressive Hybridization                        159
    *Am. J. Bot.* **25**:396–402 (1938)

13  HEISER, C. B., JR.:   Hybridization Between the Sunflower Species
    *Helianthus Annuus* and *H. Petiolaris*                                   166
    *Evolution* **1**:249–262 (1947)

14  ALSTON, R. E., and B. L. TURNER:   Natural Hybridization among Four
    Species of *Baptisia* (Leguminosae)                                       180
    *Am. J. Bot.* **50**:159–173 (1963)

15  CARSON, H. L., P. S. NAIR, and F. M. SENE:   *Drosophila* Hybrids in Na-
    ture: Proof of Gene Exchange Between Sympatric Species                    195
    *Science* **189**:806–807 (1975)


## PART V: HYBRID ZONES

**Editor's Comments on Papers 16, 17, and 18**                                  198

16  MOORE, W. S.:   An Evaluation of Narrow Hybrid Zones in Vertebrates       201
    *Quart. Rev. Biol.* **52**:263–277 (1977)

17  HUNT, W. G., and R. K. SELANDER:   Biochemical Genetics of Hybrid-
    isation in European House Mice                                            216
    *Heredity* **31**:11–33 (1973)

18  BLOOM, W. L.:   Multivariate Analysis of the Introgressive Replace-
    ment of *Clarkia Nitens* by *Clarkia Speciosa Polyantha*
    (Onagraceae)                                                              239
    *Evolution* **30**:412–424 (1976)

## PART VI: REPRODUCTIVE CHARACTER DISPLACEMENT

**Editor's Comments on Papers 19 Through 22**                    254

**19**   **BROWN, W. L., JR., and E. O. WILSON:**   Character Displacement        259
*Syst. Zool.* **5**:49–64 (1956)

**20**   **KNIGHT, G. R., A. ROBERTSON, and C. H. WADDINGTON:**   Selection for Sexual Isolation Within a Species        275
*Evolution* **10**:14–22 (1956)

**21**   **EHRMAN, L.:**   Direct Observation of Sexual Isolation Between Allopatric and Between Sympatric Strains of the Different *Drosophila Paulistorum* Races        284
*Evolution* **19**:459–464 (1965)

**22**   **LEVIN, D. A., and H. W. KERSTER:**   Natural Selection for Reproductive Isolation in *Phlox*        290
*Evolution* **21**:679–687 (1967)

**A Final Note**        299
**References**        303
**Author Citation Index**        313
**Subject Index**        318
**About the Editor**        321

# CONTENTS BY AUTHOR

Alston, R. E., 180
Anderson, E., 11, 20, 159
Bloom, W. L., 239
Brown, W. L., Jr., 259
Carson, H. L., 195
DeBach, P., 73
Ehrman, L., 284
Grant, K. A., 105
Grant, V., 62, 87, 105
Heiser, C. B., Jr., 166
Hubbs, C. L., 139
Hubricht, L., 159
Hunt, W. G., 216
Kerster, H. W., 290
Knight, G. R., 275

Levin, D. A., 290
Moore, W. S., 201
Nair, P. S., 195
Parker, E. D., Jr., 117
Rao, S. V., 73
Rattenbury, J. A., 31
Robertson, A., 275
Selander, R. K., 117, 216
Sene, F. M., 195
Stebbins, G. L., Jr., 20
Stephens, S. G., 51
Straw, R. M., 101
Turner, B. L., 180
Waddington, C. H., 275
Wilson, E. O., 259

# INTRODUCTION

Hybridization, or the "crossing between individuals belonging to separate populations which have different adaptive norms" (Stebbins, 1959), has been known since antiquity to occur in both plants and animals. The scientific study of hybridization and its products, especially in plants, began in the middle of the eighteenth century as naturalists became aware of the discontinuities among species. Work on plant hybridization during the subsequent century was synthesized by Focke (1881), who reported that many groups of plants hybridize with facility while others do so with difficulty if at all, and that hybridization is promoted by asymmetrical frequencies of species in flower. These observations are still valid. Darwin (1859), in his *Origin of Species*, considered the matter of hybridization principally with regard to the distinctiveness of species, pondering the observation that there is continuum between fully fertile and completely sterile species hybrids. The possibility that hybridization might play an important role in evolution was first expounded by Lotsy (1916), whose principal contribution was the demonstration of complex and rich segregation in advanced-generation hybrids. Although other botanists of the early twentieth century (e.g., Jeffrey, 1915) also assigned a dominant role to hybridization, most were reluctant to acknowledge more than the occasional occurrence of hybrids. The minor role accorded to hybrids was even more vividly expressed by zoologists, especially since animal hybrids were recorded much less frequently than plant hybrids. Even as late as 1942, leading systematic zoologists regarded the evolutionary effects of hybridization as minimal (Mayr, 1942). Only by the middle of this century had the evolutionary significance of hybridization, based on genetic and cytological criteria, been authoritatively treated in both plants and animals (Anderson, 1949; Stebbins, 1950; Dobzhansky, 1951; White, 1954).

Hybridization may be an important evolutionary phenomenon if accompanied by the selective incorporation of genes from the nonrecurrent population system (introgression) or by the stabilization of hybrid derivatives with or without polyploidy, or if population systems diverge in areas of contact to avoid hybridization and the attendant loss of gametes. A prerequisite for gene exchange or stabilization is the production of partially fertile and vigorous $F_1$ and advanced-generation hybrids. The factors that determine whether this prerequisite is fulfilled are isolating mechanisms; they are classified in distinctive ways by several authors (e.g., Dobzhansky, 1970; V. Grant, 1971; Mayr, 1963; Stebbins, 1977). Basically, there are some differences that prevent the formation of hybrid zygotes, and some that reduce the viability and fertility of hybrids. Isolating mechanisms per se should not be viewed as adaptations, for they are not properties of single populations or species. In effect, isolating mechanisms reside in the interface between populations and are a function of the interface. They result from the incongruities in the adaptive "surfaces" of populations that happen to be paired in nature or by the evolutionist.

Almost every review or chapter on isolating mechanisms written during the past half-century includes a statement to the effect that isolation is the by-product of divergent evolution. Although this statement is widely accepted, it may be interpreted variously, and many authors fail to convey explicitly what they really mean. If isolation is a by-product of divergent evolution, should we conclude that the alternate character states resulting in isolation were not the immediate targets of selection? The answer obviously is no, since differences in flowering time, floral form and allurements, habitat requirements, and so on could enhance populations' adaptedness to their local physical and biotic environments. Therefore selection for certain types of character divergence in allopatric populations may directly impede interpopulation gene flow, although this effect may not have been directly selected for (Clausen, 1951; Stebbins, 1950). On the other hand, if the targets of selection were not immediately involved in prezygotic isolation, this or other forms of isolation may arise by the hitchhiking of neutral genes linked to those under selection (Mayr, 1942; Dobzhansky, 1937; Muller, 1942). Then a chance association of genes may incidentally result in isolation. Now neither genes involved in isolation nor the isolation that results from their divergence is a direct product of selection. Finally, isolation may be a by-product of divergent evolution through the accumulation (in different populations) of novel neutral genes that later become essential constituents of the genotype (Harland, 1936; Schmalhausen, 1949). In this scheme, selection promotes genetic divergence after the process has been initiated, but not for the value of the isolation that may incidentally accrue. Wallace (1889) and later Fisher (1930) and Dobzhansky (1941, 1951) offered a contrasting hypothesis that isolating mechanisms could also be products of selection against hybridization. Selection for reproductive isolation in areas of sympatry would reinforce previously existing barriers and serve to reduce gametic wastage, hybridization, and disruptive gene flow. V. Grant (1966a) has suggested the term *Wallace effect* for this process, which also may be included in the broader concept of character displacement (Brown and Wilson, 1956). Selection against hybridization can influence only mechanisms that act in the parental generation, and thus can relate only to prezygotic mechanisms.

The questions about pre- and postzygotic isolation have shifted in emphasis from what to why within molecular and genetic contexts. This is most evident in studies on gene expression and development in species hybrids, although the work on the constituents of floral fragrance and animal sex pheromones in the prezygotic realm is quite important. A major area of study involves the compatibility of and interactions between nuclear genes and of cytoplasmic genes as they relate to the viability and development of hybrids. Parental enzyme phenotypes in hybrids have been examined using electrophoretic techniques in several animals including sea urchins (Ozaki and Whitely, 1970), fish (Hitzeroth et al., 1968; Whitt et al., 1977), amphibians (Johnson and Chapman, 1971; Etkin, 1977), birds (Meyerhof and Haley, 1975), and mammals (Epstein et al., 1972). The expression of parental genes may not be synchronous during early development, paternal genes being suppressed. Asynchronous development of parental genomes is also evident for nuclear genes in somatic cell hybrids (R. L. Davidson, 1974), and for cytoplasmic genes in sexually produced hybrids (Bogorad, 1975; Barrett and Flavell, 1977). On the other hand, in hybrids between *Xenopus laevis* and *X. mulleri* there is preferential transcription of ribosomal RNA of the former species, regardless of which species serves as the egg parent (Honjo and Reeder, 1973). In general, it seems that in animal hybrids, (1) early morphogenesis is dictated principally or exclusively by maternal components present in the egg at fertilization, and (2) the presence of catabolic enzymes depends on maternal components rather than on a new embryo transcripts. Yet the parental genome as a whole is not repressed during early morphogenesis (E. H. Davidson, 1976).

Biosynthetic processes in the development of plant hybrids is poorly understood. Genetic tumors are produced in hybrids within several genera. In *Nicotiana*, tumor formation is associated

3

with auxins in higher-than-regulatory amounts (Bayer and Ahuja, 1968). The biosynthesis of secondary products may be disrupted, with the accumulation of the precursors or the production of novel compounds (Ornduff et al., 1973; Williams et al., 1974). The molecular approach to the study of hybrids extends to chromosomes, where studies of pairing relationships are now considering the level of unique DNA divergence and the amount and divergence of redundant DNA (Narayan and Rees, 1977), and ribosomal-DNA constancy and nucleolar organizer expression (Doerschug et al., 1976; Bicudo and Richardson, 1977). If the genomes of $F_1$ hybrids have antagonistic or incompatible regulatory systems, evidence of such should be forthcoming from physiological as well as biochemical studies. Unfortunately, little attention has been given to tissue, organ or whole-organism physiology of hybrids.

Some hybrids are inviable or weak and would not be expected to perform well in nature. Others are not only vigorous but fertile as well. In contrast to the extensive study of genome compatibility, rather little attention is being given to the relative fitness of hybrids vis à vis their parents in experimental or natural populations. Botanists have all but ignored this topic (but see Hiesey et al., 1971) except with cultivated species, whereas zoologists have conducted some substantive laboratory experiments (Lewontin and Birch, 1966; Nagle and Mettler, 1969; Sokal and Taylor, 1976). The survival success of hybrids and their parents in nature has been studied only superficially (Blair, 1956; Levin, 1973).

The role of hybridization in evolution, allopolyploidy aside, remains a matter of conjecture. Hybrid swarms have been described in numerous plant and animal genera, but hybridization does not necessarily result in the selective incorporation (or infiltration) of genes from one population into another. As noted in a recent critique by Heiser (1973), introgression may be important, but there are very few convincing cases in either the wild plant or animal literature. In some instances, apparent introgression has been disproved with the implementation of new techniques; in other instances, different explanations for concordant variation patterns are more parsimonious. The stabilization of hybrid derivatives also could account for the evolution of new lineages, and the literature contains many putative examples. As noted by Gottlieb (1972), however, judgments of hybrid origin often are based on one or a few criteria that do not provide a rigorous and unequivocal means of testing the hypothesis of hybrid origin. Hybridization or the potential for such could result in a strengthening of pre-existing prezygotic barriers in areas of sympatry. Although there have been numerous suggestions of this form of character displacement, the evidence that the challenge of a related taxon is responsible for divergence is often not substantial (P. R. Grant, 1972).

Our difficulty in assessing the role of interspecific gene exchange in evolution is principally attributable to our inability to detect and quantify gene exchange. Many morphological, physiological, and chemical expressions of species are controlled by several unlinked genes (Clausen and Hiesey, 1958; Alston, 1967; Wallace et al., 1972). To transfer a diagnostic multigenic character from one species to another, all genes involved must cross the "hybrid bridge" between the species and be reunited in one species by recurrent backcrosses. The more unlinked genes involved, the less likely it is that all genes will be transferred and that gene exchange will be perceived. Not only may we not be able to recognize introgressants, but in some instances we cannot recognize $F_1$ hybrids by morphological criteria (Ornduff, 1969; Lee, 1975; Raven and Raven, 1976). The detection and quantification of alien genes may be facilitated by the study of enzyme variants (allozymes), since variants of this type are under single structural-gene control, inherited in a simple Mendelian fashion, and are identifiable in homozygous or heterozygous states (Scandalios, 1969; Ayala, 1976). However, since introgression involves the selective incorporation of genes from another species, and since some enzyme variants may be neutral or nearly so (Nei, 1975), this approach does not assure success. It is possible, however, that species-specific allozymes would be linked to genes that may be selectively advantageous in a related species and thus incidentally hitchhike from one species to another. The probability of hitchhiking increases when genes, of necessity, are transmitted as blocks in backcross hybrids (V. Grant, 1966b, 1967).

Although we need to document better the roles that hybridization may play within the evolution of a specific genus, and to do so from multiple vantage points, most authors agree that hybridization, especially if it is associated with introgression, provides populations with increased evolutionary flexibility and the opportunities to exploit resources beyond the range of the "pure" population. Indeed, Epling (1947) and others have argued that the ability to hybridize may actually be selected for in changing environments. For a contrary opinion, see Wagner (1970), who argues that hybridization is an insignificant evolutionary factor.

The ability of population systems to hybridize and exchange genes not only has been studied in its own right, but has been used in various forms to define the biological species concept. As stated by Mayr (1940, 1942, 1963), biological species are groups of actually or potentially interbreeding populations, which are reproductively isolated from other such groups. This concept is echoed in the writings of nineteenth-century botanists (Lindley, 1830; Herbert, 1837; Gartner, 1849). During the past three decades, it has been applied to many groups of organisms and has been attractive, in part because of its affinity with the area of evolutionary population genetics. The concept has raised a good deal of controversy about what species are, assuming that they are indeed real and not just taxonomic grouping categories. The notion that species should be defined on the basis of genetic or chromosomal incompatibility rather than adaptive discontinuities of which the incompatibility may be independent has been questioned by students of plant and animal evolution (e.g., Sokal and Crovello, 1970; Raven, 1976).

Evidence from the study of interspecific hybridization also has been used in conjunction with that from comparative protein chemistry to postulate two types of evolution. Mammals that can hybridize differ only slightly at the protein level, whereas frogs and birds that differ substantially at the protein level hybridize readily with their congeners (Wilson et al., 1974; Prager and Wilson, 1975). Wilson and his colleagues argue that protein evolution has proceeded at about the same rate in the three aforementioned groups of animals, but that the rate of organismal evolution has been much greater in mammals. They interpret this to mean that protein evolution may not be at the basis of organismal evolution, and that there might be two types of evolution: protein evolution, which is time-dependent, and regulatory evolution, which parallels organismal evolution.

Part I

# THE ROLE OF HYBRIDIZATION
# IN EVOLUTION

# Editor's Comments
# on Papers 1, 2, and 3

1 **ANDERSON**
*Hybridization of the Habitat*

2 **ANDERSON and STEBBINS**
*Hybridization as an Evolutionary Stimulus*

3 **RATTENBURY**
*Cyclic Hybridization as a Survival Mechanism in the New Zealand Forest Flora*

Hybridization between population systems (races to species) with different adaptive norms followed by several generations of backcrossing and selective incorporation of alien genes (namely, introgressive hybridization; see Anderson and Hubricht, 1938) may greatly enhance the level of variation and thus the evolutionary flexibility of populations. Hybridization also may set the stage for the evolution of novel lineages via stabilization of intermediate transgressive hybrid derivatives (Anderson, 1949; Heiser, 1949; Stebbins, 1950, 1959). For introgression to occur, $F_1$ hybrids must (1) grow in the proximity of one or both parental species, (2) be partially fertile, and (3) be capable of backcrossing to one or both of the parental species. For stabilization to occur, hybrids must (1) grow near one another, (2) be partially fertile, and (3) be capable of crossing *inter se*. Moreover, certain advanced-generation hybrids must be superior to their parental systems either where hybridization occurs, or in other habitats into which such hybrid derivatives may be dispersed. The novel adaptations must be under such genetic control so that they may be fixed after relatively few generations of inbreeding. Fertility must be restored if the novel lineage is to spread to other sites. The probability of a new lineage establishing itself is a function not only of its physical isolation from its progenitors, but also of its genetic isolation, because if hybridization is possible, a minority element may be "hybridized" out of existence since a larger proportion of its progeny will be hybrids than will those of a majority element, such as one of the parental taxa (Lewis, 1961).

8

Hybridization between populations is contingent on the availability of sites for F₁, backcross, or advanced-generation hybrids. These microsites probably will differ from those occupied by nonhybrid individuals because hybrids may be inferior competitors vis à vis their parents or because they have ecological amplitudes different from their parents (Anderson, 1949; Stebbins, 1950, 1959). Hybridization between species with divergent habitat requirements and hybrids restricted to narrow ecotonal zones has been described in plants (*Quercus*, Muller, 1952; *Aquilegia*, Chase and Raven, 1975) and animals (*Gasterosteus*, Hagen, 1967). Suitable microsites may be accessible when hybridization occurs at the geographical or ecological margins of one or both population systems (*Phlox*, Levin and Smith 1966; *Purshia*, Stutz and Thomas, 1964). Novel microsites also arise as a consequence of a local, short-term habitat disturbance or a broad regional disturbance that encompasses a broad time span.

Anderson (Paper 1) contends that disturbance disrupts preexisting ecological barriers and permits closer and more frequent contact of divergent populations and thus a greater potential for hybridization. He argues that hybrid swarms can survive only in disturbed or "hybridized" habitats and that in recent times humans may have been especially important in forming such habitats. In this paper and others (1953), he sites several examples for plants; additional examples are summarized by Heiser (1949a, 1973).

Recent disturbance has also enhanced the potential for hybridization in animals. The planting of alfalfa in New Hampshire has led to the partial breakdown of ecological isolation and increased hybridization in *Colias* (Hovanitz, 1948; Remington, 1954). In parts of Mexico, the clearing of forests has disrupted the original vegetation and opened up intermediate habitats that facilitate hybridization between *Pipilo erythrophthalmus* and *P. ocai* (Sibley, 1954). If recent disturbance can greatly increase the opportunities and incidence of hybridization, then it is possible that geological, climatic, and floristic shifts also may have opened new habitats. Anderson and Stebbins (Paper 2) postulate that the rapidity of evolution and bursts of innovative diversification at various times in the past may be the consequence of hybridization. Introgressants or hybrid segregates may have been preadapted for the exploitation of habitats or resources beyond the grasp of the parental taxa. These hybrid derivatives may have served as the nucleus for new species, genera, or families.

In a changing environment, the infusion of genetic variation from a related species may permit populations to respond more

readily to altered selection pressures. This is the point of Rattenbury's hypothesis in Paper 3. He argues that introgressive hybridization may have permitted the tropical forest elements in New Zealand to survive periodic cooling during the late Pliocene and early
Pleistocene, in spite of bottlenecks that species isolates experienced. Selection may have favored the ability to hybridize and may
account for the high incidence of hybridism and unusually high
levels of variation in the New Zealand flora. Weak barriers to hybridization and extensive hybridization in plants also are evident
in the Hawaiian Islands (Gillett, 1972) and other islands in the South
Pacific (Carlquist, 1974). The impact that hybridization may have
had on the evolution of the South Pacific fauna is not clear. Mayr
(1963) refers to several cases of hybridization in birds.

The outcome of hybridization (introgression or a stabilized
hybrid derivative) depends in part on the breeding system of the
participating taxa. As noted by Baker (1953) and V. Grant (1956), the
result of hybridization when the taxa are cross-fertilizing is more
likely to be introgression than the formation of new taxa. With
self-fertilization, hybrids may proceed through several generations
of inbreeding, with the resultant stabilization of certain recombinations. If genetic or chromosomal barriers are present, the factors
responsible for them may reassort and constitute a barrier between
the stabilized derivative and one or both parental taxa. V. Grant
(1971) discusses several examples.

Reprinted from *Evolution* 2:1–9 (1948)

# HYBRIDIZATION OF THE HABITAT

Edgar Anderson

*Missouri Botanical Garden, St. Louis 10, Missouri*

It has been the experience of most biologists that hybridization between species is rare in nature. Many biologists encounter interspecific hybrids in the field so rarely as to doubt if they really occur there at all or else find them under special circumstances (Epling, 1947) which raise serious doubts as to the importance of hybridization under natural conditions. After a series of investigations (Anderson, 1936b, 1936d; Anderson and Hubricht, 1938; Anderson and Turrill, 1938) it has been my own experience that clearcut out-and-out hybrids are seldom met with, even when a deliberate search is made for them, and that hybrid swarms of bizarre recombinations are found, if at all, only under peculiar circumstances.

It was once the common opinion (Zirkle, 1935) that this lack of evident hybridization was caused by the sterility of interspecific hybrids. Experimental evidence has not confirmed this judgment and modern advocates of reproductive isolation (Mayr, 1940) as a species criterion have had to phrase their definitions to permit semi-fertility between distinct species. For the higher plants there is an impressive amount of experimental evidence on this point though it is widely scattered. There are the papers of the early hybridizers (Focke, 1881; Roberts, 1929; Zirkle, 1935), much work in genetics (see for instance East, 1913, 1916) and experimental taxonomy (Clausen, Keck, and Hiesey, 1946), and the experience of numerous plant breeders. The latter, by far the largest of these three bodies of evidence, is not too accessible to most scientists since it has to be dug out piecemeal from such compendia as Rehder's Manual of Trees and Shrubs (1940). A critical summary of all this evidence is badly needed. Lacking one, the evidence for the higher plants may be roughly summarized as follows: Well-differentiated species of the same genus may or may not be interfertile when tested experimentally. On the whole it seems to vary with the genus. There are certain genera in which interspecific hybrids are difficult to make and are sterile. There are others, equally exceptional, in which the widest possible crosses within the genus will yield fertile or semi-fertile hybrids (Anderson and Schafer, 1931). The yellow trumpet Narcissi are so completely fertile with the flat, white-flowered Poets' Narcissi that the whole business of supplying new garden hybrids has been founded on it (Calvert, 1929). The Poets' Narcissi (in themselves a whole group of species and sub-species) are so interfertile with the species complex making up the long-crowned yellow daffodils that it has been possible to accomplish such recombinations as the transfer of the deep red-orange pigment from the rim of the tiny central eye in a Poets' Narcissus to the brilliant flaring orange trumpet of such modern "red trumpets" as the variety "Fortune" (Anderson and Hornback, 1946). The commonest condition among the higher plants seems to be that in which crosses within a species group are easy to make while intergroup crosses are difficult or impossible.

If, therefore, we base our explanation of the rarity of hybrids under natural conditions on the experimental facts we can make the following summary: 1. In a minority of the cases where related species occur near each other, they are completely intersterile. 2. In certain cases hybrids can be formed but are sterile. 3. In a surprising number of cases the experimental evidence shows that the species can be crossed readily, in the breeding plot, but no hybrids appear under natural conditions. 4. In many cases which have been carefully investigated, hybrids, though usually rare, do occasionally occur but leave no apparent descendants except under unusual conditions.

This experimental evidence, such as it is, justifies the generalization that species maintain themselves as recognizable units even where they are interfertile and even when there is a considerable opportunity for them to hybridize. Why should this be so? Before we can answer the question we must first have clearly in mind the known results of such hybridization under experimental conditions. If we ignore the complications due to polyploidy and other important but more-or-less specialized phenomena, the usual results of hybridization are readily summarized. They were well established by the early hybridizers and have been abundantly confirmed by modern genetic research.

1. The first hybrid generation is intermediate between the parents and is as uniform as they are, or even more so.

It is usually more vigorous than either and more robust in nature.

2. The second hybrid generation, while on the average intermediate, is extremely variable. Usually no two individuals are alike. The variation is usually bewildering, but it can be shown to have a general trend from a few individuals more or less like one of the parental species, to a great bulk more or less like the first generation hybrid, to a very few more or less like the other parental species.

3. The facts with regard to the backcrosses are equally well established but unfortunately have been so little stressed in modern times that they are not as generally familiar to biologists. If the first hybrid generation is crossed back to either parent, the first backcross is made up largely of plants which resemble the species to which they have been backcrossed. If we take one of these and cross it back a second time to the same parent, not only will the seedlings resemble this species very much indeed but some of them may be nearly or quite indistinguishable from it. In all the cases with which I am personally familiar, many, if not most, of the first and second back-crosses, if found in nature, would by taxonomists be accepted as varieties or slightly aberrant individuals of the species to which they were back-crossed (see fig. 1). Few taxonomists, even those specializing in that group of plants, would even suspect that many of these back-crosses were of partially hybrid ancestry. For the genus Apocynum we have a detailed experimental record of back-cross morphology and the taxonomic reaction to it (Anderson, 1936b).

In the genus Tradescantia, by rigidly experimental methods (Anderson, 1936a, 1936d; Anderson and Sax, 1936; Anderson and Woodson, 1935), including the production of experimental backcrosses, it has been possible to demonstrate that the principal result of hybridization, in those cases where it did occur, was a series of such backcrosses to one or to both parents. To this phenomenon I gave the

N. LANGSDORFFII

N. ALATA

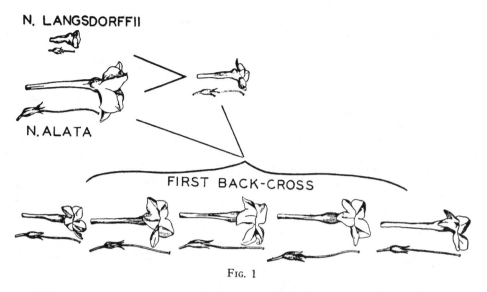

FIRST BACK-CROSS

Fig. 1

name *introgressive hybridization* (Anderson and Hubricht, 1938), since it provided a mean by which elements in the germplasm of one species might introgress into the germplasm of the other. The chief result of hybridization under such conditions is the enrichment of variation in the participating species. Such hybridization is cryptic and only by very specialized techniques can we measure its exact importance in any particular case. Since the first publications on the subject, the phenomenon of introgressive hybridization has been confirmed, with experimental verification, in numerous genera. (A comprehensive bibliography on introgression by Dr. Charles Heiser is now under way. The following papers are representative: Goodwin, 1937b; Riley, 1938, 1939a; 1939b; Dansereau, 1941; Marsden, Jones and Turrill, 1946; Epling, 1947; Stebbins, Matzke, and Epling, 1947; and Heiser, in press.) Circumstantial evidence indicates its importance in many more genera. How important is it on the whole in the higher plants and in other groups of organisms? We do not yet have any exact evidence. From those cases where experimenters have gone to the trouble of making experimental backcrosses (Epling, 1947), it is clear that many of the second backcrosses could not be recognized as of mongrel origin purely by their appearance. It is therefore clearly indicated that gene flow from one species to another may go far beyond any point which could be detected by ordinary morphological techniques. We shall not be able to assess the real importance of introgression until we can study genetically analyzed species in the field and determine the actual spread of certain marker genes. Until such data are available, any generalizations are based on mere opinions.

Introgression therefore gives us a partial answer to our original question as to why hybrids are so seldom met with in nature. It is because when hybrids do occur, they usually perpetuate themselves, if at all, in backcrosses to one or the other parental species and the mongrel nature of their descendants is not apparent to the ordinary biologist. The commonest result of hybridization is introgression, and introgression must be excessive before it will produce results conspicuous enough to impress biologists who are not making a deliberate search for such phenomena. This is only half an answer. Why do interfertile species limit themselves very largely to backcrosses when they meet under natural conditions and most particularly why is the backcrossing largely in areas where natural conditions have been very much disturbed? There are at least two main reasons; one resides in the germ plasm itself, the other in the habitat. As to the internal one, the total effect of all the forces which make for specific cohesion is very great, much greater than one would expect until he made careful calculations (Anderson, 1939a & b). Following the arguments used in these calculations it can furthermore be shown that in well-differentiated species the total effect of linkage is so strong that two well-differentiated but interfertile species, meeting in an idealized environment favorable to hybridization, would remain recognizable units in spite of their interfertility. The details of the argument are largely mathematical and are shortly to appear elsewhere, but the general conclusions can be tersely put. Linkage by itself is a force strong enough to prevent the complete swamping of interfertile species. As a factor in specific cohesion it is proportional to the differentiation of the two hybridizing entities: the greater the differentiation, the stronger the cohesive force of linkage.

The effect of the habitat, however, is also important, and it usually operates in exactly the same direction. The argument is as follows: it is now known that the physiological differences between species segregate in the same way as do the morphological ones. In Neurospora (Beadle, 1945) the mode of inheritance of scores of physiological differences are precisely understood. In yeasts the Lindegrens (1947) have demonstrated with laboratory precision the inheritance of various differences in habitat preference. The higher plants are not so amenable to precise physiological analyses of their nutritional requirements but there is abundant circumstantial evidence that a similar situation prevails and there are precise data for a few characters such as maturity (see for instance Goodwin 1937c, footnote 13, or Marsden Jones and Turrill, 1946), response to day length, etc. Nearly everyone who has grown and studied the second generation from a species cross has noted the segregation and recombination of such physiological differences as length of blooming season, resistance to diseases, dayblooming habit vs. night blooming, ease of wilting, resistance to cold, light tolerance, etc. If, therefore, we cross two species differing in their ecological requirements we may expect these physiological differences to segregate as follows: The first hybrid generation will be uniform in its requirements and on the whole they will be for conditions intermediate between those required for the two parents. The second generation will be made up of individuals each of which will require its own peculiar habitat.

Let us repeat this last statement; it is the crux of the argument. THE SECOND GENERATION WILL BE MADE UP OF INDIVIDUALS EACH OF WHICH WILL REQUIRE ITS OWN PECULIAR HABITAT FOR OPTIMUM DEVELOPMENT. As a whole the requirements of the second generation will range from a need for something more or less like one parental habitat to something more or less like the intermediate habitat of the F-1 to something more or less like the habitat of the other parent.

In nature therefore we might reasonably expect to find first generation hybrids growing in an intermediate zone between the two parental habitats. The persistence of any considerable variety of the various second generation recombinations would require a habitat such as is seldom or never met with, where various combinations of the two parental habitats are found in close juxtaposition to one another.

As a crude example let us consider the adjacent habitats in which one finds *Tradescantia subaspera* and *Tradescantia canaliculata* at home in the Ozark Plateau (Anderson and Hubricht, 1938). The former grows in deep rich woods at the foot of bluffs while the latter grows up above in full sun at the edge of the cliffs. As an over-simplified example we can list three of the outstanding differences between these two habitats as follows:

rich loam............rocky soil
deep shade...........full sun
leaf mould cover......no leaf mould cover

*Tradescantia canaliculata* and *T. sub-aspera* are well-differentiated species; neither one of them is by any means the closest relative of the other, yet Mr. Hubricht and I have found by actual experiment that not only can they be crossed readily by artificial means but they do cross abundantly when left to themselves in an experimental garden (Hubricht and Anderson, 1941). Even though both he and I were familiar with the appearance of these artificial hybrids and though we searched for them at many points where the species were growing very near one another, we found very few of the first generation hybrids. The habitats of the two species are strikingly different; in the Ozarks one seldom finds the intermediate habitat in which the hybrid is able to germinate and survive. This is a more or less intermediate condition, a gravelly soil, partial shade with some bright sunlight, and a light covering of leaf mould. Imagine, however, the habitat which must be provided if we are to see the second generation recombinations which we obtain in the breeding plot. If we consider only the three contrasting characters of the habitat which have been mentioned above, our recombinations would require the following six new habitats in addition to the parental ones (these six represent only the extreme recombinations; a whole series of intermediates will also be required):

| rich loam | rocky soil |
|-----------|-----------|
| full sun | deep shade |
| no leaf mould | leaf mould |
| rich loam | rocky soil |
| full sun | full sun |
| leaf mould | leaf mould |
| rich loam | rocky soil |
| deep shade | deep shade |
| no leaf mould | no leaf mould |

What would have to happen to any natural area before such a set of variedly intermediate habitats could be provided? It has been very generally recognized that if hybrids are to survive we must have intermediate habitats for them. It has not been emphasized, however, that if anything beyond the first hybrid generation is to pull through, we must have habitats that are not only intermediate but which present recombinations of the contrasting differences of the original habitats. If the two species differ in their response to light, soil, and moisture (and what related species do not?) we must have varied recombinations of light, soil, and moisture to grow their hybrid descendants. Only by a hybridization of the habitat can the hybrid recombinations be preserved in nature.

The actual inherent differences in ecological preference will of course be much more diverse than in the crude example given above. The number of different kinds of habitats required by the hybrids will rise *exponentially* with the number of basic differences between the species. With ten such differences, around a thousand different kinds of habitat would be needed to permit the various recombinations to find a niche somewhere as well suited to them as the original adjacent habitats were to the two parental species. With only twenty such basic differences (and this seems like a conservative figure) over a million different recombined habitats would be needed. Under natural conditions anything like such a situation is close to impossible. Ordinarily it is only through the intervention of man that it is even remotely approached. Even in these cases the new "Recombination Habitats" will largely be limited to habitats pretty much like those required by one of the parental species, but which in a few characteristics approach the requirements of the other parent. We may expect that even in such disturbed habitats there will be back-up recombinations not greatly different from one of the parents which will most readily find an ecological niche suited to them.

One of the best demonstrations yet published of the way in which man can provide strange new niches of hybrid recombinations is that given by H. P. Riley (1938, 1939a) in his analysis of the hybridization between *Iris fulva* and *Iris giganticaerulea*, two species which differ strikingly in their color, morphology, and ecological adaptations (Viosca, 1935). In one of the localities which he studied in detail on the Mississippi delta, a series of long, narrow farms run straight back from the highway side by side, in the fashion set by the French settlers, with almost the precision of experimental plots. The original environment at that point was fairly uniform, but each man has treated his farm a little differently. It was strikingly apparent from Riley's study that the numbers and kinds of hybrids varied from farm to farm. Some had few or none, while others, even when adjacent, had hybrids in great quantities; there were significantly more of them where the meadows had been pastured. In one farm in particular, the little depression which ran parallel to the highway had been subjected to a series of operations. The trees and shrubs had not been entirely removed from this area, but it had been repeatedly cut over and had in addition been heavily pastured. It had a swarm of different hybrid derivatives, almost like an experimental garden, and the hybrid area went right up to the fenceline at the border of the farm and stopped there.

Nor is this an isolated instance. Viosca (1935) and other students of Louisiana irises have worked out in considerable detail the relation between the production of hybrid swarms of these conspicuously different irises and the churning and rechurning of the habitat by ditching, pasturing, lumbering, roadbuilding, etc. It is only where man has hybridized the natural environments of the Mississippi delta that nature can find an appropriate lodging place for the hybrids she has created.

This dependence of interspecific hybridization upon the intervention of man has been described by a number of authors (Darrow and Camp,[1] 1945; Anderson and Hubricht, 1938). It was discussed in some detail by Wiegand (1935) in his paper on "A naturalist's experience with hybrids in the wild." Marie-Victorin has given a vivid description (1922, p. 32; 1935, p. 65) of its operation when the original flora of the St. Lawrence valley was largely replaced by fields and pastures. Epling, Stebbins, Dansereau, and their students have commented upon the connection in a number of different genera. Does this mean that introgression as a phenomenon is limited to the areas disturbed by man and that its results are mere artifacts and not genuine natural phenomena? I think not. Though freely admitting that nearly all the introgression which has been studied experimentally (for one exception see Dansereau, 1941) is of the nature of an artifact, I believe that at particular times, and in particular places, introgression may have been a general evolutionary factor of real importance.

Under the conditions of an experimental garden, natural selection among the progeny of a cross between species is much less severe than it is in nature. Though the optimum environments for the sister hybrids may be quite various, it is possible to raise the majority of them in one plot, providing that they are widely spaced and competition with aggressive weeds is kept at a minimum. The prevalence of iris hybrids on one or two of the farms described by Riley (1938) may have been due in part to the reduction of competition with other plants, particularly grasses, as well as to the variations in shade and moisture brought about by repeated recuttings and overpasturing.

There must have been various times, even without the intervention of man,

---

[1] Darrow and Camp also considered the reaction of hybridizing polyploid complexes with the environment. Polyploidy introduces further complications into hybridization which are beyond the scope of this paper.

when species hybridized under conditions which produced varied new habitats and when competition was not too keen, as for instance when newly colonizable areas emerged from the sea or when various floras spread out onto the northern lands denuded by Pleistocene glaciation. At such times introgression would have been an important evolutionary factor. For one area we are beginning to get actual proof that it did occur. Along the coast of California there are peninsulas which once were isolated islands but which are now united with the main land. In their studies of the California knobcone pines, fossil and living, Mason and his students are demonstrating the actual role of introgression (for a general summary see Cain, 1944, pp. 112–118) in forming these pines as we know them today and to determine in some detail how introgression operated at the time when these islands were joined to the coast.

The Edwards Plateau in central Texas is another area in which introgression may have operated on a grand scale. This comparatively small area is a center of distribution and variation for numerous genera. A mere leafing through of a series of monographs of North American genera (Larsen, 1933; Anderson and Woodson, 1935; Barkley, 1937) will demonstrate that it is one of the outstanding centers east of the Rockies. For many genera the concentration of species is higher there than at any other point and for the genus Tradescantia we know the even more significant fact that it is a center for the diploid strains of polyploid species (Anderson and Sax, 1936; Anderson, 1937). The geological evidence shows that when the Edwards Plateau came into being, it united older land masses in Mexico and in the United States. Certainly at such a time related species in many genera might have met and hybridized under conditions where competition would not have been keen and where associations of plants were in the making instead of already existing as tightly closed corporations. Tradescan-

tias from Mexico would have met species coming down from the Appalachians in an area conducive to the survival of some of the hybrid recombinations.

Woodson has recently (1947) called attention to the importance of peninsular Florida in the speciation patterns of the eastern United States. During parts of the Tertiary it was an island or group of islands which finally became attached to the mainland. Species and varieties which became differentiated during the island period must then have had unusual opportunities to hybridize with their relatives on the mainland. Giles' (1942) studies of Cuthbertia in this area have given cytological proof that such hybridization did actually take place. Careful studies of variation throughout the whole area, for a series of species, should yield data with which we could assess the general overall importance there of introgression.

## SUMMARY

1. Experimental evidence shows that sterility will not account for the rarity of hybrids under natural conditions.

2. Careful field analyses have shown that natural hybridization is largely limited to backcrosses which resemble the parental species so closely that special methods are required to detect them readily.

3. One of the factors limiting hybridization to such introgression is imposed by the habitat for the following reason: Two species differing in their habitat requirements will produce a first generation hybrid adjusted to a uniform intermediate environment. The second generation however consists of individuals each of which requires its own peculiar habitat for optimum development. Such heterogeneous habitats are seldom or never met with, the only approach to them being found in places where man has greatly altered natural conditions.

4. It is concluded that hybrid swarms can survive only in "hybridized habitats." While most of the latter result from hu-

man intervention, similar conditions have prevailed in pre-human times when new lands were opened up to colonization by diverse floras. At such times and places introgressive hybridization must have played an important role in evolution.

## LITERATURE CITED

ANDERSON, EDGAR. 1936a. A morphological comparison of triploid and tetraploid inter-specific hybrids in Tradescantia. Genetics, **21**: 61–65.

——. 1936b. An experimental study of hybridization in the genus Apocynum. Ann. Mo. Bot. Gard., **23**: 159–168.

——. 1936c. The species problem in Iris. Ann. Mo. Bot. Gard., **23**: 457–509.

——. 1936d. Hybridization in American Tradescantias. Ann. Mo. Bot. Gard., **23**: 511–525.

——. 1937. Cytology in its relation to taxonomy. Bot. Rev., **3**: 335–350.

——. 1939a. The hindrance to gene recombination imposed by linkage: an estimate of its total magnitude. Am. Nat., **73**: 185–188.

——. 1939b. Recombination in species crosses. Genetics, **24**: 668–698.

—— AND LESLIE HUBRICHT. 1938. The evidence for introgressive hybridization. Am. Jour. Bot., **25**: 396–402.

—— AND EARL HORNBACK. 1946. A genetical analysis of pink daffodils: a preliminary attempt. Jour. Cal. Hort. Soc., **7**: 334–344.

—— AND RUTH PECK OWNBEY. 1939. The genetic coefficients of specific difference. Ann. Mo. Bot. Gard., **26**: 325–348.

—— AND KARL SAX. 1936. A cytological monograph of the American species of Tradescantia. Bot. Gaz., **97**: 433–476.

—— AND BRENHILDA SCHAFER. 1931. Species hybrids in Aquilegia. Ann. Bot., **45**: 639–646.

——. 1933. Vicinism in Aquilegia vulgaris. Am. Nat., **67**: 1–3.

—— AND W. B. TURRILL. 1938. Statistical studies on two populations of Fraxinus. New Phytol., **37**: 160–172.

—— AND R. E. WOODSON. 1935. The species of Tradescantia indigenous to the United States. Contrib. Arn. Arb., **9**: 1–132.

BARKLEY, F. A. 1937. A monographic study of Rhus. Ann. Mo. Bot. Gard., **24**: 265–498.

BEADLE, G. W. 1945. Biochemical genetics. Chem. Rev., **37**: 15–96.

CAIN, STANLEY A. 1944. Foundations of plant geography. Harpers. 1–556 pp.

CALVERT, A. E. 1929. Daffodil growing. Dulau. 1–389 pp.

CLAUSEN, JENS, D. D. KECK, AND W. M. HIESEY. 1946. Experimental taxonomy. Carn. Year Book, **45**: 111–120.

DANSEREAU, PIERRE. 1941. Etudes sur les hybrides de Cistes. VI. Introgression dans la section Ladanium. Can. Jour. Research, **19**: 59–67.

DARROW, GEORGE M., AND W. H. CAMP. 1945. Vaccinium hybrids and the development of new horticultural material. Bull. Torr. Bot. Club, **72**: 1–21.

EAST, E. M. 1913. Inheritance of flower size in crosses between species of Nicotiana. Bot. Gaz., **55**: 177–188.

——. 1916. Inheritance in crosses between Nicotiana Langsdorffii and Nicotiana alata. Genetics, **1**: 311–333.

EPLING, CARL C. 1947. Natural hybridization of Salvia apiana and Salvia mellifera. Evolution, **1**: 69–78.

FOCKE, W. O. 1881. Die Pflanzen-mischlinge. 569 pp.

GILES, N. H., JR. 1942. Autopolyploidy and geographical distribution in Cuthbertia graminea Small. Amer. Jour. Bot., **29**: 637–645.

GOODWIN, RICHARD H. 1937a. Notes on the distribution and hybrid origin of X Solidago asperula. Rhodora, **38**: 22–28.

——. 1937b. The cytogenetics of two species of Solidago and its bearing on their polymorphy in nature. Am. Jour. Bot., **24**: 425–432.

——. 1937c. The role of auxin in leaf development in Solidago species. Am. Jour. Bot., **24**: 43–51.

HUBRICHT, LESLIE, AND EDGAR ANDERSON. 1941. Vicinism in Tradescantia. Am. Jour. Bot., **28**: 957.

LARSEN, ESTHER L. 1933. Astranthium and related genera. Ann. Mo. Bot. Gard., **20**: 23–44.

LINDEGREN, CARL C., AND GERTRUDE LINDEGREN. 1947. Mendelian inheritance of genes affecting vitamin-synthesizing in Saccharomyces. Ann. Mo. Bot. Gard., **34**: 95–99.

MARIE-VICTORIN, FR. 1935. Flore Laurentienne. Montreal. 917 pp.

——. 1922. Esquisse systematique et ecologique de la flore dendrologique. Contrib. Lab. Bot. de l'Univ. de Montreal, no. 1: 1–33.

MARSDEN JONES, E. M., AND W. B. TURRILL. 1946. Researches on Silene maritima and S. vulgaris. Kew Bull., **1946**: 97–107.

MASON, H. L. (see Cain, 1944).

MAYR, ERNST. 1940. Speciation phenomena in birds. Am. Nat., **74**: 249–278.

REHDER, ALFRED. 1940. Manual of cultivated trees and shrubs. Second edition. 996 pp. Macmillan.

RILEY, H. P. 1938. A character analysis of colonies of Iris fulva, Iris hexagona var. giganticaerulea, and natural hybrids. Am. Jour. Bot. **25**: 727–738.

——. 1939a. The problem of species in the Louisiana irises. Bull. Am. Iris Soc., 3–7.

——. 1939b. Introgressive hybridization in a natural population of Tradescantia. Genetics, 24: 753–769.

ROBERTS, H. F. 1929. Plant hybridization before Mendel. 374 pp. Princeton.

STEBBINS, G. L., E. G. MATZKE, AND C. EPLING. 1947. Hybridization in a population of *Quercus marilandica* and *Quercus ilicifolia*. Evolution, 1: 79–88.

VIOSCA, P. 1935. The irises of southeastern Louisiana. Bull. Am. Iris Soc., 57: 3–56.

WIEGAND, K. 1935. A taxonomist's experience with hybrids in the wild. Science, 81: 161–166.

WOODSON, R. E., JR. 1947. Notes on the historical factor in plant geography. Contrib. Gray Herbarium, 165: 12–25.

ZIRKLE, CONWAY. 1935. The beginnings of plant hybridization. 231 pp. Univ. of Pennsylvania, Phila.

# 2

Reprinted from *Evolution* **8**:378–388 (1954)

## HYBRIDIZATION AS AN EVOLUTIONARY STIMULUS

E. Anderson and G. L. Stebbins, Jr.

*Missouri Botanical Garden and University of California, Davis*

One of the most spectacular facets of the newer studies of evolution has been the demonstration that evolution has not proceeded by slow, even steps but that seen in the large there have been bursts of creative activity. Some of the evidence for these bursts is from paleontology; Simpson (1953) has recently assembled a wealth of data concerning them and has discussed in detail their possible causes. Paleobotanists are equally aware of such events as the great upsurge of angiosperms in the Cretaceous, and of primitive vascular plants in the Devonian period. Other evidence for evolutionary bursts comes from the existence of large clusters of related endemic species and genera in the modern fauna and flora of certain regions, particularly oceanic islands and fresh water lakes. The snails (Achatinellidae) and honey sucker birds (Drepanidae) of Hawaii are classical examples, as are also the Gammarid crustaceans of Lake Baikal, and the fishes of Lakes Tanganyika and Nyasa in Africa, and particularly of Lake Lanao in the Philippines (see Brooks, 1950 for a summary and discussion of the data). It is true that some of these examples may represent normal rates of evolution occurring in a restricted area which has been isolated for a very long time, but there can be little doubt that in the case of others evolution has been phenomenally rapid.

As Simpson (1944, 1953) has clearly stated, the cause of this rapid evolution is to be sought in the organism-environment relationship. Along with most authors, however, he has tended to emphasize the peculiar environment present during these evolutionary bursts, and has suggested that one need not postulate any unusual type of population structure as a contributing factor. Zimmermann (1948) has given a plausible account of the environmental factors operating in the case of oceanic islands; reduction of competition, frequent migration to new habitats, and populations repeatedly reduced to a very few individuals, giving a maximum opportunity for the operation of chance as well as for the rapid action of selection.

To the student of hybridization, however, another factor which may have contributed largely to these evolutionary bursts presents itself. Hybridization between populations having very different genetic systems of adaptation may lead to several different results. If the reproductive isolation between the populations is slight enough so that functional, viable and fertile individuals can result from segregation in the $F_2$ and later generations, then new adaptive systems, adapted to new ecological niches, may arise relatively quickly in this fashion. If, on the other hand, the populations are well isolated from each other so that the hybrids between them are largely sterile, then one of two things may happen. The hybrids may become fertile and genetically stabilized through allopolyploidy, and so become adapted to more or less exactly intermediate habitats, or they may back cross to one or both parents, and so modify the adjoining populations of the parental species through introgression. This latter phenomenon has now been abundantly documented in the higher plants, and several good examples are known in animals (see bibliographies in Anderson, 1949; Heiser, 1949; Anderson, 1953). By introgressive hybridization elements of an entirely foreign genetic adaptive system can be carried over into a previously stabilized one, permitting the rapid reshuffling of varying adaptations and complex modifier systems. Natural selection is presented not with one or two new alleles but with segregating blocks of genic material belonging to entirely different adaptive systems. A simple analogy will show the comparative effectiveness of introgression.

Let us imagine an automobile industry in which new cars are produced only by copying old cars one part at a time and then putting them together on an assembly line. New models can be produced only by changing one part at a time. They cannot be produced *de novo* but must be built up from existing assembly lines. Imagine one factory producing only model 'T' Fords and another producing model 'T' Fords and also modern station wagons. It will be clear that if changes could only be brought about by using existing assembly lines these would have to proceed slowly in the factory which had only one assembly line to choose from. In the other factory, however, an ingenious mechanic, given two whole assembly lines to work with, could use different systems out of either and quickly produce a whole set of new models to suit various new needs when they arose.

Just as in the example of the two assembly lines, hybrids between the same two species could produce various different recombinations, each of which could accommodate itself to a different niche. When a big fresh water lake was formed *de novo,* hybrids between the same two species could rapidly differentiate into various new types suitable for the various new niches created in the big new lake. A technical point of much significance is that each of the various heterozygous introgressive segments brought in by hybridization would (by crossing over) be capable of producing increased variation generation after generation for periods running into whole geological eras (Anderson, 1939). The enhanced plasticity due to crossing over in introgressed segments has been shown on theoretical grounds to be present for many generations. Such studies as those of Woodson on Asclepias (1947, 1952), of Hall on Juniperus (1952) and of Dansereau on Cistus (1941) indicate that this does actually happen and that introgressive segments may persist for geological periods and produce effects of continental magnitude.

To students of introgressive hybridization it would seem like an excellent working hypothesis to suppose that when Lake Baikal was formed, and when each new island of the Hawaiian archipelago arose from the ocean, species belonging to different faunas and floras were brought together and that physical and biological barrier systems were broken down. There were increased chances for hybridization in an environment full of new ecological niches in which some new recombinations would be at selective advantages. There is a growing body of experimental data to support such an hypothesis. These data fall largely in two groups (1) Evolution under domestication, (2) Evolution in disturbed habitats.

(1) For evolution under domestication the evidence is overwhelming that by conscious and unconscious selection, man has created forms of plants and animals which are specifically distinct from their wild progenitors. This large body of evidence demonstrates that given a habitat in which novelties (or at least some of them) are at a great selective advantage, evolution may proceed very rapidly. There is presumptive evidence that many of these domesticates originated through introgression but the process began so early that getting exact experimental evidence for the history of any one of them will entail long-continued cooperative research (see, however, Mangelsdorf and Smith (1949), Alava (1952), and Nickerson (1953) for evidence that modern Zea is greatly different from the maize of five thousand

years ago and that much of this differentiation may well be the result of introgression from Tripsacum). For some ornamentals, domestication is such a recent event that critical evidence is easier to assemble. Anderson (1952) has presented in elementary detail the case of *Tradescantia virginiana*. He shows that in four hundred years by introgression from *T. ohiensis* and *T. subaspera* (unconsciously encouraged by man) it has evolved under cultivation into a variable complex quite distinct from *T. virginiana* as a genuinely wild species.

(2) Evolution in disturbed habitats. It has been repeatedly shown (Anderson, 1949; Heiser, 1949; Epling, 1947) that species which do not ordinarily produce hybrids and backcrosses may readily do so when man or any other agent disturbs the habitat. This phenomenon was referred to as "Hybridization of the Habitat" by Anderson (1948). After citing the work of several authors who have emphasized the role of man in promoting and creating habitats favorable for the perpetuation of hybrids and hybrid derivatives, he reached the following conclusion (1948, p. 6). "Does this mean that introgression as a phenomenon is limited to the areas disturbed by man and that its results are mere artifacts and not genuine natural phenomena? I think not. Though freely admitting that nearly all the introgression which has been studied experimentally (for one exception see Dansereau, 1941) is of the nature of an artifact, I believe that at particular times, and in particular places, introgression may have been a general evolutionary factor of real importance."

The great frequency of hybrid derivatives in disturbed habitats is only in part due to the breaking down of barrier systems, allowing previously isolated species to cross. It can and does occur when the barrier systems are not broken down (see for instance Heiser, 1951). Much more important is the production of new and varying ecological niches; more or less open habitats in which some of the almost

infinitely various backcrosses and occasional types resulting directly from segregation in $F_2$ and later generations will be at a selective advantage. A particularly significant example was investigated by Anderson (unpublished) who studied *Salvia apiana* and *Salvia mellifera* in the San Gabriel mountains, confirming and extending Epling's (1947) previous studies. He found hybrid swarms not in the chaparral itself where both of these species are native but adjacent to it in cutover live oaks amidst an abandoned olive orchard. In this greatly disturbed area, new niches were created for the hybrid progeny, which are apparently always being produced in the chaparral but at a very low frequency. In this strange new set of various habitats some of the mongrels were at a greater selective advantage and the population of the deserted olive orchard was composed of hybrids and back-crosses to the virtual exclusion of *S. apiana* and *S. mellifera*.

It has been customary to dismiss the evidence of introgression under the influence of man as relatively unimportant to general theories of evolution because nothing quite like it had previously occurred. A little reflection will show that this is not so. Man at the moment is having a catastrophic effect upon the world's faunas and floras. He is, in Carl Sauer's phrase, an ecological dominant but he is not the first organism in the world's history to achieve that position. When the first land vertebrates invaded terrestrial vegetation they must have been quite as catastrophic to the flora which had been evolved in the absence of such creatures. When the large herbivorous reptiles first appeared, and also when the first large land mammals arrived in each new portion of the world there must have been violent readjustments and the creation of new ecological niches.

The last of these (the arrival of the large land mammals) is close enough to us in geological time so that we have witnessed the very end of the process. The vegetation of New Zealand had had no experience with mammals until the arrival of the Maori in the fourteenth century followed by Europeans in the 18th and 19th centuries. Man, pigs, horses, cattle, rats, sheep, goats, and rabbits were loosed upon a vegetation which had had no previous experience with simians or herbivores. The effect was catastrophic. Hybrid swarms were developed upon the most colossal scale known in modern times. A succession of New Zealand naturalists have occupied themselves with the problem and it has been treated monographically by Cockayne (1923) and by Allen (1937)

The extent to which disturbance of the habitat combined with reorganization of adaptive systems through hybridization could have been responsible for evolutionary bursts, "proliferation," "tachytely" or "quantum evolution" (Simpson 1953) can best be estimated by summarizing the geological and paleontological evidence concerning the time of occurrence of habitat disturbances, and comparing this with probable evolutionary changes in certain groups of organisms which were most likely initiated by hybridization. In such a survey, all three of the possible results of hybridizations—introgression, segregation of new types without backcrossing and allopolyploidy—must be considered. Reference to allopolyploids is particularly important, since hybrid derivatives of this type can easily be distinguished from their parental species by their chromosome numbers, and the time and place of hybridization can often be indicated with a high degree of probability (Stebbins, 1950, Chap. 9).

Preceding the advent of man, the most revolutionary event in the history of the northern continents was the Pleistocene glaciation and the contemporary pluvial periods of regions south .of the ice sheet. This involved not only radical oscillations in climate, but also great disturbances of the soil, both in the glaciated regions and in areas to the south of them. In the latter, the extensive deposits of loess immediately south of the ice margin and the masses of alluvium carried for miles down the river valleys must have disturbed these areas almost as much as the ice sheets churned up the areas which they covered.

The activity of hybridization in developing plant populations adapted to these new habitats is amply evident from the frequency of allopolyploids in them. Specific examples are *Iris versicolor* and *Oxycoccus quadripetalus* (Stebbins, 1950); the polyploid complexes of *Salix, Betula, Vaccinium, Antennaria, Poa, Calamagrostis,* and many others can also be cited. The best example of introgression among species which have invaded the ice-free areas in post-Pleistocene time is in the complex *Acer saccharophorum* (Dansereau and Desmarais, 1947). The numerous examples cited by Anderson (1953) of hybrid and introgressant types which occupy the central Mississippi Valley between the Appalachian, Ozark, and central Texas highlands probably represent late Pleistocene or post-Pleistocene invasion of these areas which were strongly affected by outwash from the ice sheet and from the postglacial lakes. The origin of *Potentilla glandulosa* subsp. *Hanseni* in the post-Pleistocene meadows of the Sierra Nevada is discussed by Stebbins (1950, p. 279).

During the Tertiary and earlier geological periods three types of changes in the inanimate environment can be singled out which probably gave rise to disturbed habitats favorable to the establishment of hybrid derivatives. These were mountain building movements, advance and retreat of epicontinental seas, and radical changes in the earth's climate.

Some of the direct effects of mountain buildings are the rapid creation of raw, unoccupied habitats (such as lava flows, for instance), in which plants belonging to very different ecological associations may temporarily mingle and gain a chance to hybridize. In central California the canyon of the Big Sur River is a typical example of the mixing together of species belonging to very different floras in a region of recent uplift which has a rugged, youthful topography. Here yuccas and redwoods grow within a stone's throw of each other. An example of hybridization in this area is between two species of *Hieracium; H. albiflorum,* which is typical of northern California, the Pacific Northwest, and the Rocky Mountains, and *H. argutum,* a Southern California endemic which here reaches its northern limit except for one known station in the Sierra foothills. Examples such as this could undoubtedly be multiplied by a careful study of any youthful mountain region.

The retreat of epicontinental seas in the latter part of the Pliocene period, plus faulting in the Pleistocene and recent times, has been largely responsible for the present topography of coastal California with its flat valleys and abrupt mountain ridges. One hybrid polyploid which appears to have spread as a result of these changes is the octoploid *Eriogonum fasciculatum* var. *foliolosum* (Stebbins, 1942); another is probably the tetraploid *Zauschneria californica* (Clausen, Keck, and Hiesey, 1940). A series of hybrid swarms which may have arisen in response to the same topographical changes is that of *Quercus Alvordi* (Tucker, 1952). *Delphinium gypsophilum* is a relatively well stabilized species, probably of hybrid origin, endemic to this same recently emerged area of California (Epling, 1947), and the species of *Gilia* considered by Grant (1953) to be of hybrid origin have the same general distribution. In the Old World, Dansereau (1941) has suggested that *Cistus ladaniferus* var. *petiolatus,* which occupies the recently emerged coast of North Africa, is a product of hybridization between typical *C. ladaniferus* and *C. laurifolius* both of which occur in the more ancient land mass of the Iberian Peninsula.

Among the radical changes in the earth's climate which occurred recently enough so that their effect on the vegeta-tion can be recorded, is the advent of the Mediterranean type of climate with its wet winters and dry summers in most of California. The time of this climatic change is now fully documented by the fossil record; it took place during the middle part of the Pliocene period. It was preceded by a general decrease in precipitation, with biseasonal maxima (Axelrod, 1944, 1948).

The effects of this climatic change on the woody vegetation of the area are also well documented by the fossil record. One very probable example of a hybrid swarm exists in a fossil flora. In the Remington Hill Flora, which was laid down in the Sierra foothills at the beginning of the Pliocene, there is a great abundance of oak leaves corresponding to the modern *Q. morehus,* a hybrid between the mesophytic, deciduous *Q. Kelloggii,* and the xerophytic, evergreen *Q. Wislizenii* (Condit, 1944). That these fossil leaves were borne by hybrid trees is evidenced not only by their very unusual and characteristic shape, but also by their great variability and the fact that no similar leaves are found in any of the numerous Miocene and Pliocene floras of California. Furthermore, the Remington Hill is the only one of these fossil floras which contains the counterparts of both parental species. At present, the *Q. Kelloggii* × *Wislizenii* hybrid is frequent in the Sierra foothills, but it usually grows as single trees in company with dense stands of *Q. Wislizenii* and *Q. Kelloggii.* The populations of the parental species growing in the vicinity of the hybrids appear little or not at all different from those occurring by themselves, far from any other species of this complex. On the other hand, *Q. Wislizenii* shows considerable geographic variation, with the more northernly and more coastal variants, i.e., those adapted to increasingly mesic climates, possessing an increasingly greater resemblance to *Q. Kelloggii* in habit, leaves, buds, and fruits. This suggests that the present variation pattern in *Q. Wislizenii* is the result chiefly of extensive introgression from *Q. Kelloggii,* which began with the hybrid swarms of Mio-Pliocene time, and has since been ordered into a regular, clinical series of variants by the selective action of the changing Pliocene and Pleistocene climates. Tucker (oral comm.) has suggested that *Q. Douglasii,* a completely unrelated oak with a similar geographical distribution, may have also originated from one or more hybrid swarms of a similar geological age. The modern variation pattern of the common chaparral species *Adenostoma fasciculatum* (Anderson, 1952) could be interpreted on the same basis, while less thorough observations by the junior author suggest that several other examples can be found in the California flora.

Conditions favorable for the origin and spread of hybrid derivatives are made not only by changes in the inanimate environment, but also by the advent and disappearance of various types of animals. Previous to man and his associated domesticates, some of these disturbances were as follows. In the Eocene and Oligocene periods, large grazing mammals made their first appearance on the earth. Their effect on the woody vegetation cannot be detected in the fossil record, and probably was not great. The herbaceous plants, however, must have been greatly affected by their inroads, and if these smaller plants had been abundantly preserved as fossils, we might be able to record a burst of evolution in them during these early Tertiary epochs. Babcock (1947, p. 132), after careful consideration of all lines of evidence, has suggested the latter part of the Oligocene as the time of origin of the genus *Crepis,* one of the larger, more specialized, and probably more recent genera of Compositae. On this basis, one might suggest that the greatest period of evolution of genera in this largest of plant families was during late Eocene and Oligocene time. The junior author, from his studies of various grass genera of temperate

North America, believes that many facts about their present distribution patterns could best be explained on the assumption that they began their diversification during the Oligocene epoch. They appear to have attained much of their present diversity by the middle of the Miocene, by which time many of the now extensive polyploid complexes, such as those in *Bromus, Agropyron,* and *Elymus,* had begun to be formed. The extensive Miocene record of species belonging to the relatively advanced tribe Stipeae (Elias, 1942) would support such an assumption.

At an earlier period, namely the beginning of the Cretaceous, the world saw for a relatively short time the dominance of the largest land animals which have ever existed, the great herbivorous dinosaurs. These monsters must have consumed huge quantities of the fern and gymnosperm vegetation which prevailed at the time, and it is difficult to see how these plants, with their relatively slow growth and reproduction, could have kept up with such inroads. It is very tempting, in fact, to speculate that over grazing on the part of giant dinosaurs contributed toward the extinction of the Mesozoic gynmospermous vegetation, as well as of the larger dinosaurs themselves, during the middle of the Cretaceous period. At the same time, shallow epicontinental seas were advancing and retreating, leaving coastal plain areas open for plant colonization; other significant events during this period were the rise of modern birds and of Hymenoptera, particularly bees.

The writers venture to suggest that these four nearly or quite concurrent events—retreat of seas, overgrazing by dinosaurs, advent of a diversified avifauna which transported seeds long distances, and rise of flower pollinating bees and other insects—all contributed to the greatest revolution in vegetation which the world has even seen; the replacement of gymnosperms by the predominant angiosperm flora of the upper part of the Cretaceous period. One should note that all of these conditions would favor hybridization and the spread of hybrid derivatives, by giving unusual opportunities for previously separated types to be brought together by wide seed dispersal, by permitting cross pollination between types previously isolated from each other, and by opening up new areas for colonization by the hybrid derivatives. The suggestion has been made elsewhere (Stebbins, 1950, p. 363) that differentation of genera and sub-families among primitive angiosperms took place partly via allopolyploidy; the time of origin of this polyploidy may well have been during the Cretaceous period. Evidence of introgression at so remote a time is probably impossible to obtain; by an analogy we should assume that in the past as now, conditions favorable for allopolyploidy also promoted introgression.

Going still further into the past, let us speculate on the events which must have taken place at the time when vascular plants and vertebrates first spread over the land. The principal geological period involved is the Devonian. At the beginning of this period comes the first extensive fossil record of vascular plants, all belonging to the primitive order Psilophytales. By the end of the Devonian, forms recognizable as club mosses (Lycopsida), ferns (primitive Filicales), and seed plants (Pteridospermae) were already widespread. We shall, of course, never know what chromosome numbers existed in these extinct groups of primitive vascular plants. But their nearest living descendants are nearly all very high polyploids, as has now been most elegantly demonstrated by Manton (1950, 1953). She has suggested that the living Psilotales, which have gametic numbers of about 52, 104, and over 200, "are the end-products of very ancient polyploid series which date back to simple beginnings. . . ." The relationship between the modern Psilotales and the Devonian Psilophytales is not clear, but to the present authors they appear to resemble each other nearly enough so that they could belong to the same complex

network of allopolyploids, which developed its greatest diversity in the Devonian period. In the genus *Ophioglossum,* generally regarded as one of the three most primitive genera of true ferns, we have the highest chromosome numbers known to the plant kingdom, namely n = ca. 256 in the northern *O. vulgatum* and n = ca. 370 in the tropical *O. pendulum.* These ferns are not preserved in the fossil record because of their soft texture, but their origin during the Devonian period is a fair inference. There is very good reason to believe, therefore, that the great proliferation of genera and families of vascular plants during this earliest period of their dominance was accompanied by allopolyploidy just as it has been in the more recent periods of very active evolution. Where allopolyploidy was widespread, we can also suspect abundant introgression.

The reader may well ask at this point whether any of this evidence contributes to the central theme of the present discussion, namely the hypothesis that these extensive hybridizations, both ancient and relatively modern, gave rise to really new types, which formed the beginnings of families, orders, and classes having different adaptive complexes from any plants previously existing. It is undoubtedly true that the results of introgression and allopolyploidy are chiefly the blurring of previously sharp distinctions between separate evolutionary lines, and the multiplication of variants on adaptive types which were already established during previous cycles of evolution. Nevertheless, the fact must not be overlooked that conditions favorable for introgression and allopolyploidy, namely the existence of widely different and freely recombining genotypes in a variety of new habitats, also favor the establishment and spread of new variants. Establishment of new adaptive systems is under any circumstances a relatively rare event; in any group of organisms we have hundreds of species and subspecies which are variants of old adaptive types to one which represents a really new departure. Hence we cannot expect to recognize introgressive or polyploid complexes which have given rise to such new types until we have carefully analyzed hundreds of those which have not. Furthermore, our methods of recognizing these complexes almost preclude the chance of identifying the new types which have arisen from them. We make the assumption that hybrid derivatives, whether introgressants or allopolyploids, have characteristics which can all be explained on the basis of intermediacy between or recombination of the characteristics of the putative parents, and then devise methods of verifying this assumption. The new types, falling outside of this assumption, would be rejected by our methods.

The junior author can suggest two examples known to him of new and distinctive morphological characteristics which may have arisen in recent hybrid derivatives. One of these is the presence in *Ceanothus Jepsonii,* a species narrowly endemic to the serpentine areas of northern California, of flowers with six and seven sepals, petals, and anthers (Nobs, 1951). This characteristic is not known anywhere else in the family Rhamnaceae or even in the entire order Rhamnales, an order which almost unquestionably dates back to the Cretaceous period. Mason (1942) has given strong evidence for the recent origin of *Ceanothus Jepsonii.* It inhabits an environment which is certainly recent, since the mountains on which it occurs were covered by a thick layer of volcanic rocks even as late as the end of the Pliocene epoch, and the serpentine formations to which it is endemic were not exposed until after the faulting which occurred at the beginning of the Pleistocene (Mason, 1942). It belongs to a complex of closely related species and subspecies, among which hybridization is still very actively taking place (Nobs, 1951). In characteristics other than sepal and petal number, it is intermediate between various ones of its relatives rather than an extreme type.

Hence there is a good reason to suspect that *Ceanothus Jepsonii* represents a species of relatively recent (i.e. Pleistocene) hybrid origin which has evolved a morphological characteristic previously unknown in its family, and in fact one which is relatively uncommon in the entire subclass of dicotyledons.

The other example is in the grass species *Sitanion jubatum.* This species is distinguished by possessing glumes which are divided into a varying number of linear segments, a characteristic not found elsewhere in the tribe Hordeae, and one which is the basis of a distinctive mechanism for seed dispersal (Stebbins, 1950, p. 141). *Sitanion jubatum* is endemic to Pacific North America, being most abundant in the coast ranges and foothills of central California. Its nearest relative is *S. hystrix,* a species found in the montane areas of the same region, and extending far eastward and southeastward. The two species are distinguished only by the degree of division of the glumes, and in fact appear to grade into each other. Field observations suggest that they actually consist of a large swarm of genetically isolated microspecies, such as has been demonstrated experimentally to exist in the related *Elymus glaucus* (Snyder, 1950).

Cytogenetically, both species of *Sitanion* are allotetraploids (Stebbins, Valencia, and Valencia, 1946). Extensive chromosome counts from various parts of the ranges of both species plus still more numerous measurements of sizes of pollen and stomata have failed to reveal any form of *Sitanion* which could be diploid. Furthermore, the chromosomes of both species are strongly homologous with those of *Elymus glaucus,* as evidenced by complete pairing in the $F_1$ hybrid. All of this evidence suggests that *S. jubatum* did not have any diploid ancestors which possessed its distinctive glumes, but has evolved out of a complex of allopolyploids which has existed in western North America for a long time, probably since the middle of the Tertiary period.

The forms of *S. jubatum* which have the most extreme division of the glumes occupy habitats which are recent, and which in some ways are intermediate between the most extreme habitats occupied by *Sitanion* and those characteristic of *Elymus glaucus*. They are known from the shore of San Francisco Bay, in northeastern Marin County, from the eastern edge of the Sacramento Valley north and east of Sacramento and from the Sierra foothills in Mariposa county. In growth habit, these races of *S. jubatum* could be regarded as intermediate between the most extreme xerophytes found in *Sitanion* on the one hand, and *E. glaucus* on the other.

Experimental evidence (Stebbins, unpublished) has now indicated that the complex of microspecies within the taxonomic species *Elymus glaucus* originated partly if not entirely through introgression. The probability is strong that *Sitanion* consists of a similar swarm of microspecies which originated also by introgression. The extensive subdivision of the glumes in some of these microspecies, therefore, may well have originated through the establishment of new mutations, or of new types of gene interaction, in genotypes produced by hybridization and introgression between morphologically very different and genetically well isolated species.

When all of this evidence has been considered, the writers can hardly escape the conclusion that hybridization in disturbed habitats has produced the conditions under which the more familiar processes of evolution, mutation, selection, and the origin of reproductive isolation barriers, have been able to proceed at maximum rates. Far from being insignificant because much of it is in habitats greatly disturbed by man, the recent rapid evolution of weeds and semi-weeds is an indication of what must have happened again and again in geological history whenever any species or group of species became so ecologically dominant as greatly to upset the habitats of their own times.

## SUMMARY

(1) It has been established by recent work in Palaeontology and Systematics that evolution has not proceeded at a slow even rate. There have instead been bursts of evolutionary activity as for example when large fresh water lakes (Baikal, Tanganyika, and Lanao) were created *de novo*.

(2) Recent studies of introgression (hybridization and subsequent back-crossing) have demonstrated that under the influence of man evolution has been greatly accelerated. There has been a rapid evolution of plants and animals under domestication and an almost equally rapid evolution of weed species and strains in greatly disturbed habitats.

(3) The rapidity of evolution in these bursts of creative evolution may well have been due to hybridization. At such times diverse faunas and floras were brought together in the presence of new or greatly disturbed habitats where some hybrid derivates would have been at a selective advantage. Far from being without bearing on general theories of evolution, the repeated demonstrations of accelerated introgression in disturbed habitats are of tremendous significance, showing how much more rapidly evolution can proceed under the impact of a new ecological dominant (in this case, Man). Such an agent may bring diverse faunas and floras into contact. Even more important is the creation of various new, more or less open habitats in which novel deviates of partially hybrid ancestry are at a selective advantage. The enhanced evolution which we see in our own gardens, dooryards, dumps and roadsides may well be typical of what happened during the rise of previous ecological dominants. The first vertebrates to enter isolated continents or islands, the first great herbivorous reptiles, the first herbiv-

orous mammals must have created similar havoc upon the biotae of their own times. Introgression must have played the same predominant role in these disturbed habitats as it does today under the impact of man. These arguments are supported by a homely analogy (page 379) and by various kinds of experimental and taxonomic data.

## LITERATURE CITED

ALAVA, REINO O. 1952. Spikelet variation in *Zea Mays L.* Ann. Mo. Bot. Gar., **39**: 65–96.

ALLAN, H. H. 1937. Wild species-hybrids in the phanerogams. Bot. Rev., **3**: 593.

ANDERSON, E. 1939. Recombination in species crosses. Genetics, **24**: 688.

——. 1948. Hybridization of the habitat. Evolution, **2**: 1–9.

——. 1949. Introgressive Hybridization. Wiley & Sons, New York, 109 pp.

——. 1952. The ecology of introgression in Adenostoma. Nat. Acad. Sci.: Abstracts of papers presented at the autumn meeting, Nov. 10–12, 1952 (Sci., **116**: 515–516).

——. 1952. Plants, Man and Life. Little, Brown & Co., Boston, 245 pp.

——. 1953. Introgressive hybridization. Biol. Rev., **28**: 280–307.

AXELROD, D. I. 1944. The Pliocene sequence in central California. Carnegie Inst. Wash. Publ., **553**: 207–224.

——. 1948. Climate and evolution in western North America during Middle Pliocene time. Evolution, **2**: 127–144.

BABCOCK, E. B. 1947. The genus *Crepis*. Part I. The taxonomy, phylogeny, distribution and evolution of *Crepis*. Univ. Calif. Publ. Bot., **21**: 1–198.

BROOKS, J. L. 1950. Speciation in ancient lakes. Quart. Rev. Biol., **25**: 131–176.

CLAUSEN, J., D. D. KECK AND W. HIESEY. 1940. Experimental studies on the nature of species. I. Effect of varied environment on western North American plants. Carnegie Inst. Wash. Publ., 520, vii, 452 pp., figs. 1–155.

COCKAYNE, L. 1923. Hybridism in the New Zealand flora. New Phytol., **22**: 105–127.

CONDIT, C. 1944. The Remington Hill flora. Carnegie Inst. Wash. Publ., **553**: 21–55.

DANSEREAU, P. 1941. Etudes sur les hybrides de Cistes. VI. Introgression dans la section Ladanium. Can. Jour. Research, **19**: 59–67.

—— AND Y. DESMARAIS. 1947. Introgression

in sugar maples. II. Amer. Midl. Nat., **37**: 146–161.

ELIAS, M. K. 1942. Tertiary prairie grasses and other herbs from the high plains. Spec. Papers Geol. Soc. Amer., **41**: 176 pp.

EPLING, C. 1947. Actual and potential gene flow in natural populations. Am. Nat., **81**: 81–113.

——. 1947. Natural hybridization of *Salvia apiana* and *S. mellifera*. Evolution, **1**: 69–78.

GRANT, V. 1953. The role of hybridization in the evolution of the leafy-stemmed *Gilias*. Evolution, **7**: 51–64.

HALL, M. T. 1952. Variation and hybridization in *Juniperus*. Ann. Mo. Bot. Gard., **39**: 1–64.

HEISER, C. B., JR. 1949. Natural hybridization with particular reference to introgression. Bot. Rev., **15**: 645–687.

——. 1951. A comparison of the flora as a whole and the weed flora of Indiana as to polyploidy and growth habits. Indiana Acad. Sci., Proc., **59**: 64–70.

MANGELSDORF, P. C., AND C. E. SMITH, JR. 1949. New archeological evidence on evolution in maize. Bot. Mus. Leaflets, Harvard Univer., **13**: 213–247.

MANTON, I. 1950. Problems of Cytology and Evolution in the Pteridophyta. 316 pp., Cambridge University Press.

——. 1953. The cytological evolution of the fern flora of Ceylon. Soc. Exp. Biol., Symp., **7**: Evolution, 174–185.

MASON, H. L. 1942. Distributional history and fossil record of *Ceanothus*. pp. 281–303 in Van Rensselaer, M., and H. E. McMinn. *Ceanothus*, publ. by Santa Barbara Bot. Gard., Santa Barbara.

NICKERSON, N. H. 1953. Variation in cob morphology among certain archaeological and ethnological races of maize. Ann. Mo. Bot. Gard., **40**: 79–111.

NOBS, M. 1951. Ph.D. Thesis, Univ. California, Library.

SAUER, C. O. 1952. Agricultural origins and dispersals. Am. Geogr. Soc., Bowman memorial lectures Ser. II, v. 110 pp., 4 pls. New York.

SIMPSON, G. G. 1944. Tempo and Mode in Evolution. xiii, 237 pp., New York.

——. 1953. The Major Features of Evolution. Columbia University Press, New York, 434 pp.

SNYDER, L. A. 1950. Morphological variability and hybrid development in *Elymus glaucus*. Amer. Jour. Bot., **37**: 628–636.

STEBBINS, G. L., JR. 1942. Polyploid complexes in relation to ecology and the history of floras. Am. Nat., **76**: 36–45, figs. 1–2.

——. 1950. Variation and Evolution in Plants. Columbia University Press, New York, 643 pp.

——, J. I. VALENCIA AND R. M. VALENCIA. 1946. Artificial and natural hybrids in the Gramineae, tribe Hordeae I. *Elymus, Sitanion* and *Agropyron*. Am. Jour. Bot., **33**: 338–351.

TUCKER, M. 1952. Evolution of the California oak *Quercus alvordiana*. Evolution, **6**: 162–180.

WOODSON, R. E., JR. 1947. Some dynamics of leaf variation in *Asclepias tuberosa*. Ann. Mo. Bot. Gard., **34**: 353–432.

——. 1952. A biometric analysis of natural selection in *Asclepias tuberosa*. Nat. Acad. Sci.: Abstracts of papers presented at the autumn meeting, Nov. 10–12, 1952 (Sci., **116**: 531).

ZIMMERMAN, E. C. 1948. Insects of Hawaii. Vol. I. Introduction. University of Hawaii Press, Honolulu, 206 pp.

# 3

Reprinted from *Evolution* **16**:348–363 (1962)

## CYCLIC HYBRIDIZATION AS A SURVIVAL MECHANISM IN THE NEW ZEALAND FOREST FLORA[1]

J. A. RATTENBURY

*University of Auckland, New Zealand*

Received October 6, 1961

The vegetation of New Zealand, more especially the tracheophyte element, has attracted the interest and aroused the curiosity of botanists since the time of Cook's voyages nearly two centuries ago. It has long been recognized (e.g., Colenso, 1844; Diels, 1896) that a number of aspects of the flora are remarkable in a temperate region, and discussion of these features has continued until the present day, especially in publications of Cockayne (1928), Allan (1940), and Millener (1960). These special features, while not unique to New Zealand, are exceptionally frequent there and concern both vegetative and reproductive parts of the plant. The high incidence of divaricating habit, microphylly, heteroblasty, epiphytism, white or inconspicuous flowers, conspicuous and edible fruits, anemophily, dioecism, hybridism, as well as the occasional appearance of cauliflory, ramiflory, and liane-habit, has led to the formulation of various hypotheses to explain some or all of the unusual aspects of the vegetation.

Cockayne (1928) attributes many of the unusual features of the flora to multiple invasions, mainly by land connections, from several quarters including a tropical element from Malaya and an Antarctic element from the south. Huxley (1942) suggests that the widespread hybridism may be correlated with the absence of herbivores. Fleming (1949) points out that the archipelagic structure of New Zealand, which has existed from early Tertiary times, could account for considerable

amounts of speciation. Millener (1960) states that many of the adaptive features of the vegetation reflect the past climatic history of the region. Oliver (1930) relates the prevalence of epiphytism to past tropical conditions. Heine (1937) shows clearly the correspondence between white and inconspicuous flowers and the native insect fauna, and points out that this is in turn related to dioecism and hybridization.

The present paper is an attempt to provide an hypothesis which will serve to explain all (or most) of the peculiarities of the flora in terms of its supposed evolutionary history from early Tertiary times. The problem is twofold, namely, to determine possible historical reasons for the existence of a diversity of adaptive types, and to explain in genetical and evolutionary terms how such diversity may have arisen and even, perhaps, been preserved by natural selection. Brief reference will be made to many genera, but a few clearcut examples will be described and illustrated in greater detail. Following a description of the kinds of variation which characterize the vegetation, an attempt will be made to relate the contrasting types to edaphic and/or climatic conditions. Then will follow a discussion of the breeding systems and modes of dissemination of the plants involved to show how diversity of this type might more readily have become established. Finally, reference to the climates of the Tertiary and Quaternary periods will be made in order to provide a basis for an hypothesis to account for the preservation of such large-scale diversity in the vegetation.

### POLYMORPHY IN THE NEW ZEALAND FLORA

The term *polymorphy* has been applied to the New Zealand vegetation (e.g., Allan,

[1] Modified from a paper presented at the Tenth Pacific Science Congress of the Pacific Science Association, held at the University of Hawaii, Honolulu, August 21 to September 6, 1961, and sponsored by the National Academy of Sciences, the Bernice P. Bishop Museum and the University of Hawaii.

31

1961) to describe the often quite astonishing extent of variation among members of the same interbreeding population, between forms of the same species from region to region or between species of the same genus where, as is frequently the case, interspecific hybridity occurs. In some cases, the polymorphy may be a manifestation of the genetic process known as *polymorphism*, as defined by Ford (1940). In other instances, particularly where there is considerable regional separation of the potentially interbreeding polymorphic forms, the process might better be described as *geographic polymorphism, regional polymorphism,* or *polytypicism*. There is ample evidence (Allan, 1961) that much of the variation described as polymorphic is environmental rather than genetic, but many of the most striking examples have been shown (Allan, 1940, and others) to be of the latter type. The term "polymorphy" as used in this paper merely describes the state of affairs that exists in the New Zealand vegetation, without implying any specific genetic mechanism.

Much of the variation that has been described as polymorphic, while it has a genetic basis, is highly variable in itself in the degree to which the polymorphs belong to the same species or occupy the same region. In many cases, the extent of the differences between the extreme types is so great that they are regarded as distinct species, despite their complete interfertility when brought together and the full viability of their offspring. *Melicope ternata* J. T. et G. Forst. and *M. simplex* A. Cunn. are both true-breeding when growing in isolation and are strikingly distinct in morphological features, *M. ternata* being an erect, openly branched tree with large, trifoliate leaves and *M. simplex* a divaricating shrub with small, unifoliate (adult) leaves. Even the naturally occurring $F_1$ hybrid is sufficiently distinct to have been given a specific name, *M. "mantelli."* Yet the segregation in the second generation produces a large hybrid swarm of intermediates and parental types (Allan, 1961). The three "species" are illustrated in fig. 1A.

Much the same situation holds for some 120 genera (30% of the tracheophyte flora —fig. 1). In a number of these, such as *Coprosma, Aristotelia, Hebe, Fuchsia, Celmisia, Lophomyrtus, Neopanax,* and *Plagianthus,* the hybrids have been given specific status before there had been a full appreciation of their origin and their degree of heterozygosity. As an extreme example, six of the seven New Zealand genera of the Tribe Inuleae (Compositae) are considered (Allan, 1961) to form natural intergeneric hybrids. The extraordinary extent of interspecific hybridization has been commented on from the time of Hooker (cf. Travers, 1869) and studied intensively by Cockayne (1929), Cockayne and Allan (1934) and Allan (1961, including bibliography of earlier studies). It may well be argued that true species are not involved here but rather subspecies in the biosystematic sense.

Many almost pure stands of these polymorphs exist in almost complete isolation, but frequently populations in nature are mixtures of these same species. Allan (1961) cites numerous examples of field observations which indicate that natural mixed populations of polymorphically varying species may have been comparatively common even prior to the widespread destruction of the forests by man (cf. for example, statements by Huxley, 1942; Anderson, 1953). The beech (*Nothofagus*) forests of Tongariro National Park in the North Island contain hybrid swarms involving six or more species of *Coprosma*, a source of considerable confusion to taxonomists, yet the effects of man have been negligible in that area since no significant destruction of the forests by human agencies has taken place.

The most obvious and spectacular difference between closely related and at least occasionally interbreeding groups of plants are those that affect the overall habit of the species and the size of its leaves. If one arbitrarily chooses the erect, openly spreading or even arborescent habit as "normal," departures from this condition may take the form of densely compacted shrubs with

32

FIG. 1. · Interfertile "species" of New Zealand plants. A. *Melicope ternata* (left—a single leaf) *M. simplex* (right) and the F₁ hybrid. B. *Lophomyrtus bullata* (left) *L. obcordata* (right) and some of the F₁ segregants. C. *Aristotelia serrata* (left) *A. fruticosa* (right). D. *Plagianthus betulinus* (left) *P. divaricatus* (right).

tortuously interlacing branchlets (the so-called "divaricating habit" shown in fig. 2) or of decumbent or even prostrate states. Some 50 species of woody plants in New Zealand are divaricating (Millener, 1960), although these are not all capable of interspecific hybridization. Of those which do interbreed with "normal" species, the best examples are in the genera *Coprosma*, *Melicope*, *Aristotelia*, *Sophora*, *Corokia*, *Plagianthus*, and *Neopanax*. In such genera as *Fuchsia*, *Coriaria*, *Pimelea*, and *Coprosma*, erect species cross with decumbent or prostrate ones.

Leaf size is, perhaps, the most frequent variable, and microphylly is always associated with divaricating habit and the prostrate state when these are present. It is, however, often the only conspicuous variable and the degree of size difference between the extreme types may be of the

order of one hundredfold. *Coprosma repens* A. Rich., for example, has leaves which may frequently be ten times as long and broad as *C. rhamnoides* A. Cunn. (see fig. 2B) yet these species hybridize in nature and produce the intermediate *C. "neglecta"* Cheesem. (Allan, 1961). Nearly half of the 45 New Zealand species of *Coprosma* form interspecific hybrids under natural conditions, leaf-size varying most frequently and conspicuously. Some 20 other woody genera show comparable size differences in their leaves between interfertile species (see fig. 1).

Another way in which leaf surface area can vary is through differences in leaflet number in plants with compound leaves. A good example is the genus *Weinmannia*. Adult specimens of *W. silvicola* Sol. are mostly trifoliate. In *W. racemosa* Linn. f. they are unifoliate. Leaflet number is

FIG. 2.   A. The divaricating habit in a *Coprosma* hybrid. B. Leaf-size differences in two *Coprosma* species (*C. repens,* above; *C. rhamnoides,* below).

greater in both species in the juvenile stages, but always two to three times as great in *W. silvicola.* The hybrids are intermediate. The same situation obtains in several other genera, notably *Melicope, Neopanax, Pseudopanax, Rubus, Sophora.* Frequently, smaller leaflet number is associated with smaller leaf size (see fig. 1A).

Total leaf surface may be reduced by varying amounts of dissection of the blade, as in species of *Neopanax* and *Clematis*; or the leaf-blade may be absent altogether as in some species of *Rubus, Clematis,* and *Carmichaelia.* Finally, the surface of the leaf may vary from species to related species in degree of pubescence (*Senecio, Metrosideros*) or extent of inrolling of the leaf margins (*Olearia*) or in the amount of cuticle present (*Coprosma*). Diels (1896) describes and illustrates many anatomical features of the leaves of New Zealand plants (e.g., thickness of the palisade layer) which suggest that comparative studies of interfertile species may demonstrate polymorphic variation in microscopic leaf and stem structures as well.

## ADAPTIVE SIGNIFICANCE OF VEGETATIVE POLYMORPHY

An examination of morphological variation exhibited by the numerous interfertile species groups referred to above, immediately suggests that the erect, open growth and the large, entire, glabrous leaves are mesomorphic features suiting the plants to a moist habitat and temperate climate; while the divaricating or sprawling habit and the small or dissected or pubescent leaves are more xeromorphic, adapting the organism to drier or cooler conditions. The presence of such adaptations in the New Zealand flora was commented on as early as 1896 by Diels who, in discussing certain derivatives of forest types (microphyllous, divaricating, prostrate forms), stated that in spite of the climate of New Zealand being less extreme in temperature and with fewer droughts than that of Central Europe, nevertheless the reduction in transpiration is no less great than among plants of the water-poor steppes. Considerable work has been done on New Zealand plants to demonstrate the existence of anatomical modifications of leaf and stem which are probably concerned with water retention (Diels, 1896; Tschirch, 1881; Finlayson, 1903; Herriott, 1906; McIndoe, 1932). So far there appears to have been no experimental work to determine to what extent those species with xeromorphic characters are better adapted to drier conditions than their genetically related mesomorphic counterparts.

Kramer (1959) states that deficiencies in the root absorption system can be very important in the production of water deficits in plants, even more so than losses by transpiration. He also points out that both aeration and temperature are factors in the operation of the root system. One would assume, therefore, that the cold climatic conditions coupled with water-logged soils and strong winds that are typical of stations at moderately high altitudes in winter in most parts of New Zealand, would greatly reduce the absorptive efficiency of the roots and at the same time cause considerable evaporation from the leaf surface. This is further accentuated, as Diels (1896) points out, by the fact that day temperature of the air may be quite high in these latitudes, even in winter. It is not surprising, then, to find a high proportion of xeromorphic forms at altitudes where winter soil temperatures are close to freezing as is the case, for example, in the beech forests of the Volcanic Plateau in the North Island.

To illustrate this correspondence between higher altitude and the preponderance of xeromorphic species, we may cite again Coprosma which, in these forests, at an altitude of about 4,000 feet, is present predominantly as small-leaved and divaricating or prostrate species. Fifteen hundred feet lower down, in milder surroundings, the large leaved, membraneous C. australis

is found. The same is true in this region for Myrsine, Aristotelia, Elaeocarpus, and others.

A rather similar picture obtains with latitude, the smaller leaved, divaricating or prostrate forms occupying the colder southern regions, while the large-leaved forms are confined in the south to protected coastal localities, many of them not extending as far as Stewart Island. In Coprosma, for instance, of 33 species with small leaves (4–40 mm in length) 28 or about 85% reach latitudes higher than 45° S while of the remaining 12 species with large leaves (55–200 mm) only three (25%) reach this latitude and only one extends to Stewart Island. Similarly the large leaved species of such genera as Hoheria, Weinmannia, Neopanax, Pseudopanax, and Corokia have lower latitudinal ranges than their more xeric counterparts. (See fig. 3.)

Evidence of this kind strongly reinforces the opinions of Diels (1896) and Millener (1960), who interpret the morphological and anatomical features of so many woody New Zealand plants as xeromorphic. Perhaps too little emphasis has so far been placed on the importance of these as adaptations to low soil temperatures and the resulting poor root absorption rather than to low soil moisture bringing about excessive transpiration losses.

While temperature, as a function of both altitude and latitude, may well be the most important factor in relating these adaptive forms to their environment, it is by no means the only deciding element. Water deficiency in the soil resulting from a number of other causes may sometimes play a part. The eastern slopes of the South Island ranges are markedly drier than the western sides. Within an 80-mile zone, mean annual rainfall may vary from 15 inches to 300 inches. The xeromorphic species of, for example, Pittosporum, Sophora, and Hymenanthera are predominantly east of the divide. Sand dunes throughout the country support the microphyllous, divaricating Coprosma acerosa A. Cunn., and the small leaved, creeping species of Fuchsia and Muelhenbeckia also are found in sandy

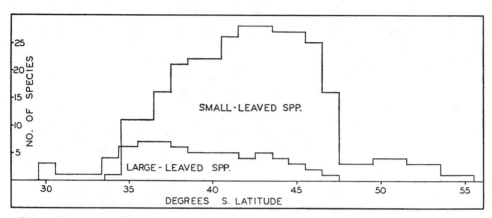

Fɪɢ. 3. North-south distribution of large- and small-leaved *Coprosma* species in New Zealand and adjacent islands. The abscissa represents the range of the species. Data from Allan, 1961.

or gravelly places. The physiologically arid salt marshes harbor the xeromorphic *Plagianthus divaricatus* J. R. & G. Forst. On the whole it appears that those species with xeromorphic adaptations have exploited those habitats where the availability of soil moisture is, for various reasons, relatively low. This is, of course, to be expected, and the evidence is presented here in support of the thesis that the widespread genetic variability is related almost exclusively to modifications affecting loss of water from the plant. Reduction in stomatal number, by decreasing the leaf size and interference with stomatal efficiency by various anatomical modifications of the leaf surface, are fairly obvious ways of reducing transpiration. The divaricating and prostrate states would appear to be effective in reducing the action of air movements, particularly strong winds, as a factor in the transpiration process. The densely compacted nature of most divaricating shrubs, especially since the leaves are often confined mostly to inner shoots, should act as a most effective windbreak. Whenever an extensive shrub layer of this type exists, as in some of the beech forests of both islands, virtually no wind movement is felt a few yards from the forest margins.

A special and quite remarkable case of polymorphy is that known as heteroblasty, in which juvenile and adult forms of a species differ strikingly in habit and leaf characters. In a number of cases (e.g., *Pseudopanax, Hoheria, Elaeocarpus*) heteroblastic and homoblastic species will hybridize naturally. The adult specimens are usually rather similar in each species pair, but in the juvenile stages the two may differ greatly, one member being much more xeromorphic in structure. Possibly the adaptation at the juvenile stage is in response to a poor root system in the early stages of growth. Heteroblasty is, however, of several kinds and there may be more than one explanation for its occurrence.

Finally, reference might be made to the considerable numbers of epiphytes in the flora. The epiphytic state may constitute the utilization of a suitable edaphic niche by xeromorphically modified species. When such species do occupy terrestrial sites, invariably they do so on well-drained or well-aerated terrain, as for example along ridge tops or on volcanic scoria. They may also become terrestrial at higher altitudes or latitudes. They are predominantly a feature of the northern part of New Zealand.

## ACHIEVEMENT OF GENETIC DIVERSITY

The wide scale polymorphy shown by the New Zealand tracheophytes, particularly the forest trees and shrubs, suggests that certain conditions for rapid evolution must have been nearly ideal. We see throughout the flora nearly isolated, almost pure-breeding populations which on the one hand have diverged so greatly in their morphological features as to be regarded by taxonomists as good Linnaean species, but which on the other hand have retained in a large number of cases (120 of the 391 genera) almost complete interfertility between some or all of the species. The problem of achieving and maintaining a satisfactory degree of isolation will be discussed in the next section. First let us examine those features of the plants which have made possible the genetic diversity.

Genetic diversity is a function both of mutation and recombination. For the latter to be at all important, sexual outcrossing must be achieved with a considerable degree of efficiency. At first glance, the New Zealand flora would seem to be poorly adapted to effect cross pollination. With very few exceptions, the angiosperms possess flowers lacking brilliant coloration. Most frequently they are white or quite inconspicuous. As Thomson (1880) and Heine (1937) have pointed out, the lack of colors, particularly reds and blues, is associated with the complete absence of long-tongued bees from the native insect fauna, most insect pollination being caried out by Diptera and to a lesser extent Lepidoptera and Coleoptera. Lack of bright color is further associated with absence of much floral specialization for hymenopterous cross pollination.

Outcrossing is, however, very effectively guaranteed by devices which reduce the frequency of selfing to a minimum. It is unlikely that self fertilization has been completely abandoned, since the establishment of isolates may have depended on its retention in the breeding system. Nevertheless, 14.5% of the seed plants are dioecious (E. J. Godley, personal communication) as against 2.0% for Great Britain (Lewis, 1952). In many other groups, protandry or protogyny effectively increase the outbreeding potential. Thomson (1880) estimates that nearly 50% of the New Zealand angiosperms are normally cross fertilized. Not surprisingly, wind pollination often accompanies the dioecious condition. Devices such as the massing of flowers into large and conspicuous inflorescences and the development of ramiflory are frequent, and serve to increase the efficiency of outcrossing both by insects and wind. Despite the absence of native long-tongued bees, cross pollination seems to be very effectively achieved and genetic diversity is correspondingly great.

Genetic recombination alone is not enough, however, for the preservation of many widely divergent types. In the absence of reproductive isolation, swamping will tend to occur with the result that the more extreme phenotypes will be rare. Diversity in the New Zealand vegetation has been achieved, probably, as a result of geographical isolation acting on a wealth of heterozygosity provided by the efficient outbreeding systems present there.

The speciation process, in the New Zealand forests at least, has resulted in the establishment, in quite a high percentage of the genera, of rather loosely knit complexes ("Rassenkreisen") whose components are perhaps more properly called "subspecies." The situation resembles the "recombinational reticulate speciation" of Huxley (1942). The genetic mechanism is probably not unlike the "homoselection" proposed by Carson (1959) where "small, ecologically marginal, inbred populations of a polytypic species [are] characterized by a tendency toward fixation of alleles or chromosome structures and the attainment of adjustment based on specific, genetically-fixed, adaptive features." Carson indicates that this kind of selection may further be characterized by random drift. Species formation is promoted. The presence in the New Zealand forests of many, partially isolated populations among which gene exchange is effectively, if infrequently, accomplished should be ideal for rapid evolution and adaptive radiation (Wright, 1940). The special conditions favoring partial isolation in New Zealand are discussed below.

While heterozygosity may be a feature of large, freely interbreeding populations, recombination in such groups does not usually produce markedly distinctive segregants. The interchange of the large groups of genes, controlling physiologically adjusted cell processes, that is necessary to effect profound changes in a population seems to be a feature of introgressive hybridization. In the words of Anderson and Stebbins (1954) "Natural selection is presented not with one or two new alleles but with segregating blocks of gene material belonging to entirely different adaptive systems." That hybridization in the New Zealand flora has been of this larger scale, introgressive type is borne out by the extent of the differences involved as well as by hybrid analyses (Cooper, 1954a, b; 1958, and unpublished student exercises at Auckland and Wellington). Introgressive hybridization requires initial isolation and differentiation of populations which ultimately come together again marginally and produce important new recombinations. The problem consists essentially of finding a system which will promote, on a fairly permanent basis for many species, the establishment of isolated, potentially interbreeding populations.

## Achievement of Isolation

If isolated segregants of a plant species are to be established and allowed to differentiate, two conditions must be satisfied. In the first place, the propagules must be capable of widespread distribution; and secondly, this distribution process must not be too frequently effective. Isolation cannot be too rare an event nor should it be complete, but swamping has to be prevented. Long range dissemination is best achieved by wind or by animals, particularly birds. Seed dispersal by water along sea coasts and river courses may also be effective in some cases. *Sophora* and *Entelea* may spread in this way to judge by their present distribution and the structure of their diaspores. *Nothofagus* also appears to be water distributed (Holloway, 1954).

If we confine our attention to those New Zealand tracheophytes in which interspecific hybridism is believed to occur (data from Allan, 1961; Cockayne and Allan, 1934), we find that, of 120 genera, 61

develop succulent or conspicuous fruits, hooked seeds, or are otherwise to some degree capable of being disseminated by birds, 39 produce large numbers of minute seeds or devices to assist in wind dispersal, and two are capable of transmission by water. Of the remaining 18 genera, 9 show little or no evidence of polymorphy. The conclusion from these data seems to be that polymorphy is linked very closely with a long range dissemination potential.

For genetic drift to be effective, isolation must be almost complete, and any mechanism for long distance dispersal cannot be so highly developed as to allow of frequent secondary colonizations of the same area. The conditions in New Zealand seem to have been almost ideal in this respect, at least up to the coming of man. The number of indigenous bird species which regularly feed on native berries is not great, and these are forest dwelling species like the pigeon, tui, and bell bird, which normally have no very extensive migratory habits. Winds within the forest are not very effectual, especially near the forest floor, except on rare occasions. Furthermore, the germination of seeds of many native species is a matter of extreme difficulty, as is evidenced by the sporadic appearance of seedlings which often seem to require special conditions for their development. Competent nurserymen have experienced great difficulty in germinating native seeds, often resorting to powerful treatments for breaking the dormancy. In many cases the viable period is very short.

Isolation must also involve barriers to cross fertilization between isolates. The fact that so many of the genera here considered have at least one and usually most of their species as inhabitants of the forest (see Cockayne, 1928) means that wind pollination over long distances occurs rarely. The insect pollinators (mostly forest inhabiting) are for the most part rather restricted in their ranges. It is only through widespread destruction of the forest and the introduction of many European and Asiatic bird and bee species by man that newly exposed marginal populations of true-breeding species of indigenous plants have come to interbreed on a large scale. This has resulted in the wholesale breakdown of barriers in such genera as *Coprosma*, *Epilobium*, and *Hebe*, which today present nearly insoluble tangles to the taxonomist (see, for example, Allan, 1940).

A further isolation-promoting circumstance has been the frequently archipelagic nature of the land-surface. During interglacial periods, particularly, many offshore islands were present. Fleming (1949) states ". . . we are justified in thinking of post-Jurassic New Zealand as being, what it is now, archipelagic in its geography" and ". . . this condition of a changing archipelago is one that would encourage 'speciation', i.e. the formation of two or more species by successive invasions, reinvasions, or back invasions of stock from one island to another, and this would account for genera with a multitude of species and for species in which complete physiological isolation . . . preventing interbreeding, had not been attained when the two daughter stocks came together again, so that they 'hybridized'." Most of the offshore islands today possess species and varieties of mainland genera which bear evidence of independent evolutionary trends. The examples of large leaved, offshore island forms is particularly noteworthy in *Hedycarya*, *Geniostoma*, *Macropiper*, *Rhabdothamnus*, and others.

### SELECTION AGAINST REPRODUCTIVE ISOLATION

The question will arise as to why genetic (reproductive) isolation has not developed subsequent to geographical separation. The answer is that it probably has done so in many genera, of which a great many component species have become extinct. Millener (1960) points out that the New Zealand angiosperm flora is on the whole a very meager one, particularly in terms of the average number of species per lowland genus (1.3 as against a world figure of 12.5). Among the New Zealand dicotyledons, there are some 75 genera whose propagules are apparently suitable for wind or bird dissemination but which are not suspected of hybridism. In two thirds of these only one, rather uniform species exists today in New Zealand and it is tempting to suggest that these represent survivors of a more ancient polymorphy. In others, such as *Alectryon*, *Myoporum*, *Solanum*, it would be surprising if hybridization were not possible when the species are brought together. In the remainder, speciation is by now probably complete.

Mather (1955) suggests that isolation (rather than polymorphy) results if "different forces of selection be effective [although] genetic exchange between the populations [need not be] completely absent, but [must] not rise above a certain maximum." It would appear that a kind of recombination threshold exists between polymorphy and isolation in the evolution of New Zealand plants, and that the maximum survival point lies just below this threshold. It is of considerable evolutionary interest that so many (30%) of the genera of vascular plants in New Zealand have apparently remained below the threshold of complete isolation but have nevertheless continued to be highly polymorphic.

It would seem likely, then, that reproductive isolation has not developed in this large segment of the vegetation because selection favored those groups that remained interfertile and, therefore, polymorphic. The rather surprising fact of an insular forest flora of relatively high latitude possessing a high incidence of diploidy as against the polyploid status of closely related species in lower latitudes and more continental areas to the north has been pointed out by Rattenbury (1961). The explanation is probably not (as that publication suggests) one of direction of migration, but rather that the situation is the consequence of natural selection in the New Zealand flora for the more interfertile diploid types.

### CYCLIC HYBRIDIZATION AS A SURVIVAL MECHANISM

A number of unusual features in the New Zealand flora have been described.

The high incidence of microphylly, divaricating habit, heteroblasty, and epiphytism appear to be adaptive features relating the plants to edaphic and climatic conditions of cold or aridity. The floral features promote outbreeding, and thus genetic variability, in the absence of long-tongued bees. The mobile diaspores and usually poor seed germination serve, in a forest environment, to promote isolation and genetic drift. There remain unaccounted for the exceptional amount of hybridism in the flora and the extent to which interbreeding taxa are differentiated.

The genetic variation in the New Zealand vegetation might be considered a transient feature associated with the speciation process, and that in time reproductive isolation would develop and less adaptive combinations be eliminated. If this were the case, why then is the amount of interspecific hybridization so great in New Zealand? Many of the genera involved date back to early Tertiary times at least. Why should speciation have only recently become widespread and divergent, or, if this is not the case, why has reproductive isolation not developed long since?

The alternative is to assume that continuous interfertility of morphologically very distinct species is a fairly stable feature of the vegetation, and that there has been an adaptive advantage to those organisms that have evolved in a manner permitting retention of the capacity to interbreed. It is the main purpose of this paper to present a hypothesis which will account for the persistence of such a breeding system within a flora over considerable periods of geological time. The hypothesis is based on the assumption that allelic combinations involving many genes have persisted in suitable habitats for long periods as distinct (Linnaean) species. Nevertheless, not infrequently, recombination has occurred and a fund of heterozygosity has been maintained in the overall breeding group. The evidence in support of this assumption is presented below.

In the simpler system of balanced polymorphism, a change in the selective action of the environment on the allelic pairs results in a shift in the balance such that the proportions of these alleles reach a new equilibrium. The same should apply to the larger scale variation in the New Zealand forests. As climatic changes occur, so should the proportions of adaptive ecotypes in the vegetation, provided always that the rate of gene exchange can keep pace with the climatic changes.

Climatic variation during the Pleistocene glacial and interglacial periods was sufficiently great to cause profound changes in the forest cover of New Zealand (see fig. 4). During the cooler parts of the cycle, *Nothofagus* forests covered most of the North Island while the South Island was almost devoid of forests. Couper and McQueen (1954) state that the Ohuka Creek (37°30′ S) mid-Pleistocene plant microfossils are "almost identical in species and relative abundance of species with plant microfossils of a recent core sample from Milford Sound (44°30′ S)." Since both stations are at sea level, this represents an equivalent latitudinal displacement of seven degrees. In the North Island, *Nothofagus* is confined at present to small areas at altitudes of 2,500 feet or more.

The frequent association of xeromorphic species with these beech forests today strongly suggests a wider distribution of these forms in the cooler parts of the Pleistocene. Similarly, the disappearance of the broad leaved *Nothofagus* species of the *brassi* group from the flora in early Pleistocene (Couper, 1960) indicates that mesic forms were more abundant in the milder Tertiary. There may, then, have been a kind of recurrent variation extending over periods of time of the order of magnitude of the duration of glacials and interglacials, but the presence of sufficient diversity of habitat, even during the maximum extent of Pleistocene cooling (to judge by the survival, apparently, of such frost-tender species as *Agathis australis* Salisb., *Ackama rosaefolia* A. Cunn., and others which do not now regenerate south of about 38° S)

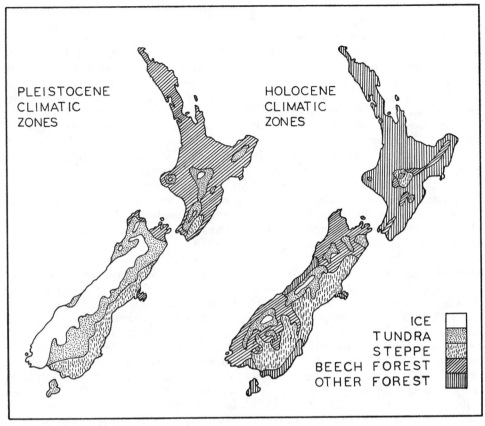

FIG. 4. New Zealand vegetation during the maximum Pleistocene glaciation (after Willett, 1950) and in pre-European times (after McLintock, 1959).

means that a wide range of polymorphic forms has been capable of simultaneous existence throughout Quaternary times. Only the proportions of these will have changed.

The term *cyclic hybridization* is proposed to describe the situation in which there is alternating fluctuation in the proportions of allelic combinations (and therefore of characters) in a group of interbreeding organisms, which corresponds with an approximately simultaneous alternation in environmental conditions such as, for example, the climatic fluctuations which occured in the Pleistocene. Wide variations are made possible by genetic contributions to the interbreeding group from different and often ecologically or geographically segregated forms. Cyclic selection could

apply equally well to the smaller scale of variation within a single population, where it might more appropriately be called *cyclic polymorphism* or cyclically variable selection. Cyclic polymorphism differs only in the duration of the cycles from the seasonally variable polymorphism known in various insect species having short life cycles. It differs both from "balanced polymorphism," where stability is present as a kind of genetic equilibrium, and from "transient polymorphism," in which certain alleles or combinations of alleles are gradually reduced to almost or quite negligible proportions in the population (cf. Ford, 1953).

The process of cyclic hybridization should not be confused with the concept of "differentiation–hybridization cycles" proposed by Ehrendorfer (1959). Ehrendorfer describes a cycle in which the early stages are characterized by differentiation through mutation, resulting in isolation, the differentiation later breaking down through hybridization. The resulting hybrid complex then begins a new cycle of differentiation–hybridization. He further suggests that allopolyploidy may be an important mechanism for overcoming barriers built up in the diploid differentiating stages. In cyclic hybridization, on the other hand, once adaptive forms have arisen they are perpetuated by an essentially *continuous* process of hybridization or recombination which is cyclic only in its consequences, i.e., in giving rise to environmentally controlled fluctuations in the proportions of adaptive types present in the interbreeding group. Polyploidy will generally be unfavorable to such a system, since it will interfere with both the efficiency of the outbreeding system and the expression of the recombinant types. It is important for cyclic hybridization that barriers be only partial, that is, ecological or geographical rather than reproductive.

From the foregoing discussion it is clear that the cyclic or recurrent variation described will act as a buffer to changes in the selective action of the environment, especially if the latter are themselves of an alternating or cyclic nature. Why then should such a genetic system have become so highly developed in New Zealand? The answer lies in the history and geography of the islands. New Zealand is isolated from the nearest sizable land masses (Australia to the west, New Caledonia to the north) by about 1,000 miles of ocean. This isolation has existed from early Tertiary times, and climatic fluctuations since the more or less tropical Miocene period must have imposed a severe strain on a flora of predominantly tropical structure. Unlike floras of the Northern Hemisphere, which were able to migrate in advance of the moving ice sheets of the Pleistocene, the New Zealand vegetation was faced in considerable degree with evolution or extinction.

Those lowland forest species that were able to evolve morphological variants suited to colder conditions have been successful in persisting until the present time; these variants are now more or less restricted to higher latitudes, higher altitudes, and special edaphic niches. That warm temperate forms also exist today, in dominant numbers in much of the forest region, is not likely to be the result of subsequent reverse mutation or new mutations so much as of recombinations of genetic materials that were present throughout the Quaternary. In other words, the forest forms have been reconstituted through a continuing process of hybridization and subsequent selection.

The evidence in terms of megafossil records during the Pleistocene is still, and may always be, inadequate to demonstrate a cyclic fluctuation, corresponding to glacial and interglacial periods, in the proportions of any given pair of polymorphic forms. Nevertheless, it seems reasonable to suppose (in view of the present distribution of genera like *Coprosma*, coupled with the knowledge that many of these genera have apparently continuous pollen records extending back at least into mid-Tertiary times) that some form of climatically induced alternation has indeed been present in the past, permitted perhaps by the process of cyclic hybridization, and that this mechanism, in some cases at least, has been important for the survival of the group.

The problem can, moreover, be approached indirectly by an examination of the fossil record of the Cenozoic in terms of relative predominance of small and large leaved species of *Nothofagus*. Couper (1960) suggested three infrageneric categories for *Nothofagus*: the *fusca, menziesii,* and *brassi* groups, distinguishable by their pollen morphology. All three groups were present continuously in New Zealand from mid-Eocene until early Pleistocene, when the *brassi* group disappeared along with a number of other more tropical elements. The present day New Zealand representatives of the first two groups are small

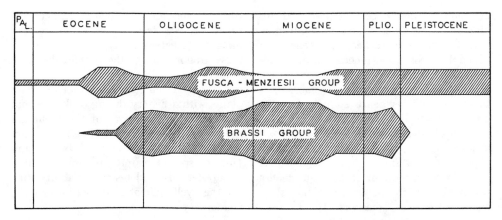

Fig. 5. Relative abundance of large-leaved (*brassi*-group) and small-leaved *Nothofagus* species during the Cenozoic. (See text for discussion.)

leaved (5–20 mm in length). The *brassi* group exists today in New Caledonia and New Guinea, and the leaves are considerably large (25–100 mm) (Langdon, 1947). These leaf size differences appear in macrofossil records.

Acording to Couper, abundant pollen of species of the *brassi* group is found in the Miocene and Pliocene beds, while the *fusca* and *menziesii* groups are rare in the Miocene and (with one exception) in the Pliocene. In the lowest Pleistocene, however, the proportions of these large and small leaved groups completely reverses, the former becoming very rare and soon disappearing, the latter remaining abundant until the present day. Fig. 5 is based on Couper's (1960) data, from which it will be seen that there is approximately an inverse abundance relationship between the large and small leaved groups. Unfortunately, no other groups of New Zealand plants are known to have a correspondence between leaf size and recognizable fossil pollen characters, and as climates are not as well understood for the Tertiary as for the Pleistocene, correlations are not as satisfactory as might be desired.

While megafossils of the *brassi* group occur throughout the period under discussion, there are no fossil leaves of the other groups recorded until the Pleistocene (information from a paper delivered by C. A.

Fleming, 10th Pacific Science Congress, Honolulu, 1961). This fact suggests either that the leaves of all *Nothofagus* species in the Tertiary were large, and indistinguishable from the *brassi* group, or alternatively that members of the small leaved groups were very restricted in their range, a conclusion not borne out by the pollen data.

No evidence exists that the large and small leaved groups of *Nothofagus* formed part of a single interbreeding system. All species in the *fusca* group do, however, hybridize freely today (Allan, 1961), and it is to this group that the *brassi* type is most nearly related (Langdon, 1947; Poole, 1953) both in morphology and pollen type.

### NONPOLYMORPHIC VASCULAR PLANT GROUPS

No theory can be said to be satisfactory unless exceptions can be shown not to negate it. The apparent exceptions to the hypothesis of evolution of New Zealand tracheophytes outlined above are more numerous than the cases in support. In other words, the hypothesis explains only the unusually high incidence of polymorphy, hybridism, and certain diaspore types. Even assuming that the 75 dicotyledon genera (together with many monocotyledons and vascular cryptogams) which produce wind or bird disseminated propagules (but which provide no present evidence for polymorphy) may once have possessed this kind of genetic diversity, we are confronted with 100 or more tracheophyte genera for which little or no evidence exists, either past or present, of polymorphic variation. It could well be argued that the survival of these groups is evidence enough that the process of cyclic hybridization is not a prerequisite to survival under New Zealand conditions even for those types which appear to show it.

Of the 80 or more dicotyledonous genera which neither are polymorphic nor show adaptive features (diaspore types) suggestive of possible past polymorphy, 80% are herbs almost exclusively of open, often alpine, areas and (from their distribution pattern) wide-ranging in their climatic tolerances. Only 10% are forest trees or forest dwelling shrubs. Many of these 80 genera, including all the true forest types, consist of one native species each, often of almost cosmopolitan distribution. The distribution of some of them along the shore lines and inland marshy areas and river beds suggests that migratory shore or water birds were responsible for their dispersal. This may mean that they are recent immigrants. Almost all are absent from the known microfossil record (Couper, 1960), and the fact that virtually none of them is dioecious favors their establishment by recent transoceanic dispersal (Baker, 1955). Much has been written about the importance of transoceanic dispersal (see Davis, 1950; Gordon, 1949). In New Zealand, the evidence favoring this method of plant distribution comes largely from strand plants and the predominantly herbaceous floras of the sub-Antarctic islands, where ocean currents, migratory birds and high winds (largely unimpeded by forest formations) seem to be mainly responsible.

Certainly the present day mountain species, numbering about 1,000, of which half are not found at lower levels (Millener, 1960), should have been capable of surviving without modification, since the combined opportunities provided by latitude (13 degrees, exclusive of the smaller islands) and altitude at the higher latitudes (10,000 feet or more at present) offer far greater chances for survival of these groups than of the present day lowland species.

The conclusion reached is that, of those plants with a history dating back well into the Pleistocene, survival of the mountain species has been mainly through migration within New Zealand and of the lowland (forest) species mainly through retention of interfertility and hence variability during the cyclic climatic fluctuations of the Quaternary.

## ALTERNATIVE EXPLANATIONS FOR POLYMORPHIC VARIATION IN THE NEW ZEALAND FLORA

As alternatives to the hypothesis of cyclic hybridization, it can be supposed (1) that the polymorphy is a transient aspect of speciation of ancient New Zealand plant groups which has occurred only in Holocene times; or (2) that the polymorphy represents adaptive radiation of plant groups recently introduced into New Zealand and which have exploited a wide variety of newly available edaphic niches.

The New Zealand geological record is sufficiently well known now (see, for example, Couper and McQueen, 1954) to permit us to see that climatic conditions comparable to the present have existed several times in the past and have followed or preceded periods of greater climatic extremes (tropical in the Miocene; cool-temperate to cold in the Pleistocene). There is every reason to suppose that similar evolutionary trends occurred in the past as a result of similar climatic changes. It is less justifiable to conclude, however, that evolutionary reversals have taken place, and one might assume that survival of each newly evolved form depended on its adaptation to a suitable, continuously available habitat and its subsequent propagation with little modification throughout the Pleistocene.

How then are we to explain the persistence of interfertility in so many of these highly variable groups? This is the real crux of the problem. It is difficult to explain the high incidence of hybridism unless we assume either that the segment of the flora possessing it is very young or that natural selection has favored the retention of interfertility.

The arguments against the recent introduction of plant groups, with subsequent rapid radiation, concern mainly their presence or absence in neighboring land masses. Naturally, one looks to Australia as an important source, especially as Australian dust and forest-fire smoke have been observed over New Zealand, and birds have been blown across the Tasman Sea from that continent. Oliver (1953) lists many plants, particularly from Australia, considered by him to be Tertiary and post-Tertiary introductions. Even in the earlier group, a large number of the genera still possess a single New Zealand species, and few exhibit polymorphy. Of the supposedly post-Tertiary introductions, only a few form hybrids and these are confined to nonforest species. The evidence is strong then, that new arrivals have not undergone rapid speciation resulting in widespread polymorphic variation.

Further arguments against the recent introduction of species which have given rise to polymorphic complexes are the high degree of functional unisexuality in these groups which would not be considered conducive to ready establishment of chance colonizers (Baker, 1955), and the low viability and poor germination qualities of the seeds.

## Summary

According to the hypothesis of cyclic hybridization advanced above, we may picture the history of the New Zealand forest vegetation somewhat as follows. Faced with the climatic cooling of the late Pliocene and early Pleistocene, the predominantly tropical flora of the northern forest communities, being unable to migrate to lower latitudes, was in danger of large scale extinction. Those groups with highly developed outbreeding systems and mobile diaspores were able, by a system of partial isolation favoring genetic drift, to evolve forms whose adaptations permitted them to survive under conditions equivalent to an increase in altitude of not less than 2,500 feet (800 m) or an increase in latitude of at least four degrees. Such modified species and varieties would be capable of survival by utilization of special ecological niches or by migration alone during the subsequent climatic changes of the remainder of the Pleistocene.

Nevertheless, that segment of the vegetation with a tropical facies has survived from Miocene times and is generally dominant today in lowland forests of most of the North Island and parts of the South Island. This may be construed as indicating that warm temperate species were able to survive even the severest of the glacial regimes in isolated pockets in the north. The fact, however, that at the present time very distinct species of a large segment of the forest flora and its derivatives can, and do, hybridize freely to produce fully viable and widely segregating offspring from their crosses, strongly suggests that survival of the more tropical elements has depended in a large number of cases on their cyclic reappearance as a result of genetic recombination. Certainly, there is no reason to suppose that many of them have re-established themselves from neighboring land masses since the last glacial advance, or to consider them in the early adaptive radiation stages of speciation.

### Literature Cited

Allan, H. H. 1940. Natural hybridization in relation to taxonomy. (The New Systematics, ed. J. S. Huxley) Oxford.
——. 1961. Flora of New Zealand, Vol. 1, Wellington.
Anderson, E. 1953. Introgressive hybridization. Biol. Rev., **28**: 280–307.
—— and G. L. Stebbins, Jr. 1954. Hybridization as an evolutionary stimulus. Evolution, **8**: 378–388.
Baker, H. G. 1955. Self-compatibility and establishment after "long-distance" dispersal. Evolution, **9**: 347–348.
Carson, H. L. 1959. Genetic conditions which promote or retard the formation of species. Cold Spring Harbor Symposia on Quant. Biol. XXIV: 87–105.

COCKAYNE, L. 1928. The vegetation of New Zealand (Die Vegetation der Erde, 14) 2nd Ed. Leipzig.
——. 1929. Hybridism in the forests of New Zealand. Acta For. Fenn., 34: 1–23.
—— AND H. H. ALLAN. 1934. An annotated list of groups of wild hybrids in the New Zealand flora. Ann. Bot., 48: 1–55.
COLENSO, W. 1844. Journal of a naturalist in some little known parts of New Zealand. Lond. J. Bot., 3: 1–62.
COOPER, R. C. 1954a. Pohutukawa × Rata. Variation in Metrosideros (Myrtaceae) on Rangitoto Island, New Zealand. Rec. Auckland Inst. Mus., 4: 205–212.
——. 1954b. Variation in Hebe (Scrophulariaceae) at Huia and Blockhouse Bay, New Zealand. Rec. Auckland Inst. Mus., 4: 295–308.
——. 1958. Pohutukawa × Rata No. 2. Variation in Metrosideros (Myrtaceae) in New Zealand. Rec. Auckland Inst. Mus., 5: 13–40.
COUPER, R. A. 1960. New Zealand Mesozoic and Cenozoic plant microfossils. N. Z. Geol. Surv. Palaeo. Bull. no. 32. Wellington.
—— AND D. R. McQUEEN. 1954. Pliocene and Pleistocene plant fossils of New Zealand and their climatic interpretation. N. Z. J. Sci. Tech. B., 35(5): 398–420.
DAVIS, J. H. 1950. Evidences of trans-oceanic dispersal of plants to New Zealand. Tuatara, 3: 87–97.
DIELS, L. 1896. Vegetations-Biologie von Neu-Seeland. Engl. Bot. Jb., 22: 202–300.
EHRENDORFER, F. 1959. Differentiation-hybridization cycles and polyploidy in Achillea. Cold Spring Harbor Symposia on Quant. Biol., 24: 141–152.
FINLAYSON, A. C. 1903. The stem-structure of some leafless plants of New Zealand, with especial reference to their assimilatory tissue. Trans. N. Z. Inst., 35: 360–372.
FLEMMING, C. A. 1949. The geological history of New Zealand. Tuatara, 2: 72–89.
FORD, E. B. 1940. Polymorphism and taxonomy. In The New Systematics (Ed. J. S. Huxley) pp. 493–513. Oxford.
——. 1953. The genetics of polymorphism in the Lepidoptera. Adv. in Genet., 5: 43–87.
GORDON, H. D. 1947. The problem of Sub-Antarctic plant distribution. Report of the 27th Meeting of Aust. & N. Z. Assoc. Adv. Sci., 27: 142–149.
HEINE, E. M. 1937. Observations on the pollination of New Zealand flowering plants. Trans. Roy. Soc. N. Z., 67: 133–148.
HERRIOTT, E. M. 1906. On the leaf-structure of some plants from the Southern Islands of

New Zealand. Trans. N. Z. Inst., 38: 377–422.
HOLLOWAY, J. T. 1954. Forests and climates in the South Island of New Zealand. Trans. Roy. Soc. N. Z., 82: 329–410.
HUXLEY, J. S. 1942. Evolution: The Modern Synthesis. London.
KRAMER, P. J. 1959. The role of water in the physiology of plants. In Water and its relation to soils and crops. Adv. in Agronomy, 11: 51–70.
LANGDON, L. M. 1947. The comparative morphology of the Fagaceae I. The genus Nothofagus. Bot. Gaz., 108: 350–371.
LEWIS, D. 1942. The evolution of sex in flowering plants. Biol. Rev., 17: 46–67.
MATHER, K. 1955. Polymorphism as an outcome of disruptive selection. EVOLUTION, 9: 52–61.
McINDOE, G. 1932. An ecological study of the vegetation of the Cromwell District, with special reference to root habit. Trans. Roy. Soc. N. Z., 62: 230–250.
McLINTOCK, A. H. (ed.) 1959. A descriptive atlas of New Zealand. Wellington.
MILLENER, L. H. 1960. Our plant world. N. Z. Junior Encyclopaedia. pp. 310–336, Melbourne.
OLIVER, W. R. B. 1930. New Zealand epiphytes. Jour. Ecol., 18: 1–50.
——. 1953. Origin of the New Zealand flora. Proc. 7th Pac. Sci. Congr., 5: 131–146.
POOLE, A. L. 1953. New data concerning the distribution of Nothofagus. Proc. 7th Pac. Sci. Congr., 5: 159.
RATTENBURY, J. A. 1961. Origins of the New Zealand flora: Cytogeobotanical observations on the 'Malayan element.' Darwin Centenary Vol. Roy. Soc. Victoria. Melbourne.
THOMSON, G. M. 1880. On the fertilization, etc., of New Zealand flowering plants. Trans. N. Z. Inst., 13: 241–288.
TRAVERS, W. T. L. 1869. On hybridization, with reference to variation in plants. Trans. N. Z. Inst., 1 (ed. 2, 31–35).
TSCHIRCH, A. 1881. Uber einige Beziehungen des anatomischen Baues der Assimilationsorgane zu Klima und Standort, mit specieller Berucksichtigung des Spaltoffnungsapparates. Linnaea, 43: 139–252.
WILLETT, R. W. 1950. The New Zealand Pleistocene snow line, climatic conditions and suggested biological effects. N. Z. Jour. Sci. & Technol. B., 32: 18–48.
WRIGHT, S. 1940. Breeding structure of populations in relation to speciation. Amer. Nat., 74: 232–248.

Part II

# EXPERIMENTAL HYBRIDIZATION STUDIES

# Editor's Comments
# on Papers 4, 5, and 6

**4**   **STEPHENS**
*The Cytogenetics of Speciation in* Gossypium. *I. Selective Elimination of the Donor Parent Genotype in Interspecific Backcrosses*

**5**   **GRANT**
*The Origin of a New Species of* Gilia *in a Hybridization Experiment*

**6**   **RAO and DeBACH**
*Experimental Studies on Hybridization and Sexual Isolation Between Some* Aphytis *Species (Hymenoptera: Aphelinidae). III. The Significance of Reproductive Isolation Between Interspecific Hybrids and Parental Species*

Given the potential of hybridization to alter the genetic composition of populations, it is important to recognize the relative paucity of recombination products from species crosses. Anderson (1939) demonstrated that in crosses between *Nicotiana langsdorfii* and *N. alata*, the hindrance to recombination was attributed to gametic elimination, zygotic elimination, pleiotropism, and linkage. The restrictive effect of linkage on recombination was found to be severe. However, tight coherences present in $F_2$ and $F_3$ *Nicotiana* hybrids may be broken to release variation in later generations, as shown in crosses between *N. langsdorfii* and *N. sanderae* (Smith and Daly, 1959). Strong character correlation is also evident in interracial hybrids within *Potentilla, Layia,* and *Madia* (Clausen and Hiesey, 1958), and subspecific hybrids in *Solidago* (Charles and Goodwin, 1943), *Gilia* (V. Grant, 1950), and *Antirrhinum* (Bauer, 1932).

Another expression of restricted recombination in species crosses lies in the large proportion of subvital and inviable segregates in the $F_2$ population. Stephens (Paper 4) finds that the viable fraction of $F_2$ hybrids of *Gossypium hirsutum* × *G. barbadense* consist primarily of plants resembling the parental species or the $F_1$ hybrid. In several other plant genera, $F_2$ and later-generation hybrids have frequency distributions of phenotypes skewed toward a parental type (e.g., *Lycopersicon*, Rick, 1963; *Tragopogon*, Winge, 1938; *Phaseolus*, Lamprecht, 1941, 1944; *Rubus*, Vaarama, 1954).

The differential elimination of donor parent genome in interspecific backcrosses is known in several plant genera. In backcross progeny of *Gossypium hirsutum* × (*G. hirsutum* × *G. barbadense*), Stephens (Paper 4, 1950) found significant elimination of the donor germplasm through both male and female gametes. Rick (1963) reported that backcrosses of the F₁ hybrid of *Lycopersicon esculentum* × *L. chilense* to *L. esculentum* contained an excess of *L. esculentum* markers and that backcrosses to *L. chilense* were deficient in such markers. Backcrossing maize × *Tripsacum* hybrids to maize is accompanied by the rapid elimination of maize genomes contaminated through crossing over with *Tripsacum* (Mangelsdorf, 1958), although there is evidence that inbred lines of maize may be modified and improved by the receipt of certain *Tripsacum* genes (Reeves and Bockholt, 1964). Studying two forms of leucine aminopeptidase, Wall (1968) reported that reciprocal backcrosses of *Phaseolus vulgaris* × *P. coccinea* F₁'s to both parents show a highly significant deviation from the expected 1:1 ratio when the donor allele is transmitted through the pollen but not through the egg. In *Lycopersicon*, the most probable cause of the skewed segregation ratios is a selective elimination of zygotes in the backcrossed generation, whereas in the other examples cited, the probable cause is selective abortion or nonfunctioning of gametes. Rick (1969) obtained hybrids between *Solanum pennellii* and *Lycopersicon esculentum* that were successively backcrossed to the latter, but monogenic ratios departed from 1:1, the deficiency of *esculentum* alleles (recurrent parent) being most frequent. This trend is contrary to the majority of observations in species hybrids, as exemplified in backcrosses of F₁ *Gossypium barbadense* × *G. hirsutum* (Paper 4). In Rick's (1969) study, linkage favors certain alleles but also preserves large blocks of the introgressant species. Such restriction of recombination would reduce the assortment of the germplasm of the two species and impede the transference of small gene blocks between species.

V. Grant (1967) formulated a concept of morphology-viability linkage to explain the skewedness noted above and in advanced-generation hybrids. The basic supposition is that genes controlling the phenotype and development form multifactorial constellations that differ allelically from species to species. These constellations must be transmitted intact if gametophytes or sporophytes produced during advanced-generation hybridization or backcrossing are to be viable. Morphology-viability linkage, or what Clausen and Hiesey (1960) call genetic coherence, preserves parental character combinations in spite of hybridization. Linkage (by whatever mechanism) delays the emergence of new character ensembles that may be superior to those of the parental taxa. With character cohesion, the results of advanced-generation hybridization may be much like those of introgression.

Several evolutionists have hypothesized that two taxa differing by two or more reciprocal translocations, and thus mostly sterile, could yield upon hybridization, recombination, and inbreeding new population systems that are on the same ploidal level as the parental systems but reproductively isolated from them. This process may be called recombinational speciation. The hypothesis has been experimentally verified in progeny of *Crepis tectorum* × *C. nova* (Gerassimova, 1939), *Nicotiana langsdorfii* × *N. sanderae* (Smith and Daly, 1959), and *Elymus glaucus* × *Sitanion jubatum* (Stebbins, 1957). The most comprehensive confirmation of recombinational speciation was provided by V. Grant (Paper 5) with derivatives of *Gilia malior* × *G. modocensis*. The *Gilia* study shows the emergence of a novel fertile, reproductively isolated lineage from a sterile hybrid by straight inbreeding for ten generations, with selection for vigor and fertility. In a companion paper on *G. malior* × *G. modocensis*, Grant (1966a) discusses the results of artificial selection in twenty-eight inbred lines derived from the $F_2$ generation. Eight lines passed through the bottleneck of sterility in the early generations and emerged with full vigor and fertility by the $F_{10}$ generation.

The possibility of recombinational speciation in animals has received less attention than in plants, but it has been noted with reference to *Drosophila* (Sears, 1947) and *Euschistus* (Sailer, 1953, 1954). The demonstration of fertile advanced-generation hybrids and reproductive isolation between hybrids and parental species in *Aphytis* by Rao and DeBach (Paper 6) clearly demonstrates the potential for recombinational speciation in animals.

# 4

Reprinted from *Genetics* **34**:627–637 (1949)

## THE CYTOGENETICS OF SPECIATION IN GOSSYPIUM. I. SELECTIVE ELIMINATION OF THE DONOR PARENT GENOTYPE IN INTERSPECIFIC BACKCROSSES

S. G. STEPHENS

*Texas Agricultural Experiment Station,[1] College Station, Texas*

Received February 14, 1949

THE cultivated amphidiploid species, *G. hirsutum* L. and *G. barbadense* L., cross readily and the $F_1$ hybrid shows regular bivalent pairing and is apparently fully fertile. $F_2$ progenies show considerable depression in vigor, and the net effect of inbreeding is the establishment of types practically indistinguishable from the parent species. All intermediate types are at a great selective disadvantage and fail to establish themselves. A similar situation is found in the hybrid progenies of the Asiatic cultivated diploid species, *G. arboreum* L. and *G. herbaceum* L.

HARLAND's interpretation of these phenomena (see HARLAND 1933, 1936, 1939) is that the species differences are mainly attributable to differences in "genetic architecture." Through natural selection an integrated system of modifier complexes is built up which is characteristic of each species. When different species are intercrossed, modifier segregation occurs in $F_2$ and later generations, with a consequent mutual disruption of the two internally balanced parental systems. Species differences on this basis would be independent of structural differentiation of the chromosomes, and the respective genomes could remain cytologically homologous throughout. The regular bivalent pairing and full fertility of the $F_1$ hybrid has been usually considered to support HARLAND's hypothesis (SILOW 1941, 1944; DOBZHANSKY 1941; MATHER 1943). For a contrary viewpoint see GOLDSCHMIDT (1940).

In a recent review (STEPHENS 1949b) the writer has suggested an alternative interpretation, and has cited lines of evidence from cytological, genetic and plant breeding sources, which show collectively that something more than multiple gene substitution has been responsible for the differentiation of these species. It has been pointed out that regular bivalent pairing is not a valid criterion of structural homology when the chromosomes are small and the chiasma frequency low. Preferential pairing in synthetic allotetraploids, selective elimination of genes from the donor parent in certain interspecific backcrosses, block transference of linked complexes in others, and the frequency of pseudo-allelic complexes, all suggest that the chromosomes of different species of *Gossypium* have manifold small scale structural differences which are not easily detected by cytological methods. For this type of differentiation, STEBBINS (1945, 1947) has used the convenient term "cryptic structural differentiation."

The evidence for cryptic differentiation is at present only to be found in data

---

[1] Department of Agronomy, Cotton Investigations Section. Technical Article No. 1193.

accumulated for other purposes, and it is important that tests now be set up for the specific purpose of determining its extent and importance in the speciation mechanism. Three possible tests have been briefly indicated (STEPHENS, *loc. cit.*), and it is the purpose of this paper to consider the results of only one of these.

## METHODS

If interfertile species have cryptic structural differences it is likely that crossing over between corresponding but only partially homologous chromatids in the $F_1$ hybrid will lead to the production of gametes containing several small deficiencies and duplications. In Gossypium species the typical situation during meiosis is for two chiasmata to be produced (one per chromosome arm)

TABLE 1

*Genetic constitution of parents and tester stocks*

| | Parents | | Backcross Testers | |
|---|---|---|---|---|
| hirsutum | | barbadense | 7 independent loci | 5 independent loci |
| $R_1$ Red plant body | $r_1$ | Green plant body | $r_1$ Green plant body | $r_1$ Green plant body |
| $R_2{}^{AF}$ Weak spot | $R_2{}^{AS}$ | Full spot | $R_2{}^{AO}$ Spotless | $R_2{}^{AS}$ Full spot |
| $p$ Cream pollen | $P$ | Yellow pollen | $p$ Cream pollen | $P$ Yellow pollen |
| $y$ Cream petal | $Y$ | Yellow petal | $y$ Cream petal | $Y$ Yellow petal |
| $K$ Brown lint | $k$ | White lint | $k$ White lint | $k$ White lint |
| $N$ Naked seed | $n$ | Not Naked (tufted) | $n$ Not Naked (fuzzy) | $n$ Not Naked (tufted) |
| $L^O$ Narrow leaf | | seed | seed | seed |
| $Cr^H$ "Low" Normal | $L^E$ | Intermediate leaf | $l$ Broad leaf | $L^E$ Intermediate leaf |
| | $Cr^B$ | "High" Normal | $Cr^H$ "Low" Normal | $cr^D$ Crinkled |

resulting in a ring bivalent at Metaphase I. In interspecific hybrids several of the paired chromosomes only form one chiasma and a rod bivalent results. A type of selection may therefore be envisaged in interspecific backcrosses which would tend to eliminate gametes carrying crossover chromatids, and favor gametes containing non-crossover chromatids. It is likely, too, that non-crossover chromatids of the *recurrent* parent type would have a selective advantage over non-crossover chromatids of the *donor* parent type, since KEARNEY and HARRISON (1932) have shown that selective fertilization occurs in favor of *barbadense* pollen when mixed *barbadense* and *hirsutum* pollen is applied to *barbadense* stigmas, and in favor of *hirsutum* pollen when the mixed pollen is applied to *hirsutum* stigmas. The net effect to be expected, therefore, would be a selective elimination of the donor parent genotype. This may be tested by introducing suitable marker genes into the two species under investigation, making an interspecific hybrid heterozygous for several independent gene pairs, and testing their segregations in reciprocal backcrosses. On the basis of random recombination none of the segregations obtained should differ significantly from a 1:1 ratio but if preferential elimination of gametes occurs the ratios should be skewed due to deficiencies in the classes containing genes from the donor parent.

**52**

The stocks chosen for experiment were (1) a line of Upland cotton (*G. hirsutum*) carrying six dominant and two recessive independent marker genes which has been maintained in the genetic collection at this station and (2) an inbred line of Seaberry Sea Island (*G. barbadense*) which carries contrasting alleles to those carried by the Upland stock at the same eight loci. The detailed constitutions of these stocks are shown in table 1. The interspecific hybrid obtained by crossing these two stocks was thus heterozygous for eight marker genes. Ideally the gametes of this hybrid should have been tested by crossing to multiple recessive stocks of both *barbadense* and *hirsutum*. Unfortunately such stocks were not available, and the *hirsutum* stock used (Deltapine 14) only tested seven out of the eight loci. The *barbadense* stock (Sea Island Crinkle) only tested five loci. The detailed constitutions of these tester stocks are also shown in table 1. Backcrosses, *i.e.* crosses to the tester stocks were carried out reciprocally ($F_1 \female \times$ tester $\male$ and tester $\female \times F_1 \male$)

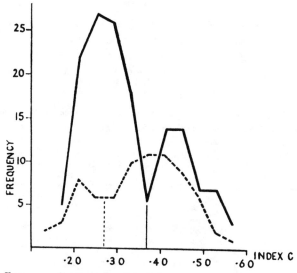

FIGURE 1.—Frequency distributions of leaf shape indices in the backcross, $F_1 \times$ *hirsutum* (solid line), and the backcross, $F_1 \times$ *barbadense* (broken line). For additional information see text.

No difficulty was encountered in scoring the backcrosses, as the alternative classes were sharply distinct and with the exception of the leaf shape segregations could be scored visually. In order to classify the leaf shape segregations a climax leaf from each plant was measured and its Index C value calculated according to SILOW's (1939) method. Frequency distributions of Index C values for both backcrosses (data from reciprocal backcrosses being pooled) are shown in figure 1. It can be seen that in both cases the frequency distributions are bimodal corresponding to the alternative leaf shape classes expected.

The point of minimum frequency was taken as the point of demarcation between genotypes, and the segregations were classified on this basis as follows:

|        |                                         | Class     | Index C        |
|--------|-----------------------------------------|-----------|----------------|
| (1)    | $F_1$ ($L^O/L^E$) × hirsutum ($l$)      | $L^O/l$   | $\cdots 0.16-0.36$ |
|        |                                         | $L^E/l$   | $\cdots 0.38-0.56$ |
| (2)    | $F_1$ ($L^O/L^E$) × barbadense ($L^E$)  | $L^O/L^E$ | $\cdots 0.12-0.26$ |
|        |                                         | $L^E/L^E$ | $\cdots 0.28-0.56$ |

A word of explanation is necessary about the Crinkle ($Cr$) locus. The alleles, $Cr^B$ and $Cr^H$ both give a normal phenotype but differ in dominance potency (HUTCHINSON 1946a). The heterozygote, $Cr^B/cr^D$, is phenotypically normal, while $Cr^H/cr^D$ is weakly Crinkled.

## RESULTS

The complete data for the backcrosses ($F_1$×hirsutum tester, $F_1$×barbadense tester and the corresponding reciprocals) are given in table 2. The segregations of the different allelomorphic pairs are listed in double columns, the left hand column in each case representing the gene from the hirsutum parent.

TABLE 2

*Monofactorial segregations in reciprocal backcrosses of the interspecific hybrid hirsutum×barbadense. Alleles from the hirsutum parent are listed in the left hand column in each case.*

|                                              | $R_1$ | $r_1$ | $R_2^{AF}$ | $R_2^{AS}$ | $p$ | $P$ | $y$ | $Y$ | $K$ | $k$ | $N$ | $n$ | $L^O$ | $L^E$ | $Cr^H$ | $Cr^B$ |
|----------------------------------------------|-------|-------|------------|------------|-----|-----|-----|-----|-----|-----|-----|-----|-------|-------|--------|--------|
| (a) h:rsutum ♀ ×$F_1$ ♂                      | 34    | 23    | 39         | 12         | 31  | 25  | 32  | 24  | 22  | 35  | 24  | 33  | 34    | 23    | —      | —      |
| F ♀ ×hirsutum ♂                              | 54    | 47    | 50         | 38         | 47  | 44  | 48  | 43  | 25  | 63  | 45  | 44  | 69    | 31    | —      | —      |
| (a) Combined data                            | 88    | 70    | 89         | 50         | 78  | 69  | 80  | 67  | 47  | 98  | 69  | 77  | 103   | 54    | —      | —      |
| Heterogeneity $\chi^2$                       | 0.555 |       | 4.988      |            | 0.191 |   | 0.268 |   | 1.436 |   | 0.994 |   | 1.270 |       | —      | —      |
| P (1)                                        | .30–.50 |     | .02–.05    |            | .50–.70 | | .50–.70 | | .20–.30 | | .30–.50 | | .20–.30 |     | —      | —      |
| $\chi^2$ (1:1)                               | 2.051 |       | 10.942     |            | 0.551 |   | 1.150 |   | 17.940 |  | 0.438 |   | 15.293 |      | —      | —      |
| P (1)                                        | .10–.20 |     | v. low     |            | .70–.80 | | .20–.30 | | v. low |   | .50–.70 | | v. low |       | —      | —      |
| (b) barbadense ♀ ×$F_1$ ♂                    | 3     | 10    | —          | —          | —   | —   | —   | —   | 6   | 6   | 5   | 7   | 6     | 7     | 4      | 9      |
| $F_1$ ♀ ×barbadense ♂                        | 37    | 27    | —          | —          | —   | —   | —   | —   | 21  | 36  | 28  | 29  | 13    | 51    | 22     | 42     |
| (b) Combined data                            | 40    | 37    | —          | —          | —   | —   | —   | —   | 27  | 42  | 33  | 36  | 19    | 58    | 26     | 51     |
| Heterogeneity $\chi^2$                       | 5.215 |       | —          | —          | —   | —   | —   | —   | 0.686 | | 0.221 |   | 2.887 |       | 0.056  |        |
| P (1)                                        | .02–.05 |     | —          | —          | —   | —   | —   | —   | .30–.50 | | .50–.70 | | .05–.10 |     | .80–.90 |        |
| $\chi^2$ (1:1)                               | 0.117 |       | —          | —          | —   | —   | —   | —   | 3.261 | | 0.130 |   | 19.753 |      | 8.117  |        |
| P (1)                                        | .70–.80 |     | —          | —          | —   | —   | —   | —   | .05–.10 | | .70–.80 | | v. low |       | .001–.01 |       |

Significant differences at 5 percent level shown in bold type. For evidence of allelomorphism of genes listed see HARLAND 1929a ($R_1$=R and $R_2$=S in his nomenclature), 1929b ($P$), 1929c ($Y$), 1935 ($K$), 1938 ($N$); STEPHENS 1945 ($L$); HUTCHINSON 1946a ($Cr$).

*The backcross, hirsutum tester×$F_1$, and its reciprocal.* The results of this backcross are shown in table 2 (a). Of the seven independent monofactorial segregations tested only one ($R_2^{AF}:R_2^{AS}$) shows a significant difference between reciprocal crosses. In this case $\chi^2$ for heterogeneity is just significant at the 5 percent level of probability. Inspection shows, however, that in both crosses there is a deficiency in the $R_2{}^{AS}$ class, and since the object of the experiment was to test the direction rather than degree of skewness, it seems permissible to pool the data from the two crosses. Considering, therefore, the combined data it can be seen that in five out of the seven segregations tested, the class containing the gene from the *hirsutum* parent is in excess of that expected on a 1:1 basis and in two of these five cases the deviation is highly significant, *viz.* in the segregations $R_2{}^{AF}:R_2{}^{AS}$ and $L^O:L^E$. In the remaining two segregations, a reverse situation is found—the classes containing genes from the *barbadense* parent are in excess and in one case ($K:k$) the deviation is highly significant.

*The backcross, barbadense tester* $\times F_1$, *and its reciprocal.* The results are shown in table 2 (b). Unfortunately only a small family of the cross *barbadense* tester ♀ $\times F_1$♂ was available for study, so that no adequate test could be made for possible reciprocal differences. One significant difference between reciprocal crosses was found, in the segregation $R_1:r_1$, where the ratios are skewed in opposite directions. However, the numbers in the first family are so small that they cannot materially affect the ratios obtained in the second family, so that the combined data may be considered as in the case of the *hirsutum* backcrosses in table 2 (a). Of the five segregations recorded, four show an excess in the class containing the gene from the *barbadense* parent and in two of these the deviation from a 1:1 ratio is highly significant, *viz.* in the segregations $L^O:L^E$ and $Cr^H:Cr^B$.

*Comparison between reciprocal crosses in the two backcrosses.* Reciprocal differences are only tested adequately in the backcross to *hirsutum*. Nevertheless within the limits of the material studied there seems to be no over-all tendency for selective elimination to be affected by the way in which the cross is made. More precisely there is no consistent tendency for selective elimination to be greater in the $F_1$ pollen than in the $F_1$ ovules. This does not imply that gametic selection is unimportant, and in fact there is some evidence that selective elimination in both male and female gametes rather than zygotic selection is mainly responsible for the skewed backcross ratios. It was noticed in making the backcrosses that when the $F_1$ was used as female about eight to ten percent of the seeds were abortive ("motes"), while when the $F_1$ was used as male not more than two percent abortive seeds occurred *i.e.* probably not more than would occur by chance in the self fertilized parental species. It is clear that if elimination occurred in the zygote the same proportion of motes should occur whichever way the cross was made, so the most likely situation is that elimination occurs both in the male and female gametes. Elimination in the former cannot be tested directly since there is no adequate technique available for testing the viability of cotton pollen (IYENGAR 1939; STEPHENS 1942) but the absence of heterogeneity in most of the reciprocal backcross segregations does suggest that elimination may be of the same order in male and female gametes.

*Comparison between backcross to hirsutum and backcross to barbadense.* In comparing the segregations obtained in backcrosses to *hirsutum* and back-crosses to *barbadense* it will be convenient to consider each allelic pair separately. The segregation $L^O:L^E$ (tables 1 (a) and (b)) gave highly significant deviations from a 1:1 ratio in both backcrosses, and it can be seen that in both cases there is a selective elimination of the allele carried by the donor parent. Thus in the backcross to *hirsutum* the $L^O$ class is in excess of the $L^E$ class while the reverse situation is found in the backcross to *barbadense*. This reversal shows that the deviations cannot be due to the effects of the alleles themselves, and suggests strongly that there is a selection against the chromosome segment from the donor parent which is marked by the particular locus. In other words it is consistent with the hypothesis that the corresponding leaf shape chromosomes of *hirsutum* and *barbadense* are structurally differentiated. This interpretation is supported by independent evidence (STEPHENS 1949b) that the leaf shape genes are apparently transferred with a block of linked genes in backcrosses of *hirsutum* to *barbadense* and vice versa.

In the case of the $Cr^H:Cr^B$ segregation (table 2 (b)) there is a marked deficiency of the gene from the donor parent in the backcross to *barbadense*. Owing to the absence of a suitable tester stock no comparable data could be obtained in the backcross to *hirsutum*. However, HARLAND's data (1932) do provide this information. They show that in a first backcross of *barbadense* to *hirsutum* of the constitution $Cr^H/cr^D(F_1) \times Cr^H$ (*hirsutum*) the class containing the gene from the donor parent is markedly deficient. It seems likely that the Crinkled chromosomes of *hirsutum* and *barbadense* may also be structurally differentiated. Again this interpretation is supported by independent evidence, as the Crinkled "locus" and the closely linked Corky "locus" are suspected to constitute a complex of pseudo-alleles (STEPHENS 1949a), and the strength of the linkage between Green lint and Crinkled is altered after transference from *barbadense* to *hirsutum* (STEPHENS 1949b).

In the backcross to *hirsutum* (table 1 (a)) the segregation $R_2^{AF}:R_2^{AS}$ shows selective elimination of the gene $R_2^{AS}$ from the donor parent. This situation is in agreement with HARLAND's early data (1929a) where a 4th backcross involving the same locus, *viz. hirsutum* spotless $\times$ (*hirsutum* spotless $\times$ *barbadense* full spot), segregated 579 spot : 706 spotless—a highly significant deficiency of the donor parent class. Unfortunately no comparable data are available for the backcross to *barbadense*.

The only significant deviation from a 1:1 ratio which cannot be interpreted as the result of selective elimination of a gene from the donor parent, occurs in the segregation $K:k$ in the backcross to *hirsutum* (table 1 (a)). Here there is a highly significant deficiency of $K$ types, *i.e.* in this case the gene from the donor parent has an apparent selective *advantage* over its allele from the recurrent parent. No explanation can at present be offered for this phenomenon. The segregating types were readily classified and there appears to be no suggestion in previous literature (WARE 1932; HARLAND 1935; HUTCHINSON 1946b) that the white lint gene *per se* has any selective advantage over its brown lint alleles. In the corresponding backcross to *barbadense*, there is a tendency which almost reaches significance for the $K$ class from the donor parent to be deficient.

In none of the other segregations studied were there significant deviations from the expected monofactorial ratios, though in the case of $R_1:r_1$ and $y:Y$ in the backcross to *hirsutum* there is a tendency for the donor parent class to be deficient.

*The cumulative effect of selective elimination of the donor parent genotype.* Inspection of tables 1 (a) and (b) shows that out of the total of twelve backcross ratios tested nine show a deficiency in the class containing the gene from the donor parent, though in only four of these does the deficiency reach significance. If selective elimination were dependent on structural differentiation, it would seem that only the larger or more extensive structural differences would

TABLE 3

*Frequencies of plants carrying various numbers of hirsutum "marker" genes in the backcrosses (a) $F_1 \times hirsutum$, (b) $F_1 \times barbadense$.*

| NUMBER OF *hirsutum* MARKERS | 7 | 6 | 5 | 4 | 3 | 2 | 1 | 0 | TOTAL |
|---|---|---|---|---|---|---|---|---|---|
| (a) Expected $(\frac{1}{2}+\frac{1}{2})^7$ | 1.09 | 7.60 | 22.80 | 38.01 | 38.01 | 22.80 | 7.60 | 1.09 | 139 |
| Actual | 3 | 9 | 24 | 49 | 32 | 13 | 8 | 1 | 139 |
| $\chi^2$ | 3.35 | 0.26 | 0.06 | 3.18 | 0.95 | 4.21 | 0.02 | 0.01 | 12.04 (P(6) =0.05–0.10) |

N.B. Pooling classes with frequencies less than 5 gives $\chi^2 = 9.67$, P(4) =0.02–0.05

| (b) Expected $(\frac{1}{2}+\frac{1}{2})^5$ | — | — | 2.16 | 10.78 | 21.56 | 21.56 | 10.78 | 2.16 | 69 |
| Actual | — | — | 1 | 4 | 16 | 25 | 18 | 5 | 69 |
| $\chi^2$ | — | — | 0.62 | 4.26 | 1.43 | 0.55 | 4.84 | 3.73 | 15.43 (P(4) =0.001–0.01) |

N.B. Pooling classes with frequencies less than 5 does not alter the probabilities appreciably.

lead to a significant disturbance in any particular monofactorial segregation In other cases one might expect that the structural differences might be so small that their selective effect would be swamped by random fluctuations, so that they would not be detected in any particular single gene segregation. Nevertheless the cumulative effect of many slight structural differences could be of considerable importance. The easiest way to demonstrate a cumulative effect is to compare the proportion of plants in which the parental genotypes *as a whole* are recovered with the proportion expected on the basis of random recombination. The latter is given by the binomial expansion of $(1/2+1/2)^n$ where n is the number of genetically marked loci. Thus in the backcross to *hirsutum* in which seven loci are marked the number of plants with 7, 6, $\cdots$ 0 independent genes from the *hirsutum* parent should be given by the successive terms in the expansion of the binomial $(1/2+1/2)^7$. The corresponding formula for the backcross to *barbadense* is $(1/2+1/2)^5$. The actual numbers obtained are compared with those expected on the basis of random recombination in table 3 (a) and (b) and in figure 2. It can be seen that in both backcrosses there are significant deviations from the expected values, and inspection shows that in the backcross to *hirsutum* the genes from the *hirsutum* parent are recovered *more* frequently than expected. In the backcross to *barbadense* the genes from the *hirsutum* parent appear much *less* frequently than expected. The distributions are, therefore, skewed in opposite directions instead of being symmetrically distributed around the median class. Since the loci studied which mark only seven out of twenty-six chromosome pairs exhibit such a marked cumulative effect on the recovery of the parental genotypes, it is clear that the total cumulative effect of the differential segments present throughout the chromosome set must be of a very high order indeed, and certainly sufficient to account for the rapid recovery of the recurrent parent genotype in successive backcrosses which is observed in breeding experiments (KNIGHT 1945).

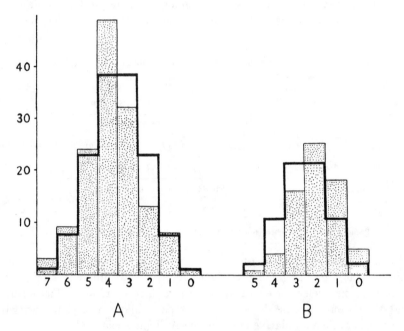

FIGURE 2.—Frequency histograms showing the numbers of plants carrying 0, 1, 2 etc. *hirsutum* marker genes (A) in the backcross to *hirsutum*, (B) in the backcross to *barbadense*. Dotted areas represent the observed frequencies, while the areas enclosed by heavy lines indicate frequencies expected on the basis of random recombination. Data are the same as those shown in table 3.

## DISCUSSION

The results of this experiment show that there is a selective elimination of the donor parent genotype in interspecific backcrosses, to whichever parent the backcross is made. The fact that this phenomenon can be demonstrated in small populations in which so few of the chromosomes are marked suggests strongly that the cumulative selective effect on the parental genotypes as a whole must be very great. This agrees with KNIGHT's (1945) statement that after only two or three backcrosses the progenies closely resemble the recurrent parent. Two obvious questions arise from a consideration of the data. When is the selection operative—in the gametes or zygotes? What is the mechanism responsible for the elimination?

The fact that very few abortive seeds were found when the $F_1$ was used as male parent and that practically a complete stand of plants was obtained from the seed collected, shows that zygotic elimination is not an important factor in the first backcross. It might, however, be very important in later backcross generations. About eight percent of all the first backcross plants failed to produce flowers but remained in a vegetative condition. Among the rest of the plants there were considerable differences in time of flowering—some of the plants were harvested before flowering occurred in others. The range of flowering period was much greater than occurs in either parent species. The late flowering types, therefore, in addition to those which did not flower at all, would probably contribute no progeny to the following generation, *i.e.* considerable zygotic elimination would be expected between first and second backcross generations. Returning to the situation in the first backcross, it appears that gametic elimination was of prime importance, and this was confirmed on the female side by the observation that abortive seeds were found when the $F_1$ was used as female parent, but not when the $F_1$ was used as male.

With regard to the mechanism responsible for selective elimination, it is clear that the multiple gene substitution hypothesis in its original form is not sufficient to explain the facts. It does not seem to have been realized that the postulation of unlimited numbers of modifiers in an attempt to explain the blurred segregations and depressed vigor in $F_2$, carries with it the responsibility of explaining their rapid elimination following inbreeding or backcrossing. The more modifiers postulated and the more minute their individual effects, the slower should be their rate of elimination on any selective basis. Another difficulty which does not seem to have been realized is the sharp contrast between $F_2$ and first backcross segregations. Usually it has been supposed that the relatively slight "modifier disturbance" in the backcross is due to the buffering effect of the balanced genome from the recurrent parent. If, however, the individual modifiers were additive and independent, we should expect that the range of expression of any particular gene in the $F_2$ would be only halved in the corresponding backcross. In order to obtain a greater contrast between $F_2$ and backcross it would be necessary to postulate multiplicative interaction between the individual modifiers. This in turn would lead to a violent skewing of $F_2$ segregations for which there appears to be little evidence. Finally it is difficult to understand how numerous freely assorting modifiers could segregate in such a way as to impart a significant skewness to a monofactorial segregation without blurring the segregating classes.

The first step in the elucidation of these difficulties is the recognition that the "modifiers" are not freely assorting but are carried in internally balanced blocks in each chromosome or chromosome segment. That is, we arrive at MATHER's (1943) important conception of "polygenic balance," though without being committed to the arbitrary classification of genes into "oligogenes" and "polygenes" which in this connection at least, is not an essential feature of his theory. It is clear that owing to the low chiasmata frequency in cotton chromosomes these internally balanced complexes could not be broken up effectively in only one generation of backcrossing, and that the chromosomes from the donor parent which carried the complexes (including certain marker genes) might be at a selective disadvantage. Further, any crossing over which did occur in the marked regions would disrupt the balanced complex, and crossover chromatids would also be at a selective disadvantage. Without independent evidence to the contrary, the skewed backcross ratios could therefore be explained by differences based on internally balanced "polygenic complexes" without its being necessary to postulate structural differences. Against this purely genetic interpretation, however, is the independent evidence cited earlier that two of the loci which show the strongest selective elimination are carried on chromosomes which are suspected to be structurally differentiated, and, more generally, the extraordinary rapidity with which the genotype of the donor parent as a whole is eliminated in backcrosses. It is clear that the mechanism of selective elimination would be all the more effective if the "polygenic complexes" were structurally differentiated (*i.e.* were differential segments) since these would have the effect of "locking" the parental gene combinations by reducing crossing over, and any rare crossovers which occurred in their neighborhood would be expected to result in deficiencies and duplications—potentially of far more selective consequence in the gametes than simple genic recombinations. More precise tests for structural differentiation will be considered in later papers in this series.

### SUMMARY

(1) Evidence is produced that there is considerable selective elimination of the donor parent genotype in interspecific backcrosses involving *G. hirsutum* L. and *G. barbadense* L. In the first backcross the elimination is primarily gametic and is operative both in pollen and ovules.

(2) The selective elimination can be detected by the significant skewness of specific monofactorial segregations and also by the cumulative tendency for the recurrent parent genotype to be recovered more rapidly than expected as a result of random segregation and recombination.

(3) Out of four loci which showed selective elimination, two are suspected on independent grounds to be carried on structurally differentiated chromosomes.

(4) The results are not explicable by interspecific differentiation based on freely assorting modifier systems, but require some form of internally balanced "polygenic complexes."

(5) Reasons are given for believing that the so-called "polygenic complexes" may actually be structurally differentiated chromosome segments.

### ACKNOWLEDGMENT

The writer wishes to thank MR. H. D. LODEN for assistance in collecting the data and for helpful suggestions as to their mode of presentation.

### LITERATURE CITED

DOBZHANSKY, TH., 1941  Genetics and the origin of species. xviii+446 pp. Columbia University Press, New York.

GOLDSCHMIDT, R., 1940  The material basis of evolution. xi+436 pp. Yale University Press, New Haven.

HARLAND, S. C., 1929a  The genetics of cotton. I. J. Genet. **20:** 365–385.

1929b  The genetics of cotton. II. J. Genet. **20:** 387–399.

1929c  The genetics of cotton. III. J. Genet. **21:** 95–111.

1932  The genetics of cotton. V. J. Genet. **25:** 261–270.

1933  The genetical conception of the species. Mem. Acad. Sci. USSR. No. 4.

1935  The genetics of cotton. XIV. J. Genet. **31:** 27–37.

1936  The genetical conception of the species. Biol. Rev. **11:** 83–112.

1938  The genetics of cotton. 193 pp. Jonathan Cape, London.

1939  Genetical studies in the genus *Gossypium* and their relationship to evolutionary and taxonomic problems. Proc. Int. Genet. Congr. Edinburgh. 138–143.

HUTCHINON, J. B., 1946a  The crinkled dwarf allelomorph series in New World cottons. J. Genet. **47:** 178–207.

1946b  The inheritance of brown lint in New World cottons. J. Genet. **47:** 295–309.

IYENGAR, N. K., 1939  Pollen tube studies in Gossypium. J. Genet. **37:** 69–105.

KEARNEY, T. H., and HARRISON, G. J., 1932  Pollen antagonism in cotton. J. Agr. Res. **44:** 191–226.

KNIGHT, R. L., 1945  The theory and application of the backcross technique in cotton breeding. J. Genet. **47:** 76–86.

MATHER, K., 1943  Polygenic inheritance and natural selection. Biol. Rev. **18:** 32–64.

SILOW, R. A., 1939  The genetics of leaf shape in diploid cottons and the theory of gene interaction. J. Genet. **38:** 229–276.

1941  The comparative genetics of *Gossypium anomalum* and the cultivated Asiatic cottons. J. Genet. **46:** 259–358.

1944  The genetics of species development in Old World cottons. J. Genet. **46:** 62–77.

STEBBINS, G. L., 1945  The cytological analysis of species hybrids. II. Bot. Rev. **11:** 463–486.

1947  Types of polyploids: their classification and significance. Advances in Genetics **1:** 403–429.

STEPHENS, S. G., 1942  Colchicine produced polyploids in *Gossypium* I. J. Genet. **44:** 272–295.

1945  A genetic survey of leaf shape in New World cottons. J. Genet. **46:** 313–330.

1949a  The genetics of "Corky" II. J. Genet. (in press).

1949b  The internal mechanism of speciation in *Gossypium*. Bot. Rev. (in press).

WARE, J. O., 1932  Inheritance of lint colors in Upland cotton. J. Amer. Soc. Agron. **24:** 550–562.

# 5

Reprinted from *Genetics* **54**:1189–1199 (1966)

## THE ORIGIN OF A NEW SPECIES OF GILIA IN A HYBRIDIZATION EXPERIMENT[1]

VERNE GRANT

*Rancho Santa Ana Botanic Garden, Claremont, California*

Received June 20, 1966

THE formation of a new species isolated by a sterility barrier is perhaps the evolutionary change of largest magnitude that can be retraced experimentally. In plants this process has been extensively studied in the special case of allopolyploidy. Allopolyploidy is in fact the only mode of speciation, in either plants or animals, which has been thoroughly studied as a process in the laboratory or breeding plot. It is very desirable to investigate by similar methods the formation of new species without change in ploidy.

The purpose of the present paper is to describe such a case of speciation under experimental conditions in the plant genus Gilia. This case involves a derivative of the sterile hybrid *Gilia malior* × *G. modocensis*. It will be shown that this hybrid derivative, designated Branch III, has recovered full fertility without change in ploidy, has attained a new combination of morphological characters, and is isolated by a strong sterility barrier from the parental species and from the other hybrid derivatives.

### MATERIALS AND METHODS

The materials for this experiment are the autogamous annual tetraploid (2n = 36) plants, *Gilia modocensis* and *G. malior*, and their hybrid derivatives. These plants have been grown over a period of 16 years at the Rancho Santa Ana Botanic Garden, Claremont, California. The experiment to be described here was begun in 1956 and concluded in 1966.

The artificial hybrid of *Gilia malior* ♀ × *modocensis* ♂ had been produced and analyzed earlier. Three significant background facts about this cross were known by 1956 (but published later). (1) The $F_1$ hybrid is highly sterile with an average of 2% well formed pollen grains and a seed fertility of 0.007% (GRANT 1964, 1966a). (2) Chromosome pairing is much reduced in the $F_1$, the number of bivalents per pollen mother cell (PMC) ranging from 1 to 10 and averaging 6.0, where 18 would represent complete pairing (GRANT 1964). The available cytogenetic evidence suggests strongly that the observed reduction in pairing is due mainly, though not entirely, to structural differences between the genomes of the parental species (GRANT and GRANT 1960; DAY 1965; GRANT 1966a). (3) The few $F_2$ progeny of this chromosomally sterile hybrid were not, contrary to expectation, doubled in chromosome number and fertile, but instead were essentially tetraploid with one to four extra chromosomes and highly sterile (GRANT 1966a).

The cross of *Gilia malior* × *modocensis* thereupon appeared to furnish suitable material for testing the hypothesis of MÜNTZING, STEBBINS and others (see DISCUSSION) that a chromosomally sterile hybrid can give rise to new fertile types on the homoploid level by recombination of the preexisting sterility factors. The new, structurally homozygous, recombination types are expected to be fertile themselves but intersterile with their parents and most of their siblings.

---

[1] This study has been supported in part by grant GB-3620 from the National Science Foundation.

The first and hardest step was to produce by inbreeding and selection an array of fertile and meiotically normal lines. For this purpose the plants were grown in an insect-proof screenhouse, selected artificially for fertility in each generation, allowed to set seeds autogamously, and propagated as a series of inbred lines. With one generation per year, this phase of the operation required quite a few years to complete. The details have been described elsewhere (GRANT 1966a). Three fertile lines, designated as branches, were obtained in the advanced generations of selection from three different $F_2$ individuals (see Figure 1).

The next and final step was to cross the fertile lines with one another and with the parental species, and to analyze the outcross and backcross hybrids cytologically. Fertile plants in the $F_6$ to $F_9$ generations were used as parents in these crosses. The meiotic behavior of the plants was studied in squash preparations of PMC's by phase and/or bright-field microscope.

RESULTS

*Fertility and cytology of Branch III:* The pedigree of Branch III is shown in Figure 1. The chromosome number of the parental species and $F_1$ hybrid was $2n = 36$. The $F_2$ plant used as the parent of Branch III was not counted, but four

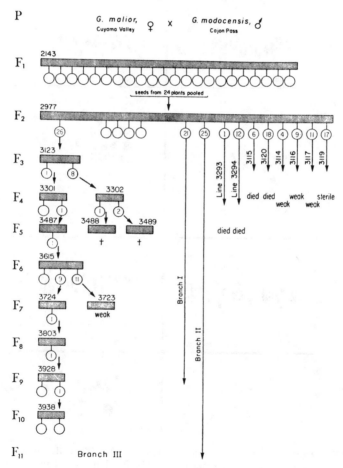

FIGURE 1.—Pedigree of Branch III.

sister plants in the $F_2$ generation ranged from 2n = 37 to 40. Two plants in the $F_3$ generation of Branch III had 39 and 40 chromosomes. The 39-chromosome individual (Plant 3123–1 in Figure 1) gave rise to the main surviving line of Branch III. Its $F_4$ descendant (Plant 3301–1) which became the parent in turn of all subsequent members of Branch III had 2n = 38. This latter chromosome number was the only one encountered in periodic checks in the $F_6$, $F_7$ and $F_{10}$ generations. The main line of Branch III evidently became stabilized at 2n = 38 from $F_4$ on.

The percentage of well formed and well stained pollen grains, taken as a fairly close measure of gametic fertility, showed the following range among sister plants in successive generations. The number of plants tested is given in parenthesis.

| | | | | | | |
|---|---|---|---|---|---|---|
| $F_1$ | 1–5 % | (11) | | $F_6$ | 8–74% | (4) |
| $F_2$ | 10–13% | (3) | | $F_7$ | 8–82% | (5) |
| $F_3$ | 9–30% | (5) | | $F_8$ | 46% | (1) |
| $F_4$ | 13–69% | (6) | | $F_9$ | 84–91% | (4) |
| $F_5$ | 15% | (1) | | $F_{10}$ | 35–89% | (9) |

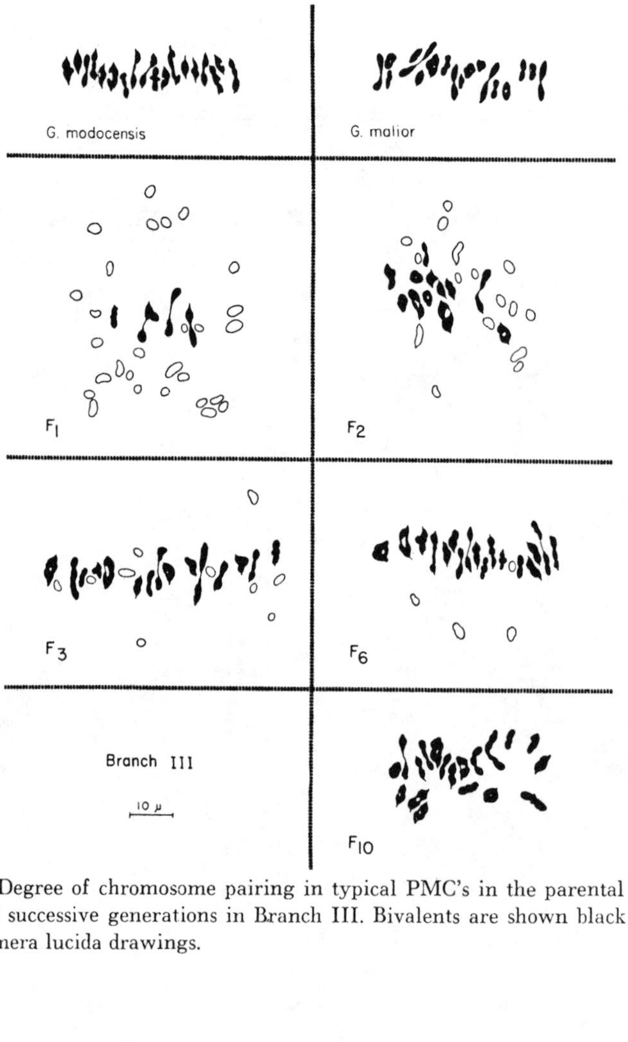

FIGURE 2.—Degree of chromosome pairing in typical PMC's in the parental species, $F_1$ and $F_2$ hybrids, and successive generations in Branch III. Bivalents are shown black and univalents white. From camera lucida drawings.

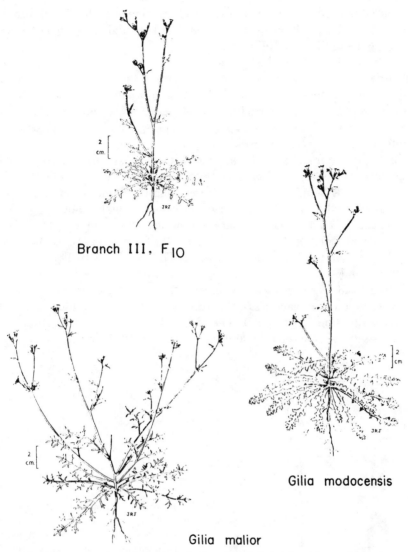

Branch III, F$_{10}$

Gilia modocensis

Gilia malior

FIGURE 3.—The parental species and their Branch III derivative. Drawn from plants grown in Claremont.

By F$_7$ some individual segregates appeared with nearly normal pollen fertility, and by F$_9$ these fertile types had become characteristic of whole families. This trend can be attributed to the artificial selection for pollen fertility.

The same selective process brought about, as expected, a correlated improvement in meiotic behavior in PMC's. The trend is shown graphically in Figure 2 In the early (F$_3$ and F$_4$) generations in Branch III there was considerable reduction in pairing and some lagging. Chains were occasionally seen in these generations. In F$_6$ meiosis was almost normal with 17 to 19 bivalents per cell, and in subsequent generations became quite normal.

Seed production was low in the early generations ($F_2$ to $F_5$) of Branch III. The plants in the $F_6$ to $F_8$ ranged from semifertile to highly fertile as to seeds. All vigorous plants in $F_{10}$ were fully seed fertile.

*Vigor:* From $F_3$ to $F_8$ inclusive the families of Branch III consisted predominantly or entirely of weak individuals. The more robust individuals were selected as parents during these generations. Response to this selection for vigor has become apparent in recent years. In $F_9$, one small family contained five vigorous and four stunted individuals. A larger family in $F_{10}$ contained 68 vigorous plants, 32 runts, and 18 plants of intermediate vigor.

As of $F_{10}$, therefore, Branch III appears to have broken through the subvitality barrier, but has not attained uniformly high vigor in the experimental environment. The line is being continued from a vigorous parental individual in order to determine whether it will continue to segregate for vigor or not.

*Morphological characters:* The parental species differ in at least 16 quantitative characters in all parts of the plants. Eleven of these proved to be useful in practice for scoring the parents and hybrid progenies. They are listed in Table 1 and shown in Figures 3 and 4.

The $F_1$ hybrid was intermediate between the parental species in all characters measured (no data on earliness). There was segregation for these characters in the early generations (before $F_5$). The parental individuals used to propagate the lines were selected for fertility and vigor, but not for any morphological traits

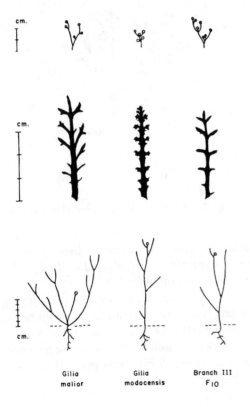

FIGURE 4.—Branching pattern, leaf form, and inflorescence in Branch III and the parental species. The circles represent each flower in a typical inflorescence (above), and the first flower to develop in the habit diagrams (below). Traced from live specimens grown in Claremont.

Gilia malior    Gilia modocensis    Branch III $F_{10}$

TABLE 1

*Morphological and developmental characters of* Gilia malior, G. modocensis *and Branch III*

| Character | G. malior | G. modocensis | F$_{10}$, Branch III |
|---|---|---|---|
| (1) Length of central leader | short | erect and long | erect and long |
| (2) Length of lateral branches | long | short | intermediate |
| (3) Orientation of lateral branches | spreading | ascending | ascending |
| (4) Point of departure of lateral branches | arising near base of plant | arising in upper part of plant | arising in upper part of plant |
| (5) Thickness of stem | slender, 1 to 2 mm diam; wiry | stout, 2 to 3 mm diam; succulent | intermediate, <2 to 2.5 mm diam; succulent |
| (6) Length of lobes of lower leaves | long | short | long |
| (7) Angle of divergence of leaf lobes | pointing forward | at right angles to rachis | intermediate |
| (8) Inflorescence | loose; first flower of a cluster pedicelled | glomerate; first flower of a cluster sessile | intermediate |
| (9) Flower size | small; corolla limb 3 to 4 mm diameter | large; corolla limb 6 to 7 mm diameter | small; corolla limb 3 to 4 mm diameter |
| (10) Pigmentation of corolla | pale violet | deep violet | deep violet |
| (11) Earliness | early; in full bloom ± 76 days after seed sowing | late; in full bloom ± 102 days after seed sowing | early; in full bloom ± 87 days after seed sowing |

per se. The families belonging to the later inbred generations (after F$_5$) were relatively uniform morphologically, and differ markedly from one another (GRANT 1966b).

The characters of vigorous plants in the F$_{10}$ of Branch III are shown and compared with those of the parental species in Table 1 and Figures 3 and 4. It can be mentioned here that the character combination in this line has not changed in any important way since F$_5$.

Branch III is like *G. modocensis* in four of the 11 characters scored, like *G. malior* in three, and like the F$_1$ hybrid in four. It has, for example, some of the branching characters of *G. modocensis* combined with certain leaf and floral features of *G. malior*, and is intermediate in various other features. Branch III thus possesses a new combination of the parental characteristics.

*Fertility relationships of Branch III:* Two fully fertile individuals in the F$_9$

generation of Branch III were crossed successfully with *Gilia malior*. The crosses were Plant 3928–1 (Br. III) ♀ × *G. malior* ♂ and 3928–2♀ × *G. malior* ♂ (see pedigree in Figure 1). The first of these replicate crosses yielded one $F_1$ hybrid and the second yielded three hybrid plants.

The backcross hybrids were highly sterile. The percentage of well formed pollen on the four hybrid plants was 4%, 10%, 13%, and 18%. The hybrids flowered and self-pollinated for seven weeks and produced four or five plump seeds per plant. The seed fertility is estimated to be 0.4%.

One hybrid individual was analyzed cytologically. It had $2n = 37$ as expected from parents with $2n = 38$ and 36 respectively. Chromosome pairing was greatly reduced at metaphase in PMC's (Figure 5). In 23 cells the number of bivalents ranged from 10 to 15 and averaged 12.9 per cell. The remaining chromosomes appeared as univalents (Figure 5). Lagging chromosomes were seen in cells at anaphase and telophase.

Branch III was outcrossed to Branch II using $F_7$ individuals with normal fertility in each line as parents. The cross was Plant 3720–94 (Branch II) ♀ × 3723–15 (Branch III) ♂. The Branch III pollen parent was fertile (82% good pollen) but weak. A single hybrid plant was obtained from this cross.

The $F_1$ hybrid was vigorous but highly sterile. It had 9% well formed pollen. A few flowers out of many on the plant produced a total of 16 plump seeds,

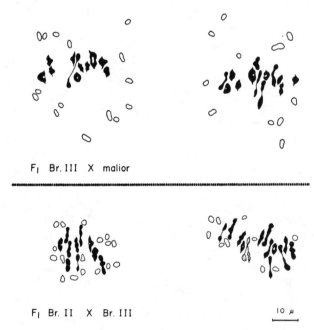

F₁ Br. III X malior

F₁ Br. II X Br. III

10 μ

FIGURE 5.—Chromosome pairing in PMC's in backcross and outcross hybrids of Branch III. Bivalents shown black, univalents white, and chains stippled. $2n = 37$ in these hybrids.

giving an estimated seed fertility of 1.3%. The one vigorous $F_2$ plant obtained from these seeds had 29% well formed pollen and was thus semisterile.

The $F_1$ hybrid of Branch II × III had 2n = 37 chromosomes as expected. Chromosome pairing was strongly reduced at metaphase (Figure 5). Sixteen PMC's had between 10 and 14 bivalents per cell (average 11.6). A chain of three chromosomes was seen in some cells (Figure 5). Some of the univalents often lagged at anaphase.

The backcross hybrid of Branch III × *G. modocensis* has not yet been available for analysis. This hybrid is being produced currently by MISS RUTH WILSON. However, Branch II has previously been shown to be interfertile and chromosomally homologous with *G. modocensis* (GRANT 1966b). The sterility and reduced pairing in the outcross hybrid between Branch III and Branch II is therefore indicative of the fertility relationships of Branch III with *G. modocensis*.

## DISCUSSION

Several authors have proposed and developed the hypothesis that two plant species differing with respect to two or more independent segmental rearrangements, and chromosomally intersterile on this account, can give rise by hybridization, recombination and inbreeding to new fertile types which are on the same ploidy level as the parental species but are separated from these by sterility barriers (MÜNTZING 1929, 1930, 1938; GERASSIMOVA 1939; STEBBINS 1942, 1950, 1957, 1959; GRANT 1956, 1958, 1963, pp. 469–481). It is convenient to refer to this hypothetical process here as recombinational speciation.

The simplest case would be that of two parental types differing by two independent translocations, and having the genomic constitutions AA BB CC DD and $A_bA_b$ $B_aB_a$ $C_dC_d$ $D_cD_c$. Their $F_1$ hybrid is partially sterile with an expected 75% of abortive pollen if the deficiency-duplication products are inviable in the gametophyte stage. The progeny of this hybrid can be expected to include some new homozygous recombination types for the translocations (i.e., AA BB $C_dC_d$ $D_cD_c$) which are fully fertile themselves but partially intersterile with both parental types. Parental plants differing in two independent sets of transpositions or re-inversions will produce an $F_1$ hybrid with a similar degree of sterility, if crossing over occurs regularly between the transpositions or within the reinversions on each chromosome pair, and can likewise give rise to new fertile recombination types in $F_2$ or later generations.

A larger number of heterozygous segmental rearrangements will of course give rise to chromosomal sterility barriers comparable in strength to those commonly found between well isolated plant species. For example, two species differing by six independent rearrangements of the types and under the conditions mentioned above will yield an $F_1$ hybrid with a gametic fertility of 1.56%. This sterile hybrid can then go on to produce fertile recombination types in later generations which are isolated by strong sterility barriers from both parental species. The most common type of fertile recombination product expected from this hybrid would yield backcross hybrids with either parental species which have a gametic fertility in each case of 12.5%.

69

This hypothesis has been developed and advocated in the hope that it would provide a partial solution for the common occurrence in many plant groups of morphologically similar species on the same ploidy level separated by chromosomal sterility barriers. A single pair of genomically differentiated species can theoretically give rise to one or more new species by hybridization and recombination without change in ploidy.

There have been three previous attempts to verify this hypothesis experimentally. and each of these has been successful in one respect or another. Two additional hybridization experiments in Erophila and Phaseolus confirm similar but somewhat different hypotheses of hybrid speciation (WINGE 1940; LAMPRECHT 1941).

GERASSIMOVA (1939) was able to derive fertile homozygous types of *Crepis tectorum* differing from one another as well as from the parental population by single translocations and semisterility barriers. These two types then produced by recombination a third fertile form designated *Crepis nova* which contained both translocations in homozygous condition and was even more intersterile with the original parental type of *Crepis tectorum*.

The $F_1$ hybrid of *Nicotiana langsdorffii* × *sanderae* is semisterile with 45 to 55% well formed pollen and slightly irregular meiosis. SMITH (1954) and SMITH and DALY (1959) derived several inbred lines from this hybrid. The diploid derivatives with short corollas (which are the ones of main interest to us here) became almost fully fertile in later generations (76 to 80% good grains in $F_6$ to $F_9$). These short-flowered derivatives produced semisterile hybrids with both parental species; the backcross hybrids with *N. langsdorffii* had 46 to 58% good pollen and those with *N. sanderae* had 60 to 61%. This fertile hybrid derivative is thus as intersterile with the two parental species as these are with one another (SMITH and DALY 1959).

The $F_1$ hybrid of *Elymus glaucus* × *Sitanion jubatum* is highly sterile with < 1% good pollen. Chromosome pairing is fairly normal in the hybrid, but structural differences are inferred to exist from the behavior of the allopolyploid derivatives (STEBBINS and VAARAMA 1954). By backcrossing the hybrid with *Elymus glaucus* and selfing the $B_1$ plant, STEBBINS (1957) obtained fully fertile homoploid progeny in $F_3$ and $F_4$. The derived line formed a hybrid with *Elymus glaucus* which was highly sterile with 0 to 3% good pollen (STEBBINS 1957).

The Crepis and Nicotiana experiments confirm the hypothesis in terms of weak sterility barriers. The Elymus experiment is the first one involving strong sterility barriers. Here the fertile hybrid derivative was obtained by backcrossing, which, in these predominantly autogamous plants, might or might not occur in nature. The Gilia experiment now provides a case of a new, fertile, strongly isolated type which has emerged from a highly sterile hybrid by the normal pathway of straight inbreeding.

Our final task is to attempt to assess the importance and extent of recombinational speciation in nature in the light of the experimental evidence. Theoretically this process can be compared with allopolyploidy which represents an alternate pathway from a chromosomally sterile hybrid to a new fertile species. Allopolyploid speciation is known to be of common occurrence in plants on the basis of experimental and cytotaxonomic evidence, and therefore we might logically expect recombinational speciation to have occurred frequently too, but to remain undetected cytotaxonomically.

The possibility that recombinational speciation may be common in plants has been in my mind during the years of my experimental work on this problem. Similar views have been expressed by STEBBINS (i.e., 1966, pp. 122–123). I am now inclined to think that recombinational speciation involving strong sterility barriers is a real and interesting but relatively rare process in nature.

The Gilia experiment was conducted with one aspect of the situation, hybrid sterility, mainly in mind. Allopolyploidy is a way out of the impasse of chromosomal sterility in a hybrid plant, as DARLINGTON (1932, 1958) and others have pointed out; and recombination of separable chromosomal sterility factors could be considered an alternative way out of the same impasse. A second important aspect of the situation, namely hybrid breakdown, was, however, neglected in predicting the experimental results. For allopolyploidy is a way around both chromosomal sterility *and hybrid breakdown;* but recombinational speciation circumvents only the first of these barriers.

In the event, the Gilia experiment was needed to reveal the full force of the hybrid breakdown barrier. The vast preponderance of $F_2$ to $F_6$ progeny of *Gilia malior* × *modocensis* were subvital, sterile, or both (GRANT 1966a). Two of the three vigorous and fertile lines derived from the hybrid turned out to be reversions to one parental type (GRANT 1966b). The only fertile derivative that possessed a new combination of sterility factors was weak during many generations and could be kept alive only with difficulty.

Among the numerous sterile species hybrids of Gilia raised in our experimental garden over the years, seven have doubled spontaneously to produce fertile or semifertile allopolyploid progeny, and some of these hybrids have doubled repeatedly in replicate cultures (GRANT and GRANT 1960; GRANT 1965; DAY 1965). The new allopolyploid plants showed good general vigor from the start. Other workers have had similar experiences in other plant groups. One gets the definite impression from such experiences that speciation accompanied by strong sterility barriers may take place fairly easily by the allopolyploid route, but only with difficulty and under exceptional circumstances by the recombinational route.

The cross-pollinations were made by DR. ALVA DAY and MRS. KAREN A. GRANT. The latter also helped with the scoring of morphological characters, and critically read the manuscript. JEANNE R. JANISH made the drawings for Figure 3, and CHARLES POPP the chart for Figure 1. The help of these workers and the financial assistance of the National Science Foundation are gratefully acknowledged.

SUMMARY

One of the selection products of a chromosomally sterile species hybrid in Gilia became fully fertile without change in ploidy. This fertile hybrid derivative has its own distinctive character combination and is intersterile with both parental species. It represents an experimental case of the hybrid origin of a new isolated species without chromosome doubling.

## LITERATURE CITED

DARLINGTON, C. D., 1932 *Recent Advances in Cytology*. Edition 1. Churchill, London. —— 1958 *The Evolution of Genetic Systems*. Edition 2. Basic Books, New York.

DAY, A., 1965 The evolution of a pair of sibling allotetraploid species of Cobwebby Gilias (Polemoniaceae). Aliso **6**: 25–75.

GERASSIMOVA, H., 1939 Chromosome alterations as a factor of divergence of forms. I. New experimentally produced strains of *C. tectorum* which are physiologically isolated from the original forms owing to reciprocal translocation. Compt. Rend. Acad. Sci. U.R.S.S. **25**: 148–154.

GRANT, V., 1956 Chromosome repatterning and adaptation. Advan. Genet. **8**: 89–107. —— 1958 The regulation of recombination in plants. Cold Spring Harbor Symp. Quant. Biol. **23**: 337–363. —— 1963 *The Origin of Adaptations*. Columbia Univ. Press. New York. —— 1964 Genetic and taxonomic studies in Gilia. XII. Fertility relationships of the polyploid Cobwebby Gilias. Aliso **5**: 479–507. —— 1965 Species hybrids and spontaneous amphiploids in the *Gilia laciniata* group. Heredity **20**: 537–550. —— 1966a Selection for vigor and fertility in the progeny of a highly sterile species hybrid in Gilia. Genetics **53**: 757–776. —— 1966b Linkage between viability and fertility in a species cross in Gilia. Genetics **54**: 867–880.

GRANT, V., and A. GRANT, 1960 Genetic and taxonomic studies in Gilia. XI. Fertility relationships of the diploid Cobwebby Gilias. Aliso **4**: 435–481.

LAMPRECHT, H., 1941 Die Artgrenze zwischen *Phaseolus vulgaris* L. und *multiflorus* Lam. Hereditas **27**: 51–175.

MÜNTZING, A., 1929 Cases of partial sterility in crosses within a Linnean species. Hereditas **12**: 297–319. —— 1930 Outlines to a genetic monograph of the genus Galeopsis. Hereditas **13**: 185–341. —— 1938 Sterility and chromosome pairing in intraspecific Galeopsis hybrids. Hereditas **24**: 117–188.

SMITH, H. H., 1954 Development of morphologically distinct and genetically isolated populations by interspecific hybridization and selection. Proc. 9th Intern. Congr. Genet. (Caryologia **6** (suppl.): 867–870.

SMITH, H. H., and K. DALY, 1959 Discrete populations derived by interspecific hybridization and selection in Nicotiana. Evolution **13**: 476–487.

STEBBINS, G. L., 1942 The role of isolation in the differentiation of plant species. Biol. Symp. **6**: 217–233. —— 1950 *Variation and Evolution in Plants*. Columbia Univ. Press. New York. —— 1957 The hybrid origin of microspecies in the *Elymus glaucus* complex. Cytologia, suppl. vol., 336–340. —— 1959 The role of hybridization in evolution. Proc. Amer. Phil. Soc. **103**: 231–251. —— 1966 *Processes of Organic Evolution*. Prentice-Hall, Englewood Cliffs, New Jersey.

STEBBINS, G. L., and A. VAARAMA, 1954 Artificial and natural hybrids in the Gramineae, tribe Hordeae. VII. Hybrids and allopolyploids between *Elymus glaucus* and *Sitanion spp*. Genetics **39**: 378–395.

WINGE, Ö, 1940 Taxonomic and evolutionary studies in Erophila based on cytogenetic investigations. Compt. Rend. Lab. Carlsberg **23**: 41–74.

# 6

Reprinted from *Evolution* **23**:525–533 (1969)

## EXPERIMENTAL STUDIES ON HYBRIDIZATION AND SEXUAL ISOLATION BETWEEN SOME *APHYTIS* SPECIES (HYMENOPTERA: APHELINIDAE). III. THE SIGNIFICANCE OF REPRODUCTIVE ISOLATION BETWEEN INTERSPECIFIC HYBRIDS AND PARENTAL SPECIES

SUDHA V. RAO[1] AND PAUL DEBACH

*Department of Biological Control,*
*University of California, Riverside 92502*

Received April 2, 1969

Interspecific hybridization is of rare occurrence in animals, especially in comparison to plants, obviously because of their very different modes of reproduction, ecology, and perhaps especially, of behavioral isolation (Mayr, 1963). Within the animal kingdom, hybridization frequency differs with the class, being more characteristic of animals which reproduce by means of external fertilization, e.g., fishes and amphibia. To illustrate this rarity of animal hybrids in nature, Mayr (op. cit.) states that perhaps one out of 60,000 wild birds is a hybrid.

In animals with internal fertilization, the rarity of hybrids may frequently be attributed to the presence of ethological isolating mechanisms. Where ethological isolating mechanisms are overcome, either under laboratory or natural conditions, and successful interspecific copulation occurs, hybrids may occasionally be produced. The steps in hybrid production subsequent to mating, such as fertilization of the egg by the foreign sperm, development of the embryo, etc., should proceed normally. These may be prevented by death of the sperm in the female genital tract, inability of the sperm to fertilize the egg, or by subsequent death of the zygote due to genetic imbalances.

Even when all these isolational barriers are circumvented and hybrids are produced, poor genetic fit (brought about as a result of deleterious translocations, duplications, inversions, etc., in the chromosomes), resulting in hybrid sterility or inviability of the $F_1$ generation, usually prevents the successful establishment of a continuing hybrid line between two species (Mayr, 1963).

In rare cases in nature where reproductive isolation mechanisms are overcome and hybrids are produced, one of the following three phenomena usually operate eventually to eliminate the hybrids:

1) Total sterility of the $F_1$ hybrids prevents them from reproducing.

2) In the event that the $F_1$ hybrids are fertile, they usually have lower fitness than either parent species and may be eliminated due to competitive displacement. In other words, natural selection and survival of the fittest operates to eliminate them (see discussion in DeBach, 1966, p. 191). Frequently, although the $F_1$ hybrids may be fertile, fertility is reduced in subsequent generations and thus the hybrid line fails to become established. For example, see Patterson and Stone, 1952, p. 451.

3) The hybrids may back-cross with the parental species to produce inferior genotypes, which again are eliminated as a result of competitive displacement. Mayr (1963) says that introgressive hybridization is rare in animals because only a small fraction of hybrids will back-cross to either of the parental species (see also Ford, 1964, p. 38, 281). In none of these cases has hybridization caused the integrity of the species to be broken down completely although continuous gene exchange may occur.

[1] Now Mrs. (Dr.) Sudha Nagarkatti. Present address: Commonwealth Institute of Biological Control, Indian Station, Bellary Road, Bangalore (6) India.

In the laboratory, however, the possibility of obtaining interspecific hybrids between animals is greater, since ecological, behavioral, and spatial barriers can be more easily overcome. Also, competition can be eliminated in the laboratory and thus greatly enhance the establishment of hybrid lines. Study of the behavior of the hybrid in the laboratory, with respect to its fitness in comparison to the parent species, its crossability with the parent species, and other characteristics, helps to determine whether or not the hybrid would be capable of establishing a successful, continuing line in the field.

We know of no previous demonstration of an interspecific animal hybrid that has been able to coexist with the two parental species and maintain its integrity in nature as a third biological species. In plants, interspecific hybridization is established as one of the methods of speciation, numerous allopolyploids being known, which, by virtue of being reproductively isolated from the parent species, qualify as a third species. In animals, however, such a phenomenon apparently never has been definitely established. In laboratory experiments, however, the possibility of obtaining a viable interspecific hybrid which would qualify technically as a biological species has at least been indicated.

Among the more intensive laboratory studies on interspecific hybridization in insects may be cited the work of Smith (1959, 1962, 1965) who has studied hybrids between species of *Chilocorus* (Coccinellidae) and other Coleoptera and has found that all native North American species of *Chilocorus* can be hybridized one way or another. In many cases, however, he has encountered a high degree of sterility; in others little is known of the fertility of the $F_1$; and in some others, where hybrids were readily produced, he believes that the normal geographic isolation between them in nature has "rendered the development of a sexual isolation barrier unnecessary" (see especially Smith, 1959).

More species hybrids of *Drosophila* have

been studied than of any other animal genus. Wharton (1944) reported many cases of cross-fertility in the Repleta group of *Drosophila* in laboratory experiments but says that "in nearly every case where the production of fertile hybrids is possible in the laboratory, potent isolating mechanisms operate to prevent such gene exchange in nature."

There are few cases reported involving animals wherein fertile $F_1$ hybrids were obtained that showed resistance to back-crossing. Foot and Strobell (1914), in reporting their studies involving hybridization between the stink bugs, *E. servus* and *E. variolarius*, stated that the hybrids were fertile among themselves but exhibited almost the same resistance to back-crossing as did the parent species to the original cross. Sears (1947) found that the cross *Drosophila munda* females × *D. occidentalis* males produced hybrids that were fertile when inbred, but the hybrid males were sterile when back-crossed to either parent type. Hybrid males from the reciprocal cross were fertile when back-crossed to *munda* females but not to *occidentalis* females. Here again gene exchange between the hybrids and parental species was not entirely restricted.

The possibility of obtaining an interspecific hybrid, reproductively isolated from both parent species, but fertile when selfed, has been investigated by Sailer (1953, 1954) who concluded from work on hybridization of the stink bugs *Euschistus servus* (Say) and *E. variolarius* (P. de B.) as follows: "The possibility that an inbred hybrid population may eventually become genetically isolated from its parent species has not been disproved, and is in fact supported by evidence that eggs from back-cross matings show a lower fertility than do selfed matings. The objective of further research on this problem is the production of a synthetic species."

This idea was kept in mind during the present bio-systematical studies on some of the species of *Aphytis* which are parasitic on diaspine scale insects. The results of these studies on experimental hybridization and sexual isolation have already been reported (Rao and DeBach, 1969a, 1969b). The purpose of the present paper is to discuss the nature of the hybrids that were obtained, with respect to the varying degrees of reproductive isolation that they exhibited from their parent species and to investigate the possibility of developing in the laboratory a synthetic biological "species."

## MATERIALS AND METHODS

All species of *Aphytis* are obligatory external hymenopterous parasites of diaspine (or armored) scale insects, and are credited with being among the most effective entomophaga responsible for regulation of population densities of diaspine scale insects. A variety of broad biological, systematic and ecological studies have been carried on with them in this laboratory for over 15 years. In many ways they are ideal laboratory animals. They and their various diaspine scale hosts are easily cultured. The *Aphytis* laboratory life cycle is short —usually about 2 to 3 weeks. The principal disadvantage in working with *Aphytis* is their small size (usually 1 mm or less in length).

The following hybrids obtained in the *Aphytis* crossing experiments were studied with respect to back-crosses. The hybrids are denoted by using the first letter of the specific name or code number of the female parent, followed by that of the male parent as a subscript. The parents used in the original cross are shown below at right (for background information, see Rao and DeBach, 1969a and 1969b):

$L_2$ = female *A. lingnanensis* Compere × male *A.* "2002"
$L_k$ = female *A. lingnanensis* × male *A.* "khunti"
$2_k$ = female *A.* "2002" × male *A.* "khunti"
$C_2$ = female *A. coheni* DeBach × male *A.* "2002"
$2_c$ = female *A.* "2002" × male *A. coheni*
$M_h$ = female *A. melinus* DeBach × male *A. holoxanthus* DeBach
$H_{\text{"R-66-19"}}$ = female *A. holoxanthus* × male *A.* "R-66-19"

The names given above in quotation marks are code names first used to designate acquisitions (live cultures) of unknown taxonomic status. For reasons discussed by Rao and DeBach (1969a), *lingnanensis* and "2002", as well as "khunti" and *coheni*, are considered to be semispecies (not species) with respect to each other.

The original collection sites of the parent cultures are listed below. These sites are not necessarily part of their indigenous areas, since the exact extent of their natural distribution is not known and accidental ecesis is common in the genus: *A. lingnanensis*—South China (Hong Kong); *A.* "2002"—Puerto Rico (San Juan); *A.* "khunti"—Northwestern India (near New Delhi); *A. coheni*—Israel (Ashkelon); *A. melinus*—Northwestern India (New Delhi and Gurgaon) and West-Pakistan (Lahore and Saidpur); *A. holoxanthus*—South China (Hong Kong); *A.* "R-66-19"—Mexico (La Paz, Baja California).[2]

For details concerning field and laboratory hosts, morphological similarities and dissimilarities, methods for setting up the original species crosses, etc., the reader is referred to the earlier paper by Rao and DeBach (1969a).

For the back-cross tests, virgin females and males of the hybrids as well as of the parent species were needed. Pupae were, therefore, isolated during the green-eyed stage (which is the stage just prior to emergence of the adults) in individual vials and held at 80 F (±2) and approximately 75% R.H. A high humidity is required for pupae removed from the host. On emergence, four to five virgin hybrid females were placed in a 3-dram vial together with three to four males of one of the parental species. Similarly, four or five females of the parental species were placed with three or four males of the hybrid, in the reciprocal cross. Thus, four different combinations were set up with each hybrid.

---

[2] This culture was acquired during the latter phases of this study in 1966 by the second author and was reared from cactus scale, *Diaspis echinocacti* (Bouche). The adults are morphologically very similar to the other species and the pupae differ only slightly in pigmentation.

Mating usually takes place immediately upon meeting of the sexes, but a 24-hour mating period was allowed during which the *Aphytis* individuals were left undisturbed in the vials. A streak of honey was provided in each vial for food.

At the end of the 24-hour period the females were anesthetized with carbon dioxide and carefully placed on a lemon bearing mature oleander scale, *Aspidiotus hederae* Vallot, 45–50 days old. This provided a good laboratory host. Each lemon was placed in a one-pint mason jar covered with a piece of muslin held tight by a screw-top lid. The *Aphytis* adults were transferred every 11th day until the female parasites died, in order to prevent development of a second generation on the same lemon. Progeny emerging from the parasitized scale on the lemons were collected every 4th day and the sex-ratio and total numbers recorded. Cactus scale, *Diaspis echinocacti*, was used only in experiments with "R-66-19" as the parasite females showed a pronounced host specificity.

## Characteristics of Reproduction in *Aphytis*

The *Aphytis* species used reproduce by arrhenotoky, which is the phenomenon whereby unfertilized (haploid) eggs give rise to males parthenogenetically, while fertilized (diploid) eggs give rise to females zygogenetically. This being the case, the production of males alone in any cross indicates the failure of the species to mate or at least of the egg to be fertilized, while the production of female progeny indicates successful fertilization, however limited such production may be. The ultimate success or failure of the "interspecific" (at least, heterogamic) cross is, of course, further determined by testing the fertility of the hybrid females over subsequent generations. Such a measure of the genetic fitness of the hybrid, therefore, involves determination of the ratio of females to males, as well as total progeny production in the $F_2$ and subsequent generations.

In interspecific crosses, therefore, only

the females in the $F_1$ generation are hybrids, while the males are not. In order to start a 50:50 hybrid culture, a few of the $F_1$ hybrid females were allowed to oviposit as virgins for 1 or 2 days and were then held in cold storage along with the others at 65 F, (with daily transfers to 80 F for 1 or 2 hours to enable feeding), until such time as their male hybrid progeny emerged after a period of 13–14 days. When the hybrid males emerged, they were placed for mating in vials at 80 F with hybrid females and the females then allowed to oviposit for production of 50:50 hybrid progeny.

Under standardized, controlled, and near-optimal laboratory conditions of 80 ($\pm 2$) F and 50% R.H., adults of all *Aphytis* species used in this study produce characteristic "standard" or "normal" sex-ratios. It was necessary to know the average ratio of females for each "species" in order to compare this with that obtained in interspecific crosses. These values were obtained for each culture from 10 randomly collected samples of about 100 individuals each from cultures held under optimal laboratory conditions. Age of females, host densities, etc., were the same for all tests. The "standard" per cent female progeny for each species is: *coheni*, 74.7; "khunti", 73.1; *lingnanensis*, 66.1; *melinus*, 64.1; "2002", 64.0; *holoxanthus*, 63.6; and "R-66-19", 60.0.

## Experimental Results

A summarized version of the experimental results is presented pictorially in Figures 1 and 2. Each case is, however, discussed individually, since important differences occur between them. In the figures, the values in each square represent the proportion of female progeny produced in a particular cross, which as mentioned earlier, measures the relative success or failure of a cross.

The per cent female progeny when *lingnanensis* and "2002" were selfed was 66.1 and 64.0, respectively (Fig. 1a). When the $L_2$ hybrid was selfed, the per cent female progeny was 60.1. This value was determined in the $F_4$ generation, by which time

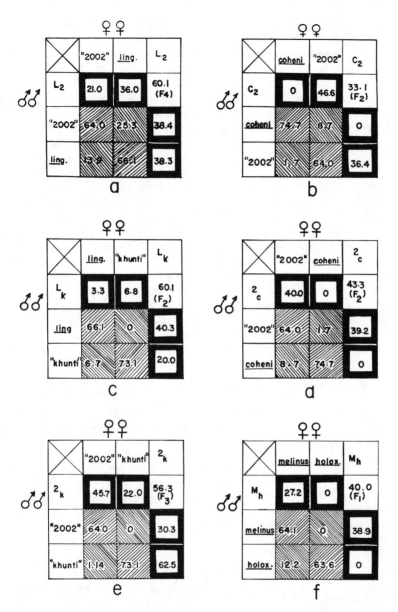

FIG. 1. Back-cross experiments using *Aphytis* hybrids. Numbers represent per cent female progeny. Dark bordered squares represent back-cross results and plain squares represent per cent female progeny in hybrid cultures. Striped squares represent results from "interspecific" and conspecific matings. Explanation of the code designations of the various crosses is given in "Materials and Methods."

the sex-ratio had definitely improved from that in the F₁ generation. The back-crosses, which were performed using the F₁ generation hybrids, showed definite reduction in the per cent female progeny. This represents a lack of compatibility be-

tween the hybrid and its parent species. A back-cross between L₂ females (F₁) and *lingnanensis* males yielded 38.3% female progeny and that between L₂ males (F₁) and *lingnanensis* females 36.0% female progeny. In back-crosses between L₂ females ($F_1$) and "2002" males, 38.4% female progeny resulted, while the reciprocal cross yielded 21.0% female progeny. However, there appears to be less reproductive isolation in general between the hybrids and the parents than between the parents themselves. Since the hybrids when selfed showed high fertility and sex-ratio more favorable than in the back-crosses, one might be prone to predict that, given sufficient time, reproductive isolation between the hybrids and the parents would reach completion and the hybrids would become a distinct third species. However, in the present case, back-crosses (using $L_2$ females and "2002" and *lingnanensis* males) made once again in the 52nd hybrid generation indicated that although reproductive isolation between $L_2$ and *lingnanensis* had strengthened, that between $L_2$ and "2002" had weakened. The per cent female progeny produced in these crosses as compared to the $F_1$ were as follows:

|  | Per cent female progeny | |
|---|---|---|
|  | 52nd gen. | 1st gen. |
| $L_2$ ♀♀ × *lingnanensis* ♂♂ : | 11.2 | 38.7 |
| $L_2$ ♀♀ × "2002" ♂♂ : | 49.0 | 38.3 |

This indicates that $L_2$ may be approaching the genetic constitution of "2002" and diverging from *lingnanensis*. There is not enough evidence to predict whether a stage would be reached when $L_2$ and "2002" would mate freely and produce normal numbers of female progeny, while $L_2$ would become completely isolated from *lingnanensis*. This possibility, however, is at least indicated.

The situation is not very different in the case of the hybrids $L_k$ and $2_k$, with some differences in the per cent female progeny in back-crosses. Inasmuch as the $L_k$ culture was discontinued after a few generations, no back-crosses could be attempted in subsequent generations. However, in the case of $2_k$, the degree of reproductive isolation from both parents appeared to be relatively unchanged even after about 40 generations.

Three of the hybrids, i.e., $C_2$, $2_c$, and

$M_h$, showed another interesting feature. The hybrids $C_2$ and $2_c$ showed complete reproductive isolation from *coheni* but only partial isolation from "2002". Whether or not this degree of isolation was maintained as such or was modified in subsequent generations could not be tested since both the hybrid cultures were lost.[3]

In the case of the hybrid $M_h$, back-crosses to *holoxanthus* produced no female progeny whatsoever, indicating complete reproductive isolation. On the other hand, back-crosses to *melinus* indicated partial reproductive isolation. Back-crosses in subsequent generations showed no change.

The most interesting of all the hybrids was $H_{\text{"R-66-19"}}$ (see Fig. 2) despite the fact that investigations on this were incomplete. In the original interspecies cross, using 25 parental *holoxanthus* females and "R-66-19" males, a single hybrid female was obtained out of a total of 385 progeny, indicating very strong reproductive isolation between the parental species. This female was allowed to oviposit for 2 days and she laid 20 hybrid male eggs. She was then held in cold storage awaiting emergence of the hybrid male adults. Unfortunately, the female died on the 10th day, 2 to 3 days before any of her hybrid male progeny emerged. This prevented the development of a 50:50 hybrid line. Subsequent attempts to repeat the interspecific cross and get hybrid females were unsuccessful.

Nevertheless, back-crosses were tried between the $H_{\text{"R-66-19"}}$ hybrid males which emerged after the hybrid female died and females of *holoxanthus* as well as females of "R-66-19". It was found that neither cross produced any female progeny whatsoever, thus indicating complete reproductive isolation from both parents. Since no hybrid females were available for back-crosses

---

[3] The loss was due to their inability to breed on oleander scale reared on banana squash, a fact that was discovered too late to prevent loss of the culture. This indicates an interesting change in host preference or host suitability to the hybrids. A change in host-preference of the hybrids would be especially important.

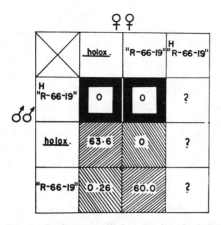

FIG. 2. Back-cross experiment using the hybrid H$_{\text{"R-66-19"}}$. Numbers represent per cent female progeny. Dark bordered squares represent back-cross results and shaded squares represent results of interspecific and conspecific matings. Explanation of the code designations is given in "Materials and Methods."

with the parental males, complete verification could not be obtained.

### DISCUSSION

Problems of speciation in sexually reproducing higher animals have been subject to much discussion. Most arguments propose geographic isolation as a prerequisite for speciation to occur, but some consider that species can evolve sympatrically. Although no consensus has been reached, most evidence appears to favor the proponents of geographic isolation. Assuming that geographic isolation is necessary, the accumulation of significant gene differences between the two geographically separated populations is not likely to occur very rapidly. At the risk of appearing trite, it is generally agreed that the hypothesis of macrogenesis is not tenable in animals; in other words, that mutations would not be able to provide the necessary degree of genetic differences and hence give rise to a viable new species in one step. Speciation must therefore take place gradually, in a series of steps; and, of course, this involves natural selection acting on genetic variants in a population (see Patterson and Stone, 1952, p. 503).

In furtherance of the preceding general-ization, instantaneous speciation through polyploidy is not known in sexually reproducing animals. It has only been proved in a few self-fertilizing or asexually reproducing lower animals. In a recent paper, Lewis (1966) has discussed speciation by "saltation" in plants. With the help of illustrations, he explains that adjacent populations, which are very similar in morphological and ecological adaptation, may differ greatly in chromosome arrangement and on occasion in basic chromosome number. The multiple chromosomal differences between such populations cause the hybrids between them to have such low fertility as to be generally considered sterile. According to Lewis, the relationship between two such species is that of parent and offspring. Although the best evidence of saltation is found in annuals, he believes that there is indirect evidence of it in woody plants.

Interspecific hybridization between two animal species leading to the formation of a distinct third "species" has apparently never been established, in view of the lack of knowledge of fertile and successful animal hybrids.

From the experimental results presented in this paper, one sees that at least in three of the cases the hybrids were completely reproductively isolated from one parent and partially from the other (e.g., the hybrids $C_2$, $2_c$, and $M_h$). In the case of the hybrid $M_h$, successful back-crossing with *holoxanthus* did not occur even after 20 generations, although $M_h$ did back-cross to a fair extent with *melinus*.

From the evolutionary or phylogenetic standpoint, since $C_2$, $2_c$, and $M_h$ show complete reproductive isolation from at least one parent, they could be considered to be semispecies. If complete reproductive isolation had appeared with respect to both parents, the hybrids could be considered artificially created biological species. Another unique feature observed in this case is that complete reproductive isolation between the hybrids and at least one of the parents appeared in a single step, i.e., in one generation as a result of hybridization, whereas the development of isolating mechanisms is generally considered to be the result of long-term evolutionary processes.

In the case of the hybrid $H_{\text{"R-66-19"}}$, the chances are that the hybrid would have been completely reproductively isolated from both parents, since this was at least partly indicated by the back-crosses in which males of the hybrid and females of the parent species were used; unless, of course, the males were sterile, of which we have no evidence. Because sterility of one of the sexes among *Aphytis* hybrids was not observed in previous crosses, it would seem safe to assume that the hybrid males were fertile. Since the single hybrid female showed good fecundity, it is very likely that a successful perpetuating hybrid line would have been established, if only the female had survived a little longer. Such a hybrid then would have represented a new biological species.

It seems unlikely that hybrids, such as those obtained in the laboratory, would develop in nature, since the laboratory conditions were made to favor hybridization but much more especially to favor survival of the hybrids. All hybrid cultures were initially poorly viable and it is questionable that they would persist in nature in competition with the parental species. Introgressive hybridization is therefore unlikely.

However, suppose that the hybrids had been produced in nature. The fact, for example, that mating between $C_2$ and *coheni* and $C_2$ and "2002" occurs less readily and fewer female progeny result in such crosses than when $C_2$ hybrids are selfed, would indicate one advantage in favor of the $C_2$ hybrid for natural selection to work on. This means that reproductive isolation between $C_2$ and *coheni* and between $C_2$ and "2002" gradually would be strengthened. In other words, $C_2$ individuals showing a higher propensity for mating with their own kind would be selected for and hence might increase in numbers, particularly if they happened to have changed host preferences or other favorable essential differences from the parental species. This could conceiv-

ably lead eventually to complete reproductive isolation between $C_2$ and the parent species. Hence, a pure $C_2$ stock would be left with a species status, unless it was eliminated through competition with the parental *coheni* and/or "2002". Such a possibility seems even more likely if we are to accept the theory of sympatric speciation in Chalcidoidea proposed by Askew (1968). According to him, an unusually high degree of inbreeding results from sib-matings in isolated populations. This, coupled with the haplo-diploid method of reproduction, which enables more rapid selection, seems to provide chalcids with features promoting a fast rate of evolution. Also, if parts of a population within the range of the whole population utilize different hosts or the same host on different host plants, sympatric speciation would be further facilitated.

The field liberation of such laboratory reared hybrids in an entirely new area where neither parent species occurs would be tantamount to releasing a third species altogether. The results of our hybridization studies are therefore significant from an applied standpoint. *Aphytis* species, as mentioned earlier, are parasitic on diaspine scale insects which are serious plant pests worldwide. All, except two of the species used herein, primarily attack the California red scale, *Aonidiella aurantii* Maskell. *A. holoxanthus* attacks the Florida red scale, *Chrysomphalus aonidum* (L.), while "R-66-19" attacks cactus scale, *Diaspis echinocacti*. Since the hybrids show resistance to back-crossing and have at least in one case proven to be biologically different (having higher heat tolerance), their use in biological control might be almost equivalent to introducing new exotic species. It will be especially interesting to study the field host-preferences of the $M_h$ hybrid since one of its parents, *A. melinus*, attacks the California scale and the other parent attacks the Florida red scale. From laboratory observations it was found that the hybrid would attack the California red scale quite readily, but no tests have been made using the Florida red scale because this species does not occur in California.

## Summary

It is well known that successful interspecific hybridization in animals is an extremely rare phenomenon. In most cases in nature, where it has been observed in animals, it takes the form of introgressive hybridization wherein the two species involved maintain their integrity while continuously exchanging genes. In the course of hybridization experiments using *Aphytis* spp. (Hymenoptera: Aphelinidae) which are parasites of diaspine scale insects, the authors found that in some cases the hybrids were partially reproductively isolated from both parent species, while in others the hybrids were partially reproductively isolated from one parent and completely isolated from the other parent species. Even more interesting was one interspecific hybrid in which the hybrid males probably showed reproductive isolation from females of both parent species. Crosses involving the hybrid females in this particular case could not be carried out as the culture was lost.

These observations point out the possibility of production of new biological species through interspecific hybridization in a single step, which to our knowledge is a phenomenon hitherto unobserved in animals. From an applied standpoint, such hybrids would conceivably represent the equivalent of new exotic species for introduction against a given pest in a biological control program.

## Acknowledgments

The authors are grateful to Mr. Stanley Warner and Mr. Walter White for providing insects and otherwise assisting in this study. Support of this study by NSF Grants G-20870 and GB-7444, and financial assistance given to the senior author by the Dry Lands Research Institute, University of California, Riverside, during the latter part of the study, is gratefully acknowledged. Thanks are due to Drs. Timothy Prout and Fred Legner for reading and criticizing the manuscript, and to Dr. Dean R. Parker for suggestions.

## Literature Cited

Askew, R. R. 1968. Considerations on speciation in Chalcidoidea (Hymenoptera). Evolution 22(3):642–645.

DeBach, P. 1966. The competitive displacement and coexistence principles. Ann. Rev. Entomol. 11:183–212.

Foot, K., and E. C. Strobell. 1914. Results of crossing *Euschistus variolarius* and *E. servus* with reference to inheritance of an exclusively male character. Linn. Soc. London J. Zool. 32:337–373.

Ford, E. B. 1964. Ecological genetics. Methuen & Co., London & John Wiley & Sons, New York, 2nd ed. 1965. 335 p.

Lewis, H. 1966. Speciation in flowering plants. Science 152:167–172.

Mayr, E. 1963. Animal species and evolution. Bellknap Press, Harvard Univ. Press, Cambridge, Massachusetts. 797 p.

Patterson, J. T., and W. S. Stone. 1952. Evolution in the genus *Drosophila*. Macmillan Co., N.Y. 610 p.

Rao, S. V., and P. DeBach. 1969a. Experimental studies on hybridization and sexual isolation between some *Aphytis* species (Hymenoptera: Aphelinidae). I. Experimental hybridization and an interpretation of evolutionary relationships between the species. Hilgardia 39(19):526–553.

——. 1969b. Experimental studies on hybridization and sexual isolation between some *Aphytis* species (Hymenoptera: Aphelinidae). II. Experiments on sexual isolation. Hilgardia 39(19):555–567.

Sailer, R. I. 1953. Significance of hybridization among stink bugs of the genus *Euschistus*. Yearbook Amer. Phil. Soc. 146–149.

——. 1954. Interspecific hybridization among insects with a report on crossbreeding experiments with stink bugs. J. Econ. Entomol. 47(3):377–388.

Sears, J. W. 1947. Studies in the genetics of *Drosophila*. VII. Relationships within the Quinaria species group of *Drosophila*. Univ. Texas Publ. No. 4720:137–156.

Smith, S. G. 1959. The cytogenetic basis of speciation in Coleoptera. Proc. 10th Internatl. Congr. Genetics (1956) 1:444–450.

——. 1962. Cytogenetic pathways in beetle speciation. Canad. Entomol. 94(9):941–955.

——. 1965. *Chilocorus similis* Rossi: disinterment and case history. Science 148(3677):1614–1616.

Wharton, L. T. 1944. Studies in the genetics of *Drosophila*. V. Interspecific hybridization in the Repleta group. Univ. Texas Publ. No. 4445:175–193.

Part III

# STABILIZATION OF HYBRID DERIVATIVES

# Editor's Comments
# on Papers 7 Through 10

7   **GRANT**
    *The Role of Hybridization in the Evolution of the Leafy-
    stemmed Gilias*

8   **STRAW**
    *Hybridization, Homogamy, and Sympatric Speciation*

9   **GRANT and GRANT**
    *Dynamics of Clonal Microspecies in* Cholla Cactus

10  **PARKER and SELANDER**
    *The Organization of Genetic Diversity in the Parthenogenetic .
    Lizard* Cnemidophorus Tesselatus

The stabilization of hybrid derivatives in nature is more likely in plants than in animals owing to their greater penchant for inter-specific hybridization, asexual reproduction, and polyploidy. (Poly-ploidy will be the subject of a separate volume by Stebbins.) In plants, putative stabilized diploid fertile hybrid entities include *Potentilla glandulosa* ssp. *hansenii* (Clausen, Keck, and Hiesey, 1940), many subspecies of *Gilia capitata* (V. Grant, 1950), *Gilia achil-leaefolia* (V. Grant, 1954), Corn Belt Dent maize (Anderson and Brown, 1952), *Clarkia deflexa* (Lewis and Lewis, 1955), *Achillea rosea-alba* (Ehrendorder, 1959), *Delphinium gypsophilum* (Lewis and Epling, 1959), *Lasthenia ferrisiae* (Ornduff, 1966), *Impatiens aurella* (Ornduff, 1967), *Purshia glandulosa* (Stutz and Thomas, 1964), *Phlox pilosa* ssp. *deamii* and *P. amoena lighthipei* (Levin and Smith, 1966), and *Stephanomeria carotifera* and *S. diegensis* (Gott-lieb, 1972). V. Grant (Paper 7) presents an example of complex retic-ulate evolution in *Gilia*, in which he distinguishes various types of hybrid complexes. The constituents of the homogamic complex in *Gilia* are a cluster of sympatric annual herbs, which are sexual and predominantly diploid. This complex is noteworthy because of the large number of elements involved. Stabilization of the gilias was the result of abiotic selective forces. As a point of contrast, Straw (Paper 8) describes the stabilization of hybrid derivatives by pol-

84

linators in *Penstemon.* This is one of the best-documented cases of the preadaptation of different hybrids to different pollinators not servicing the parental species, and stabilization via assortative pollination and selection. Moreover, it makes a strong case for the hypothesis of sympatric speciation. Another interesting example of this mode of evolution has been reported in *Ophrys* (Stebbins and Ferlan, 1956).

The fixation of sexual diploid hybrids derivatives does not seem to have played a significant role in animal evolution. Chapin (1948) reports that in Africa there are three species of paradise fly-catchers (*Terpsiphone*) that come into contact and hybridize along the edge of the African rain forest, many areas of which have been cleared in recent years. Three areas of hybrid populations are relatively constant and may represent incipient subspecies or species. There is the possibility that certain species of leafhoppers (*Erythroneura*) are products of hybridization, and that *Rana esculenta* is a hybrid derivative (Tunner, 1973; Berger, 1973).

Perhaps the best-documented case of the hybrid origin of several population systems from one pair of "parental species" is seen in the sparrows of the Mediterranean Basin. Johnston (1969) combines the pattern of spatial variation with knowledge of the phenotypes of synthetic hybrids, and the change in local variation patterns through time, to show that the "Italian sparrows" are derived from hybridization between *Passer domesticus* and *P. hispaniolensis.*

In both plants and animals, the stabilization of hybrids by asexual mechanisms may be more common than the stabilization of an intermediate phenotype in a sexual population(s). In plants, some hybrids can reproduce and spread asexually by agamospermous seed formation as well as by vegetative propagation. The asexual reproduction of diverse hybrid types from one pair of species may yield what Grant and Grant call a clonal complex (Paper 9). Their study of *Opuntia spinosior* and *O. fulgida* demonstrate the diversity, number, and distribution of hybrids that asexual reproduction permits. *Opuntia spinosior* also hybridizes with *O. versicolor,* but their hybrids reproduce sexually to produce hybrid swarms and segregating introgressive populations with a high degree of individual variation (Grant and Grant, 1971). Thus we have a good illustration of the controlling influence that the breeding system has on the outcome of hybridization. The vegetative multiplication of diploid semisterile or sterile hybrids is not well documented. Raven (1963) describes an intriguing case involving *Circaea* × *intermedia,* a putative hybrid between *C. lutetiana* and *C. alpina. Circaea lutetiana* is found nearly throughout the British Isles but becomes rarer northward; *C. alpina* is restricted almost exclusively to western England; and *C.* × *intermedia* has a rather broad range throughout the British Isles, but it is most common in Scotland. The hybrid taxon probably is polyphyletic, and apparently it has replaced its parents in certain areas. A similar example from the British flora is *Nuphar* × *intermedia*, the putatively sterile diploid hybrid between *N. lutea* and *N. pumila* (Y. Heslop-Harrison, 1953).

The hybrid origin of asexual, and often unisexual, animals is best documented in fish (Schultz, 1977) and lizards (White, 1978). Schultz (1977) describes an evolutionary explosion of unisexual species in the fish genus *Poeciliopsis*. During the past twenty years, six unisexual "species" have been discovered. They are of hybrid origin and depend on at least one parental species for sperm. The key to hybridity in the genus are *P. monacha* and *P. lucida*, which yield an all-female "species" upon hybridization. The reproductive mechanism of *P. monacha* × *P. lucida*, hybridogenesis, results in the elimination of the entire paternal (*lucida*) genome prior to or during meiosis, and the passing of the maternal (*monacha*) genome to the eggs of the hybrids. In each generation of *P. monacha* × *P. lucida*, a clonally inherited maternal genome combines with a paternal genome drawn anew from the *P. lucida* gene pool. Triploid hybrids involving the same genomes also are known, as are other diploid hybrid species involving *P. monacha* as one parent.

The lizards of the genus *Cnemidophorus* include several asexual "species" of hybrid origin. One of the best-known species is *C. tesselatus*, which comprises several biotypes, some diploid and some triploid. The species is considered to contain one genome of *C. tigris* and one of *C. septemvittatus* (Lowe and Wright, 1966). The study of Parker and Selander (Paper 10) greatly extended the allozyme studies of Neaves (1969), corroborating the origin of *C. tesselatus* and describing in great detail the nature and organization of variation in it, and the possible sources of variation.

# 7

Reprinted from *Evolution* 7:51–64 (1953)

## THE ROLE OF HYBRIDIZATION IN THE EVOLUTION OF
## THE LEAFY–STEMMED GILIAS

VERNE GRANT

*Rancho Santa Ana Botanic Garden, Claremont, California*

### INTRODUCTION

The Leafy-stemmed Gilias, a group of
annual herbs occurring in great diversity
on the Pacific slope of North America,
have long constituted a difficult taxonomic problem. They comprise one of
those so-called "critical groups," in which
the traditional methods of taxonomy have
failed to elucidate the divisions between
the species. It is not possible with the
aid of existing monographs and floras to
properly identify more than a fraction of
the diverse forms of Leafy-stemmed
Gilia that occur in nature. A taxo-
genetic study of the group was undertaken some years ago for the purpose of
discovering the evolutionary causes responsible for this situation. The basic
species were first delimited and described
(Grant, 1950 to 1953, hereafter cited
simply as parts I to V of a series). The
genetic relationships between the species
were next investigated; part VI contains
descriptions of the external morphology,
cytology and fertility of artificial hybrids
between the different species. The docu-
mentary evidence is set forth in this and
preceding papers of the series. The purpose of the present paper is to offer an
explanation of the characteristic pattern
of evolution of the group.

### GENERAL DESCRIPTION

The Leafy-stemmed Gilias comprise a
natural subdivision of the genus *Gilia,*
family Polemoniaceae, technically known
as the section *Eugilia.* The plants com-
posing this section are herbaceous annuals with blue-violet flowers borne in
loose cymes or in capitate heads. The
stems are more or less leafy throughout
their length, in distinction to some related sections which consist of rosette

plants. The members of the group grow
on loose sandy soils in the more open
communities of the Pacific slope of North
America; there is a secondary center of
distribution in temperate South America.
The nine or ten species of Leafy-stemmed
Gilia fall into three main assemblages.
The following brief recapitulation, together with the illustrations in figures 1
and 2 and the distribution maps of figure
3, will acquaint the reader with the general characteristics of the plants.

*The Gilia tricolor group.* These are
small plants which often form very extensive populations on the rolling foothills and plains of California behind the
coastline. The inflorescence is a loose
cyme and the corolla is campanulate in
form and often tri-colored due to the
presence of an orange tube and five pairs
of oval purple spots against the blue-
violet background of the lobes. The
plants are self-compatible, so that pollination of the flowers by bees brings about a
combination of selfing and outcrossing.
There are two species, *G. tricolor* Benth.
(with two subspecies) and *G. angelensis*
V. Grant, both of which are diploid with
9 pairs of chromosomes.

*The Gilia capitata group.* These are
plants of larger size with the flowers
borne in heads. The corolla is funnel-
form and blue-violet throughout. The
two species, *G. capitata* Sims and *G.
achilleaefolia* Benth., usually occur in disjunct local colonies in the hills and mountains of the Pacific states. *Gilia achilleaefolia* is self-compatible and includes,
among its numerous variations, some
large-flowered forms which are partially
cross-pollinated by bees and partially self-
pollinating, as well as small-flowered
forms which are automatically self-polli-

FIG. 1. Growth habit of eight kinds of Leafy-stemmed Gilia, grown under uniform conditions.

nated. *Gilia capitata* contains both self-compatible and self-incompatible races, all of which depend upon insects, mainly bees, for pollination. The eight named subspecies of *G. capitata* fall into three main series: a group of races which occur in mountainous habitats from central California to British Columbia (*G. c. capitata,* etc.); a group of races which inhabit sandy plains and dunes in the San Joaquin Valley and along the Pacific strand (*G. c. staminea,* etc.); and forms in the mountains and foothills of southern California and northern Baja California (*G. c. abrotanifolia*). Both *Gilia capitata* and *G. achilleaefolia* are diploid with n = 9.

*The Gilia laciniata group.* These are maritime plants, often with a prostrate habit of growth and very glandular herbage, which usually grow near the Pacific Ocean. Several of the species are self-compatible and automatically self-pollinated. In North America the group is represented by three species: *G. millifoliata* Fisch. et Mey., a diploid species which grows on coastal sand dunes; *G. clivorum* V. Grant, a tetraploid occurring along the coast and in the interior Coast Ranges of central California; and *G. nevinii* Gray, an endemic on three islands off the coast of southern California and Baja California. The group is represented in South America by two or three species of tetraploids and diploids which, for the time being, may all be referred to as *Gilia laciniata* Ruiz et Pavon sens. lat.

THE VARIATION PATTERN

The species of Leafy-stemmed Gilia are separated in nature by ecological differences, by differences in the period of flowering, and by differences in the form, color pattern and odor of the flowers which enable the pollinating bees to discriminate between them. The reproductive isolation is continued internally by strong incompatibility barriers imposed at various stages in the processes of fertilization and seed formation. It is completed by several kinds of hybrid sterility and by the inviability of a part of the later generation progeny of the hybrids. The species of Leafy-stemmed Gilia are, in short, very well isolated (see part VI).

Knowing these facts, one might have predicted that the Leafy-stemmed Gilias would group themselves into an array of distinct species. Yet as we have seen in the introduction this is by no means the case. A satisfactory classification of the group has hitherto eluded the best efforts of taxonomists; it is difficult, perhaps impossible, to devise a key for the certain identification of all the various forms of Leafy-stemmed Gilia even now. The reason for this situation may be sought in a peculiarity of the variation pattern of the group. The species, as delimited on the basis of crossability and hybrid sterility, are all more or less polytypic and include various intergrading local races. Some of the branches of a species then seem to intergrade with certain branches of other species.

*Gilia capitata*, with its globose flowering heads, approaches *G. millefoliata* with its few-flowered glomerules, on Cape Mendocino in northern California, where colonies exist with loose heads composed of an intermediate number of flowers on long pedicels. They are evidently members of *G. capitata* which vary in the direction of *G. millefoliata* without completely bridging the gap between the two species.

The resemblances in general habit and floral characters between the more southerly races of *G. capitata* and the large-flowered capitate races of *G. achilleaefolia*

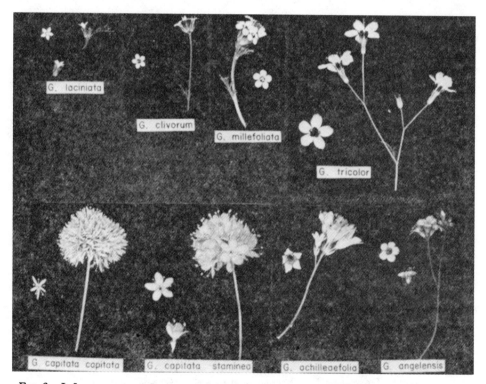

Fig. 2. Inflorescence and flowers of eight kinds of Leafy-stemmed Gilia. From the same plants as in figure 1. All to same scale.

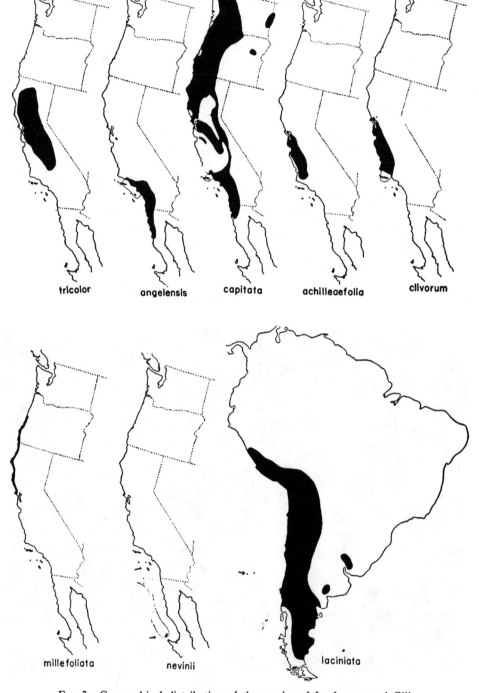

FIG. 3. Geographical distribution of the species of Leafy-stemmed Gilia.

are so striking that the two species have been united in most taxonomic treatments proposed during the past century. The large-flowered and capitate races of *G. achilleaefolia* then intergrade into small-flowered and cymose races of the same species which are so dissimilar that. the two types have been maintained in separate species for over a century.

The trailing, small-flowered races of *G. achilleaefolia* can only be distinguished with difficulty from the trailing, small-flowered types of *G. clivorum*, while the latter intergrade with larger-flowered, purple-spotted, semi-succulent forms of *G. clivorum* which closely resemble *G. millefoliata*. This particular situation is reflected in the earlier taxonomic treatments of the genus, wherein the populations of *G. clivorum* have not been recognized as constituting a distinct species, but have been assigned partly to *G. millefoliata* and partly to the small-flowered types of *G. achilleaefolia*.

The small-flowered, pale-colored, cymose races of *G. achilleaefolia* are so similar in aspect to *G. angelensis* that these two species have not hitherto been separated in taxonomic treatments. Forms of *G. angelensis* are also occasionally found which, in their dense, more capitate inflorescences and more erect habit of growth, vary towards *G. capitata*. This tendency of some populations of *G. angelensis* is matched by a counter-tendency of some types of *G. capitata* in the same area to develop loose heads with fewer flowers and longer pedicels than usual.

To summarize, partial or complete intergradation connects the species of Leafy-stemmed Gilia as follows:

(i) *millefoliata* ↔ *capitata* ↔ *achilleaefolia* ↔ *clivorum* ↔ *millefoliata*.

(ii) *capitata* ↔ *achilleaefolia* ← *angelensis* ↔ *capitata*.

Figure 4 shows what happens when we plot the variations of one species, *G. achilleaefolia*, together with those of the geographically and morphologically closest branches of two other species, *G. capitata* and *G. angelensis*, for two important di-

agnostic characters. A broad overlap is apparent in the variation patterns of the different species. Similar results have been obtained with other characters and other combinations of species.

If intergradation were to be used as a criterion of subspecific distinction, and if the limits of species were to be defined by morphological discontinuities, as is customary in taxonomic practice, the Leafy-stemmed Gilias of continental North America might logically be grouped into just two species, *G. tricolor,* and a huge and complicated assemblage composed of all the other forms. A little field work would soon bring this disposition into conflict with another established criterion of systematics, which holds that two or more natural populations coexisting in one territory are necessarily the members of distinct species, for there are innumerable sympatric contacts between the various kinds of Leafy-stemmed Gilia (see part VI). The foregoing paradox is one which actually confronted the writer at an early stage of the investigations. The genetic evidence, when it became available, was decisive and confirmed the evidence of geographical distribution, rather than that of the variation pattern. There can no longer be any doubt that the vast and complex assemblage of Leafy-stemmed Gilias in California is actually composed of several well isolated species.

EVIDENCES OF HYBRIDIZATION

An explanation is needed for the presence of the intermediate forms which obscure the morphological distinctions between the well isolated species. It is believed that the intermediate types owe their existence to occasional natural hybridization between the species. This hypothesis is supported by several lines of evidence.

In the first place, there is morphological evidence of introgressive hybridization between some of the species, as for example between *G. capitata pacifica* and *G. millefoliata* on Cape Mendocino; between *G. capitata abrotanifolia* and *G.*

*angelensis* in Santa Ana Canyon in the Santa Ana Mountains; and between *G. capitata staminea* and *G. achilleaefolia* in the Inner South Coast Range near Coalinga. In each case a population deviates in its morphological characters from the norm of its species and approaches the conditions found in some other species with which it is growing. The characteristics of the plants are such as to suggest

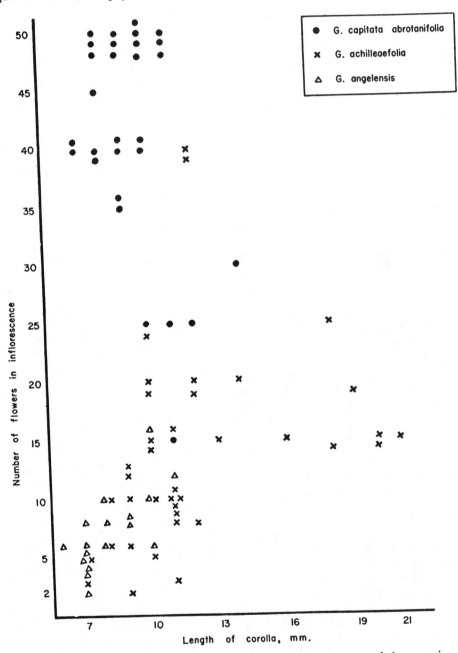

Fig. 4. Scatter diagram showing the overlapping variation patterns of three species. Each symbol represents one population.

initial crossing followed by backcrossing to one or the other parental species. The deviating populations are fully fertile in the wild and their progeny, in those cases where they have been grown, are fertile also in the experimental garden. The fertility relationships of the deviating types with the parental species have not yet been investigated. Population samples and progeny tests show that they are not excessively variable. The populations are thus probably derivatives of introgressive hybridization, which through the years have become genetically stabilized.

The introgressive hybrids show that natural hybridization does occur. The tetraploid species, *G. clivorum*, bears witness to the fact that the natural hybrids may sometimes become established as important biological entities. This species is a product of hybridization between *G. millefoliata* and *G. achilleaefolia*. It crosses with the artificial allotetraploid *G. millefoliata-achilleaefolia* to produce a fertile first generation hybrid with normal meiosis (part V).

It is significant that the taxonomic confusion in the Leafy-stemmed Gilias is confined to just those species which the crossing experiments show will occasionally hybridize (part VI). *Gilia tricolor*, which apparently forms a comparium by itself, since it is not known to hybridize with any other species either in nature or in the breeding plot (part III), presents no difficulties at all from the taxonomic point of view.

A comparison of the different species with regard to their inter-colonial variability is enlightening. The hereditary variability from population to population is very great in some of the species, whereas in others it is much less. The entities which occupy the most extensive and diverse territories, and which therefore might be expected to include ecotypes adapted to the greatest number of climates are *G. capitata capitata* and *G. tricolor*. Another wide ranging taxon is *G. capitata abrotanifolia*. Each of the

two subspecies of *G. capitata* occurs very commonly in several mountain ranges; *G. tricolor* occurs very widely on the foothills of two large mountain systems and in several intermontane valleys. These entities are largely, and in some cases entirely, cross-pollinated. Yet they are far from being the most variable species or races.

*Gilia capitata pacifica*, which occupies a narrow strip along the coast of northern California, possesses a great deal more inter-colonial variability than does *G. c. capitata* in the whole area from San Francisco Bay to Vancouver Island and from the maritime Coast Ranges to Idaho and eastern Oregon. It is more variable than *G. millefoliata*, which is equally widely diffused in the same area. *Gilia achilleaefolia*, which is restricted to a relatively small area in the South Coast Range and which reproduces partly or entirely by self-pollination, is easily the most variable species in the whole section. *Gilia clivorum*, with a more extensive area of distribution but capable of even less cross-breeding, is likewise highly variable. The wide-ranging and extensively cross-breeding *Gilia tricolor*, which occurs in the territory of *G. achilleaefolia* and *G. clivorum* and elsewhere too, cannot begin to compare with these two species in variability.

It is doubtful whether the excessive variability of *G. capitata pacifica*, *G. achilleaefolia* or *G. clivorum* can be entirely explained as a result of the accumulation of genetic changes which have occurred within these species or subspecies themselves. A far more likely explanation, in the opinion of the author, is that the variability in excess of the amount present in *G. capitata capitata* and *G. tricolor*, is a consequence of hybridization between reproductively isolated populations.

It is noteworthy in this connection that the most variable entities are also intermediate in a large number of morphological and ecological characters between certain other entities. *Gilia capitata pacifica* is intermediate between *G. capitata capitata* and *G. c. chamissonis* (part I); *G. achilleaefolia* is intermediate between *G. tricolor* and *G. c. staminea* (part IV); *G. clivorum* is intermediate between *G. millefoliata* and *G. achilleaefolia* (part V). The association of morphological intermediacy with great genetic variability is a known result of interracial or interspecific hybridization. It has therefore been suggested in earlier papers that the variable and intermediate types are derivatives of hybridization.

## THE ORIGINAL SPECIES

An obvious question concerns the identity of the original species between which the hybridizations occurred. The author regards the original species in North America as being: *G. tricolor* and *angelensis* in the *tricolor* group; *millefoliata* and *nevinii* in the *laciniata* group; and *abrotanifolia, capitata* and *staminea* in the *capitata* group.

The circumstance that the last named entities are treated taxonomically as subspecies should not be allowed to obscure the biological facts, which point to the occurrence of speciation in their past history (parts I and II). *Gilia capitata staminea* grows in the territory of *G. c. capitata* today in the Vaca Mts. of central California. Gene exchange between them is prevented by the fact that the former type inhabits the sandy washes and blooms early, while the latter grows on rocky canyon slopes and flowers later. Colonies of a maritime ecotype of *staminea* (taxonomically, *G. c. chamissonis*) and a maritime ecotype of *capitata* (*G. c. tomentosa*) occur less than 500 feet apart at Tomales Bay. This distance is within the range of seed dispersal of Gilias. Yet the sympatric populations do not hybridize today. This is shown by the fact that the *capitata* populations are 100% pure in respect to a recessive marker gene, controlling an easily scored pubescence character, the dominant allele of which is carried by the neighboring *staminea* populations. The isolated populations are separated in this case by pronounced ecological differences accompanied by weak incompatibility barriers. A foothill ecotype of *staminea* (*G. c. pedemontana*), finally, grows sympatrically with *G. c. abrotanifolia* in the Sierra Nevada, where gene exchange is barred by ecological and ethological isolation.

The existence of reproductive isolation between *G. c. capitata, G. c. staminea* and *G. c. abrotanifolia* shows that these entities have reached the stage of divergence characteristic of species. The process of speciation was reversed, however, by subsequent hybridization between *G. c. staminea* and *G. c. capitata*, in consequence of which a series of intermediate and highly variable populations arose in certain ecologically intermediate areas (part I).

The author's conception of phylogeny in the North American Leafy-stemmed Gilias is portrayed in figure 5. The skeletal phylogenetic tree in this figure, which is outlined in solid black lines, depicts the divergence of the original species just enumerated. Subsequent hybridizations between these basic species, indicated in the figure by the broken lines, have resulted in the origin of additional colonies, races and species intermediate between the former. The hybridizations have converted the phylogenetic tree into a phylogenetic net.

It is improbable that figure 5 does full justice to the actual complexity of the situation. For example, hybridization may have occurred in the remote past between *G. tricolor* and *G. millefoliata* when the former species did not constitute a separate comparium. Such hybridization would explain the presence in *G. millefoliata* of purple corolla spots identical in number and pattern, but a shade less bright in color, to those found in *G. tricolor*. Since the corolla spots are polygenically determined, as shown by the occurrence of at least four phenotypic classes of brightness in the various species and hybrids, the likelihood of this character having arisen independently in

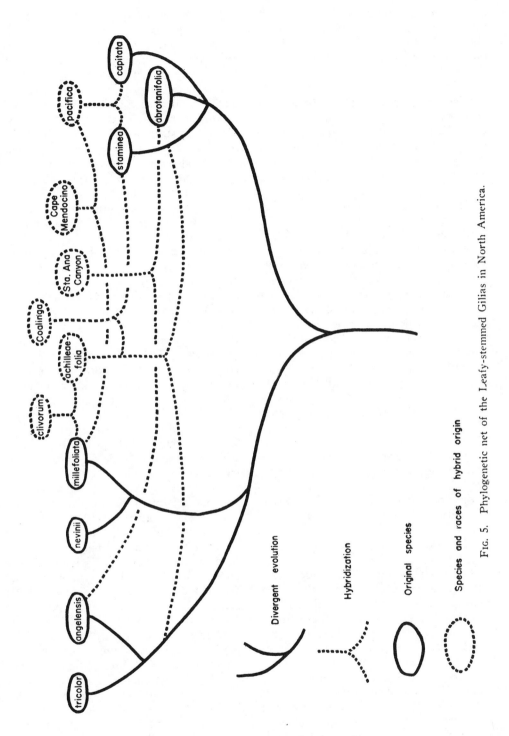

Fig. 5. Phylogenetic net of the Leafy-stemmed Gilias in North America.

two different evolutionary lines is less than the chance of its having been introduced into one of the lines by hybridization. The hypothesis of a hybrid origin of the corolla spots in *G. millefoliata,* which occupies an area contiguous to *G. tricolor,* would also explain the absence of spots in the closest relatives of *G. millefoliata,* namely *G. nevinii* and *G. laciniata,* which occur far to the south of the range of *G. tricolor* and may have never come into contact with this species.

The available evidence suggests that the pattern of hybridization was established in the Leafy-stemmed Gilias in ancient times. The age of the tetraploid *G. clivorum* is sufficiently great to have enabled it to fill a niche of its own in the oak woodland and grassland savannah of the California Coast Ranges over a longitudinal distance of 400 miles. Consideration of the amount of time required for the colonization of an area of this size imposes a definite minimum limit on the antiquity of this species. Yet *G. clivorum* could not have arisen until one of its diploid ancesters, *G. millefoliata,* had acquired the characteristic corolla spots from *G. tricolor* by introgression and until the other diploid ancestor, *G. achilleaefolia,* had originated following hybridizations between *G. tricolor* and *G. capitata.* If *G. clivorum* is old, the hybridizations of its ancestral diploid species with *G. tricolor* must be even older.

## The Hybrid Complex

The Leafy-stemmed Gilias resemble in their variation pattern some other taxonomically difficult groups, such as *Rubus* § *Eubatus,* the *Crepis occidentalis* group, *Phleum pratense* and its allies, the *Artemisia tridentata* group, *Oenothera* § *Euoenothera,* and the like, in that the species are not well separated morphologically. The resemblance was puzzling at first, because the species of Leafy-stemmed Gilia are sexual, possess a normal chromosome cycle, and are predominantly diploid. They evidently do not comprise

a polyploid complex, an agamic complex, or an assemblage of ring-forming structural heterozygotes. There is, however, an evolutionary factor common to all these complexes, which may explain their resemblances on the taxonomic level. That factor is hybridization. In some groups it has been associated with numerical or structural changes of the chromosomes or with changes in the method of reproduction. The effects of interspecific hybridization, however, are not necessarily manifested in easily recognized permanent alterations of the chromosomal mechanism. Hybridization may blur the distinctions between species without producing any easily recognizable cytogenetic features in the hybrid derivatives.

Groups of species in which hybridization has obscured the morphological discontinuities between the basic diploid types may be termed collectively *hybrid complexes.* The known types of hybrid complexes, arranged in order of decreasing taxonomic complexity, are as follows:

*Agamic complex* (Babcock and Stebbins, 1938), in which the hybrids or hybrid derivatives reproduce partially or wholly by unfertilized seeds or bulbils, which may attain a wide dispersal. Typical agamic complexes are *Crepis* § *Psilochaenia* (*op. cit.*) and *Rubus* § *Eubatus* (Gustafsson, 1942).

*Polyploid complex* (Babcock and Stebbins, 1938), in which the derivatives of hybridization are sexual polyploids. This type of hybrid complex is found in the genera *Clarkia* (H. Lewis, unpublished), *Iris* (Viosca, 1935; Anderson, 1936; Lenz, unpublished), and *Artemisia* (Keck, 1946; Ward, 1951).

*Heterogamic complex,* in which the products of hybridization are permanent structural heterozygotes. The best example of a heterogamic complex is *Oenothera* § *Euoenothera* (Cleland *et al.,* 1950).

*Clonal complex,* in which the hybrids reproduce mainly or entirely by clonal divisions and therefore tend to remain more or less localized. A good example is afforded by *Opuntia*.

*Homogamic complex,* in which the hybrid derivatives are sexual with a normal meiotic cycle and are predominantly or entirely diploid. Structural hybridity may or may not be present in the first generation hybrids; if present, it is eliminated in the later generation progeny of the hybrids by selection for sexual fertility. Homogamic complexes are developed in *Quercus* (Stebbins, 1950, chap. 2; Tucker, 1952a, 1952b; Benson, 1951), *Ceanothus* (McMinn, 1944; Nobs, 1951), *Aquilegia* (Grant, 1952a), and *Gilia* § *Eugilia.*

In their historical development, agamic, polyploid, heterogamic and clonal complexes apparently pass through a period of expansion and a period of decline. The complexes in which sexual recombination is inhibited in one way or another can expand until they have reached a climax of variability. Their fate from this point on is to gradually die out; they can give rise to no new lines of evolution (Stebbins and Ellerton, 1939; Stebbins, 1940, 1941, see also 1950, chaps. 8–11). In the homogamic complex, on the other hand, the derived forms are as fully capable of producing new gene recombinations and variations as the original species. It would seem to follow that a homogamic complex is not specially limited in its evolutionary potentialities. The original species may or may not become extinct, but the group as a whole is not predestined to decline. It may, on the contrary, continue to evolve and give rise to new evolutionary developments.

There is a second distinction between the homogamic complex and the other type of hybrid complex, which Stebbins has pointed out in a personal communication. In the agamic, polyploid, heterogamic and clonal complexes the derived types may be easily recognized by their cytological and genetic behavior. It is consequently possible to identify such complexes even after the extinction of the parental species. But the recognition of a homogamic complex as such depends upon the contemporaneous existence of both the original and the derived species. We can recognize the last relics of agamic, polyploid, heterogamic and clonal complexes which have existed for many ages and have undergone many migrations and evolutionary changes, but we can only recognize the more recent and actively evolving homogamic complexes which have not suffered widespread extinctions.

The identification of the original species in a homogamic complex is a highly uncertain procedure. How is one to decide that a supposedly original species is not really the descendant of hybrids between older species which have since disappeared from the scene? The original species of one homogamic complex could very well be the final products of pre-existing homogamic complexes. Even the most intensive study of a homogamic complex can permit only minimal estimates of the importance of hybridization in the evolution of such a group.

Anderson believes that hybridization has played a major role in the evolution of most agricultural plants and weeds. (Anderson, 1952). The Leafy-stemmed Gilias represent a group devoid of agricultural uses or weedy tendencies which has apparently been hybridizing since ancient times. It is fair to inquire how widespread the pattern of hybridization has been in the evolution of the Angiosperms as a whole. According to Stebbins (1947), approximately half of the species of flowering plants are polyploids. If, as seems probable, most natural occurring polyploids are allopolyploids (Clausen, Keck and Hiesey, 1945; Stebbins, 1947), then at least half of the species of flowering plants are of hybrid origin. This again is only a minimal estimate of the extent of hybridization.

Perhaps the most realistic appraisal of the influence of hybridization in the evolution of the Angiosperms can be gained from a consideration of the confused taxonomic structure of this class. The process of adaptive radiation has brought about fairly clear-cut divisions between the major phyletic lines in most animal groups, not excluding the almost infinitely numerous and varied class of insects. As a result zoologists have been able to devise a generally satisfactory and standardized system of classification of animals. Several generations of plant taxonomists have been unable to duplicate this achievement for the Angiosperms. A prominent plant taxonomist has recently stated that "None of [the existing classifications of higher plants] begins to answer the requirements of a truly phylogenetic system, and no system of the foreseeable future can attain such a goal" (Lawrence, 1952).

Though a primary divergence in the Angiosperms between the Monocotyledons and Dicotyledons was recognized several centuries ago, the intensive activity of the 19th and 20th centuries has not led to the recognition of major lines of evolution diverging within these two groups. Consequently there is no generally accepted classification of the Dicotyledons and Monotyledons into orders. There is apparently a reticulate, rather than a dichotomous pattern of relationships among the families and orders of flowering plants. This reticulate pattern seems to characterize the relationships of even the most primitive families of Angiosperms (Bailey and Nast, 1945; Money, Bailey and Swamy, 1950).

The reticulate taxonomic structure of the flowering plants can be explained by a modification of an hypothesis of Stebbins (1947), which holds that hybridization was of widespread occurrence during the early phases of angiosperm evolution, just as it has been in the more recent evolution of gilias, ceanothi, columbines, oaks, brambles, agricultural plants and weeds. This hypothesis is consistent with the known facts concerning the frequency of polyploidy in the Angiosperms. It accounts for the lack of clear discontinuities between the major phyletic lines by postulating the occurrence of interbreeding between those lines during the early stages of their divergence. It supposes that the Angiosperms have evolved, not along exclusively divergent lines, as seems to have been the case in most animal groups, but as a succession of hybrid complexes. The occurrence of sporadic hybridizations during the course of Angiosperm evolution may be the factor which has caused this group to grow up, not as a phylogenetic tree, but as a gigantic, snarled phylogenetic net.

Since their origin the Angiosperms have explored many new ways of life and have pursued long continued trends of specialization in both their vegetative and reproductive characters. This occurrence of an open system of evolution is probably consistent with the presence of a reticulate pattern of relationships between families and orders only if it is assumed that the Angiosperms have comprised a succession of homogamic complexes. This is because the homogamic complex is the only type of hybrid complex in which hybridization can continue to occur without resulting in the imposition of restrictions on the free recombination of genes in the derived species.

The complexity of the angiospermous phylogenetic net is no doubt increased by polyploidy, as suggested by Stebbins (1947, 1950). An unknown proportion of the families and genera may have arisen from tetraploids which have come to behave genetically like diploids. Most of the higher categories, however, have probably originated from diploid stocks. These ancestral stocks may have been hybridizing on the diploid (or diploidized) level since the earliest stages of Angiosperm evolution.

## SUMMARY

Groups of organisms in which hybridization has blurred the morphological discontinuities between the basic species are collectively designated as hybrid complexes. Hybrid complexes may be subdivided according to the mode of reproduction of the derived forms into five main types, namely agamic, polyploid, heterogamic, clonal, and homogamic complexes. In homogamic complexes the hybrid derivatives are diploid, sexual populations with a normal chromosome cycle at meiosis, rather than apomicts, polyploids, permanent structural heterozygotes, or clones. There is consequently much less restriction on gene recombination in the derivatives of homogamic complexes than in the end products of agamic, polyploid, heterogamic or clonal complexes. As a result, homogamic complexes possess an open system of evolution, as opposed to the closed system of evolution which characterizes the other types of hybrid complex.

Evolution in a homogamic complex is illustrated by the Leafy-stemmed Gilias (Polemoniaceae), a cluster of related species of annual herbs which grow sympatrically in California. The species are sexual with a normal meiotic cycle and are predominantly diploid. They are well isolated from one another in nature by a variety of external and internal isolating mechanisms. Yet they seem to intergrade in a manner which makes them exceedingly difficult to classify. There is evidence that the taxonomic complexity of the group is due to occasional hybridization between populations which under most circumstances are well isolated. The Leafy-stemmed Gilias exemplify a homogamic complex, in which hybridization could continue indefinitely without reducing the ability of the group to make progressive evolutionary advances.

The Angiosperms as a whole have undergone many progressive evolutionary changes while developing as a phylogenetic net rather than as a conventional phylogenetic tree. In the light of independent evidence for a widespread occurrence of hybridization in the past history of the Angiosperms, this combination of a reticulate pattern of relationships with progressive trends of specialization suggests that the class may have evolved as a succession of homogamic complexes.

## ACKNOWLEDGMENTS

A part of the work on the Leafy-stemmed Gilias was supported by a National Research Fellowship, for which the author is duly grateful. The manuscript was much improved by thoughtful criticisms and suggestions contributed by Drs. G. Ledyard Stebbins and Harlan Lewis. The author wishes to acknowledge his gratitude to these colleagues for their helpful interest.

## LITERATURE CITED

ANDERSON, E. 1936. The species problem in *Iris*. Ann. Missouri Bot. Gar., 23: 457–509.

——. 1952. Plants, man and life. Boston.

BABCOCK, E. B., AND STEBBINS, G. L. 1938. The American species of *Crepis*. Carnegie Inst. Washington, Publ. 504.

BAILEY, I. W., AND NAST, CHARLOTTE. 1945. The comparative morphology of the Winteraceae. VII. Summary and conclusions. Jour. Arnold Arboretum, 26: 37–47.

BENSON, L. 1951. Principles of plant classification. Mimeographed preliminary edition of a textbook, Claremont, Calif.

CLAUSEN, J., KECK, D. D., AND HIESEY, WM. M. 1945. Experimental studies on the nature of species. II. Plant evolution through amphiploidy and autoploidy, with examples from the Madiinae. Carnegie Inst. Washington, Publ. 564.

CLELAND, R. E. 1950. Studies in *Oenothera* cytogenetics and phylogeny. Indiana Univ. Publ., Science Ser., No. 16.

GRANT, V. 1950. Genetic and taxonomic studies in *Gilia*. I. *Gilia capitata*. El Aliso, 2: 239–316.

——. 1952. idem. II. *Gilia capitata abrotanifolia*. El Aliso, 2: 361–373.

——. 1952. idem. III. The *Gilia tricolor* complex. El Aliso, 2: 375–388.

——. 1953. idem. IV. *Gilia achilleaefolia*. El Aliso, 3; in press.

——. 1953. idem. V. *Gilia clivorum*. El-Aliso, 3; in press.

——. 1953. idem. VI. Interspecific relationships in the Leafy-stemmed Gilias. El Aliso, 3; in press.

——. 1952a. Isolation and hybridization between *Aquilegia formosa* and *A. pubescens*. El Aliso, 2: 341–360.

GUSTAFSSON, Å. 1942. The origin and properties of the European blackberry flora. Hereditas, 28: 249–277.

KECK, D. D. 1946. A revision of the *Artemisia vulgaris* complex in North America. Proc. Calif. Acad. Sci., 25: 421–468.

LAWRENCE, G. H. M. 1952. Morphology and the taxonomist. Phytomorphology, 2: 30–34.

LEWIS, H. 1953, Vol. III. The pattern of evolution in the genus *Clarkia* (Onagraceae). Evolution, in press.

McMINN, H. E. 1944. The importance of field hybrids in determining species in the genus *Ceanothus*. Proc. Calif. Acad. Sci., 25: 323–356.

MONEY, LILLIAN, BAILEY, I. W., AND SWAMY, B. G. L. 1950. The morphology and relationships of the Monimiaceae. Jour. Arnold Arboretum, 31: 372–404.

NOBS, M. 1951. *Ceanothus*. Carnegie Inst. Washington Year Book, 50: 117–118.

STEBBINS, G. L. 1940. The significance of polyploidy in plant evolution. Amer. Nat., 74: 54–66.

——. 1941. Apomixis in the Angiosperms. Bot. Rev., 7: 507–542.

——. 1947. Types of polyploids: their classification and significance. Advances in Genetics, 1: 403–429.

——. 1950. Variation and evolution in plants. Columbia Univ. Press, New York.

—— and ELLERTON, S. 1939. Structural hybridity in *Paeonia californica* and *P. brownii*. Jour. Genetics, 38: 1–36.

TUCKER, J. M. 1952a. Taxonomic interrelationships in the *Quercus dumosa* complex. Madroño, 11: 234–251.

——. 1952b. Evolution of the Californian oak *Quercus alvordiana*. Evolution 6: 162–180.

VIOSCA, P. 1935. The Irises of southeastern Louisiana. Bull. Amer. Iris Soc., 57: 3–56.

WARD, G. H. 1951. The *Artemisia tridentata* complex. Carnegie Inst. Washington Year Book, 50: 119–120.

# 8

Reprinted from *Evolution* 9:441–444 (1955)

# HYBRIDIZATION, HOMOGAMY, AND SYMPATRIC SPECIATION

RICHARD M. STRAW

*Rancho Santa Ana Botanic Garden, Claremont, California* [1]

Received January 18, 1955

Mayr (1947) has presented a strong case against the hypothesis of sympatric speciation in the course of supporting the universal hypothesis of geographic (or at least microgeographic) speciation. In the summary (p. 286) of that paper he made the following statement: "It is unproven and unlikely that reproductive isolation can develop between contiguous populations." Later, Grant (1949), in a discussion of pollination systems as isolating mechanisms, presented a model for sympatric speciation in angiosperms dependent on the known flower constancy of bees. This model indicated the means by which an important mutation in flower color or form might become established within a population, thus resulting in time in a new and well isolated species to which internal isolating mechanisms are unnecessary. In passing, the latter author suggested that hybridization might also provide an example of sympatric speciation. It is the purpose of the present paper to further develop this last hypothesis on the basis of a realistic model suggested by recent studies of hybridization and floral isolation in a group of *Penstemon* species. It is not the purpose of this paper to refute the general conclusions reached by Mayr (l.c.) on the importance of geographic speciation, but merely to indicate that sympatric speciation may be less unlikely than previously considered. The word, species, is here used in its biological sense.

## HYBRIDIZATION IN PENSTEMON

Four species of the section *Peltanthera* Keck of the genus *Penstemon* (Scrophulariaceae) are found in cismontane southern California. One of these, *P. cen-*

tranthifolius, belongs to the subsection *Centranthifolii* (Keck, 1937), the other three, according to Keck (l.c.), to the subsection *Spectabiles: PP. grinnellii, spectabilis,* and *clevelandii.* Despite this taxonomic disposition, these last three are quite diverse in external appearances, such as corolla color, shape, and size, and there is every indication that they belong to three different species complexes within which speciation appears to have been predominantly geographic.

*Penstemon centranthifolius* is a moderately tall (2½ ft.), erect plant with narrowly tubular bright scarlet flowers about an inch long, which are borne on lax pedicels in a semi-pendant manner. It is particularly visited and pollinated by hummingbirds (in southernmost California, mostly *Calypte anna* and *C. costae*), and is a plant of open to moderately open sites on light sandy soils in the coastal sage scrub, chaparral, pinyon-juniper, and yellow pine forest communities.

*Penstemon grinnellii* is a lower, bushier plant with pale-colored flowers (white to suffused pink or blue) characterized by short, narrow tubes, abruptly expanded, broadly ringent throats and abundant lips, and borne on stout, stiff pedicels. It is a plant of rocky and sandy soils in the yellow pine forest, where it is pollinated by large carpenter bees (*Xylocopa* spp.) whose bodies fit this corolla well.

The other two species are intermediate between these extremes of flower morphology. *Penstemon spectabilis* has flowers whose tubes are about twice the length of those of *P. grinnellii*, are only moderately expanded in the throat, and are mainly of bluish-purple colors. It is a semi-shrubby perennial pioneer in coastal sage scrub and chaparral communities, and is

---

[1] Present address, Deep Springs, California.

pollinated by the larger wasps of the genus *Pseudomasaris*, mainly *Ps. vespoides*. *Penstemon clevelandii*, the fourth member, is a smaller plant which grows among the rocks of desert-edge canyons in chaparral and pinyon-juniper communities. Its flowers are of much the same shape as those of *P. spectabilis* but are significantly smaller and of reddish-purple and intense magenta colors. The usual pollinators of this species are smaller solitary bees (Anthophoridae) and hummingbirds. All three insect-pollinated species are visited frequently by a small host of other insects, especially smaller leaf-cutter bees (Megachilidae), and *P. centranthifolius* is rarely visited by them as well.

*Penstemon centranthifolius* occurs sympatrically with all of the other species mentioned here. The hybrid *P. centranthifolius × spectabilis* (= *P. × parishii*) is fairly common in nature where the parents grow together, *P. centranthifolius × grinnellii* (= *P. × dubius*) occurs rarely (having been recorded only twice so far as I am aware), and the combination *P. centranthifolius × clevelandii* is suggested from slight evidence of introgression in the field (Straw, 1955). At the eastern edge of its range *P. spectabilis* transcends the range of *P. clevelandii*, and the appearance of bearded staminodes in colonies of *P. spectabilis* suggests gene flow into that species from *P. clevelandii*. *Penstemon grinnellii* and *P. spectabilis*, although normally ecologically separated, occasionally occur together as a result of exceptional range extensions of either species into the territory of the other. There is some inconclusive evidence of introgression of *P. spectabilis* by *P. grinnellii* also (Straw, l.c.). There is no evidence that *PP. clevelandii* and *grinnellii* grow together in nature (both species occur on Santa Rosa Mountain, Riverside County) although it is conceivable that they might rarely do so if *P. grinnellii* establishes itself at a low elevation. The hybrid *P. clevelandii × grinnellii* has, however, been successfully made in the garden (Lenz, unpublished). All of these species have

the same chromosome numbers ($n = 8$) (Keck, 1951).

Despite the facts that these species occur together in nature and apparently are all interfertile (genetic analysis is still incomplete), there is no evidence that any of the species concerned is in any danger of being swamped. There is no sign of hybrid swarms even in colonies containing very conspicuous hybrids (*PP. centranthifolius, spectabilis*, and their hybrid, *P × parishii*). Since these species all have markedly different pollinating mechanisms (hummingbirds, wasps, carpenter bees, solitary bees), it appears that hybridization is limited by their floral isolation, and that hybrids occur only rarely under special circumstances. The combination of very different floral mechanisms and high interspecific fertility relationships places *Penstemon* in a class with some other conspicuous western genera such as *Aquilegia* (Grant, 1952) and *Delphinium* (Epling, 1947; Grant, 1949: 91).

## COMPARATIVE MORPHOLOGY OF SPECIES AND HYBRIDS

The precise details of hybridization and introgression between these four species are unimportant to the present thesis and will be presented elsewhere (Straw, 1955). It is significant and suggestive, however, that two groups of hybrids, those between *P. centranthifolius* and *PP. grinnellii* and *palmeri* (a desert sister-species of *P. grinnellii* that resembles the latter in floral morphology and has the same pollinators) and between *P. centranthifolius* and *P. spectabilis*, resemble in their floral characters two other species growing in the same region, and that all of these species have distinct pollinating mechanisms.

The most extreme crosses, between *P. centranthifolius* and *PP. grinnellii* (known in nature) and *palmeri* (garden-produced hybrid by Lenz, unpublished), have flower shapes intermediate between the shapes of their parents. The tube of the corolla is longer in the hybrids than in *PP. grinnellii* and *palmeri*, while the throats of the hybrids are moderately expanded, in contrast to the almost unexpanded throat of *P. centranthifolius* and the greatly expanded throats of the other two parents. These characteristics in flower shape of the hybrids cause a striking resemblance to another, different species, *Penstemon spectabilis*, which is very common in nature. Such differences as are apparent in color of the corolla and development of a beard on the staminode are of a kind that can be easily modified by selection.

The important point in this situation is that by a single step of hybridization a new type is created that is intermediate between two other flower types in very much the same way that another related species is intermediate. All three of the species (*PP. centranthifolius, grinnellii, spectabilis*) have very different and relatively exclusive pollinators: a hummingbird, a wasp, and a bee (Straw, 1955). By an extension of this fact, it can be reasoned that should this wasp, normal to the intermediate species, happen upon a colony containing the three types it would in all likelihood visit the intermediate type, i.e., the hybrid, to the exclusion of the extreme parental types. By the same token, the normal pollinators of the parents would tend to avoid the hybrid, being excluded mechanically on the one hand and ethologically on the other.

The other important natural hybrid, *Penstemon centranthifolius × spectabilis,* resembles in shape, size, and flower color the fourth member of our series, *P. clevelandii*. This resemblance is so close that the hybrid has been and still is occasionally confused with *P. clevelandii* on herbarium sheets (Keck, 1937: 819). Here again a new type is produced by hybridization corresponding to another established species which has its own range of pollinators at least partly different from either of the hybrid's parents.

These hybrids are, furthermore, both viable and at least partly fertile. *Penstemon centranthifolius × spectabilis* has a pollen fertility averaging about 50 per cent, as tested by aniline blue in lactophenol, *P centranthifolius × grinnellii* measures 86 per cent fertile, and the $F_1$

*P. centranthifolius × palmeri* has a corresponding fertility of about 35 per cent. Second and subsequent generations, largely backcrosses, are known in two cases, with no evident breakdown so far detected. Such hybrids, even more than the mutant type suggested by Grant (1949), are available through selective pollination for the establishment of new species. There is an indication that precisely this mechanism has been effective, at least rarely, in speciation in the genus *Penstemon*. It is very tempting to postulate such an origin for both *P. spectabilis* and *P. clevelandii*, but so definite a statement must await more evidence, now being sought by genetic techniques.

That sympatric speciation by this method has not been more effective in the production of species is probably due partly to the fact that relatively few such distinct niches are available in nature, the two suggested here having already been filled by good species, and partly to a mechanism identified by Grant (1952) in *Aquilegia*. He showed first, that the intermediate types are relatively less visited by pollinators than the parental extreme types, and second, that the backcross types to either parent are successively submerged into the extreme types by selective pollination favoring the more extreme forms. A study of *Penstemon centranthifolius × spectabilis* has shown that it is infrequently visited by any pollinators, but more frequently visited by pollinators of the larger-flowered parent, *P. spectabilis*. Most backcrosses tentatively identified in the field seem to favor that parent, and are probably quickly incorporated into that rather variable species. They may even be the cause of much of its variability.

The gene flow that occurs through introgression or through backcrossing of hybrids to parents is thus, apparently, held to its minimum effectiveness by strong selection pressure due to specific and adaptive floral types and mechanisms. The selective pressure of the pollinators in keeping the species' morphological norms intact and preventing or minimizing the effects of hybridization seems to be greater than other genetic mechanisms operative here, even in the rather small populations normal to *Penstemon* species.

## HOMOGAMY AND SYMPATRIC SPECIATION

Mayr (1947: 272–273) reviewed the evidence for homogamy and concluded that " . . . the evidence in favor of homogamy is virtually nonexistent and that homogamy, where it exists, is not of the type that would lead to the establishment of discontinuities within populations." If, however, we slightly modify the lead sentence in his section on this subject to read, "The concept of homogamy states that within a population the most similar individuals *will preferentially be mated to each other,*" so that it applies better to plants, we find that this is precisely and completely the mechanism by which pollination as an isolating mechanism operates, and is the whole basis for the importance of pollinating systems to the evolution of flowering plants (cf. Grant, 1949, 1950). It is, as well, the mechanism by which swamping is prevented in the cases of *Aquilegia* and *Penstemon* mentioned above.

All that is required for sympatric speciation by means of homogamic selective pollination is that some insect not a normal pollinator of either of the parental species "adopt" a hybrid type and remain constant to it. Considering the many species of bees in many regions, this is by no means either an impossibility or an extreme improbability. Such a bee does not even have to be completely constant for its selective effect to be sufficiently intense to satisfy Hogben's calculations (1946, as quoted by Mayr, 1947: 273). That this is so is shown by the above mentioned facts that despite high inter-fertility of species and occasional inconstancy of pollinators there is no sustained tendency for discrete species of *Aquilegia* or *Penstemon,* at least, to merge into one another. Moreover, homogamy in the present case is in no way dependent on any theories of blending inheritance, upon which it was first postulated.

## SUMMARY

A model is suggested in which, through hybridization and pollination by essentially flower-constant insects, sympatric speciation in angiosperms becomes a possible and plausible mechanism. There is evidence that despite frequent hybridizations there is no tendency for distinct species of *Penstemon* to be swamped. The hybrids between some species of *Penstemon* resemble other discrete species, all of which have their own normal range of selective pollinators. Such hybrids may form, or may have formed in the past, new species. The mechanism by which this operates appears to be true homogamy as defined essentially by Mayr (1947). It is thus possible that through homogamy new species may become established without either geographic or strong genetic isolation from their parents.

## ACKNOWLEDGMENTS

The author is pleased to recognize the advice and criticism of Prof. Verne Grant throughout his research and the preparation of the present paper, but accepts full responsibility for the facts and opinions found therein. The research basic to this paper has been supported by pre-doctoral Fellowships from the National Science Foundation, whose aid is gratefully acknowledged.

## LITERATURE CITED

EPLING, C. 1947. Actual and potential gene flow in natural populations. American Naturalist, **81**: 104–115.

GRANT, V. 1949. Pollination systems as isolating mechanisms. EVOLUTION, **3**: 82–97.

———. 1950. The protection of the ovules in the flowering plants. EVOLUTION, **4**: 179–201.

———. 1952. Isolation and hybridization between *Aquilegia formosa* and *A. pubescens.* El Aliso, **2**: 341–360.

KECK, D. D. 1937. Studies in *Penstemon.* V. The section *Peltanthera.* American Midland Naturalist, **18**: 790–829.

———. 1951. *Penstemon,* in Abrams, Illustrated Flora of the Pacific States, **3**: 733–770.

MAYR, E. 1947. Ecological factors in speciation. EVOLUTION, **1**: 263–288.

STRAW, R. M. 1955. Floral ecology and evolution in *Penstemon.* Ph.D. Thesis, Claremont Graduate School, unpublished.

# 9

Reprinted from *Evolution* 25:144–155 (1971)

## DYNAMICS OF CLONAL MICROSPECIES IN CHOLLA CACTUS

Verne Grant and Karen A. Grant[1]

*Department of Botany, University of Texas, Austin, Texas*

Microspecies have been redefined recently, following long usage, as plant populations which reproduce mainly if not exclusively by uniparental methods, are morphologically uniform, occupy a definite geographical or microgeographical area, are differentiated morphologically—often slightly—from related species and microspecies, and frequently possess a hybrid constitution (see Grant, 1971, chap. 4, for discussion). Different kinds of microspecies can be recognized according to the mode of uniparental reproduction. Clonal microspecies are those that reproduce by various means of vegetative propagation.

In theory, the development of a clonal hybrid microspecies entails the simple multiplication of hybrid individuals by vegetative means, and the dispersal and establishment of the daughter individuals throughout some suitable territory. The increase in population size presumably passes through the continuous and successive stages of small clone, endemic microspecies, and geographically widespread microspecies (Grant, 1971). In practice we have very little factual information about these population units and their modes of population growth; no good studies are available of clonal hybrid microspecies in plants.

A sterile species hybrid in plants can reproduce and spread asexually by agamospermous seed formation as well as by vegetative propagation. Many cases are known in *Crepis, Taraxacum, Rubus* and other groups in which sterile interspecific hybrids have spread widely by agamospermy (Gustafsson, 1946–1947; Stebbins,

1950; Grant, 1971). But the vegetative multiplication of a sterile hybrid has been recorded in only a few instances, such as *Elymus triticoides* × *condensatus* (Stebbins, 1959), and has never been investigated or documented in any detail so far as we are aware. The purpose of this paper is to describe a case of clonal hybrid microspecies in the cholla cacti.

We initially selected the genus *Opuntia* as a potentially good group in which to study vegetative multiplication of hybrids, since these cacti are known to hybridize naturally and to propagate freely from their stem joints. The *O. phaeacantha* group in the subgenus *Platyopuntia* is currently being studied taxonomically and biosystematically by Drs. Lyman Benson, D. L. Walkington, Donald Pinkava, and their students (see Benson, 1969a, 1969b). We have concentrated our attention on the other main subdivision of *Opuntia*, the subgenus *Cylindropuntia* or cholla cacti.

The point of departure for our study was provided by an early report of natural hybridization between the cholla species, *O. spinosior* and *O. fulgida*, along the Gila River in central Arizona (Kearney and Peebles, 1942). Kearney and Peebles stated (1942, p. 616): "An apparent hybrid between *O. spinosior* and *O. fulgida* is rather abundant in the bed of the Gila River between Florence and Casa Blanca, Pinal County. The hybrid plants propagate freely by means of fallen joints." This report was confirmed and repeated by the same and other authors. Kearney and Peebles later (1964) added that the hybrid plants produce very little seed. And Benson (1940, 1969a) noted that they are intermediate between the putative

---

[1] The work reported here was carried out at the Boyce Thompson Arboretum, University of Arizona, Superior, Arizona.

Fig. 1. *Opuntia spinosior* and *O. fulgida*. (A) *O. spinosior*, Pinal Ranch, between Superior and Globe. (B) *O. fulgida*, Superior.

parental species in several characters of the joints and fruits.

We relocated the hybrid colony on the flood-plain of the Gila River near the present Indian town of Sacaton in the low desert plains of central Arizona. We discovered another larger hybrid population in desert mountains near Kelvin 45 miles east of Sacaton. This paper describes the microgeographical, the morphological, and the fertility relationships of these hybrids and their later-generation derivatives. On the basis of this evidence we can then go on to outline the clonal microspecies derived as hybrid products of O. spinosior and O. fulgida.

THE PARENTAL SPECIES

The parental species are O. fulgida, the chain-fruit cholla, and O. spinosior, the cane cholla. O. fulgida is tree-like with stout trunks topped by thick branches. Its green pear-shaped fruits hang in long branched chains from the tips of the branches (Fig. 1). O. spinosior forms shrubs with long slender branches bearing terminal yellow tuberculate fruits (Fig. 1). The character differences between these species used in the present study are listed in Table 1. There is considerable racial variation within both species. For additional details the reader is referred to Benson (1969a).

Vegetative reproduction is very common in O. fulgida. The spiny terminal stem joints detach readily from the parent plant and fall to the ground. They may take root near the parent plant, or they may be carried to considerable distances by water or animals before rooting in their final resting place. The fruits of O. fulgida also possess buds which are capable of forming new adventitious roots and shoots (Johnson, 1918). Vegetative reproduction by stem joints is occasional in O. spinosior.

O. spinosior is a good seed producer and appears to reproduce mainly by seeds. The fruits of O. fulgida, on the other hand, contain variable numbers of sound seeds, often few or none, and such good seeds as are formed frequently fail to germinate (Johnson, 1918). The breeding system of O. fulgida requires further study.

O. fulgida and O. spinosior are both widespread in central and southern Arizona and extend to northern Mexico. We are particularly concerned with the two species as they occur in the study area shown in Figure 2.

The eastern half of this area is mountainous and the western half consists of low plains and valleys. On the east the pine-covered Pinal Mts. (south of Globe) and Santa Catalina Mts. (northeast of Tucson) rise to heights of 7800 and 9100 feet, respectively. More extensive mountainous areas lie at lower elevations (3500 to 6500 feet) in the chaparral, woodland, and grassland zones. Farther west the mountains descend into the desert foothills (as at Superior and Hayden) and outwash plains (as near Florence) in the elevational range from 2800 feet to 1500 feet. Still farther west are low valleys (as at Sacaton) at 1200 to 1400 feet (see Fig. 2).

O. fulgida forms extensive populations throughout the desert plains and foothills in the creosote-bush and paloverde-saguaro communities up to 2700 feet elevation. O. spinosior characteristically occurs at higher elevations in the mountains, growing in chaparral, desert grassland, and pinyon-juniper woodland up to 6500 feet elevation; but it also extends out onto the desert plains. The lowland colonies of O. spinosior are small and spotty in the western part of our study area, as near Florence and Picacho (see Fig. 2); farther south O. spinosior becomes common in the desert plains.

The ecogeographical relationships between O. fulgida and O. spinosior differ markedly in different physiographic regions within the study area. In the southwestern part of this area the two species form extensive mixed stands. Here the lowland race of O. spinosior is biotically

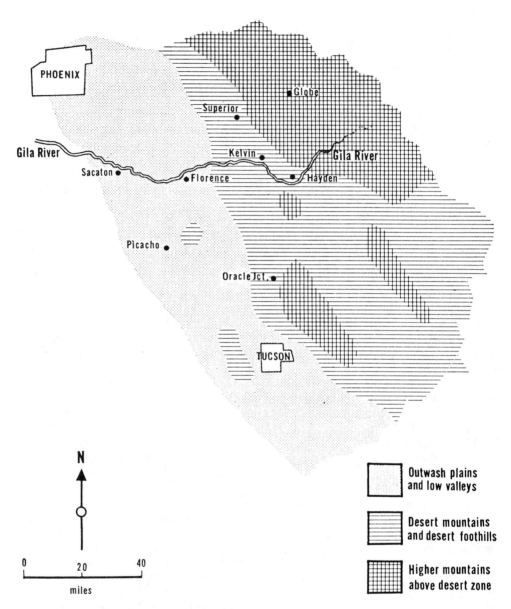

Fig. 2.   Map of study area in south-central Arizona showing main physiographic features and key localities.

sympatric with *O. fulgida* over a large area. Biotically sympatric contacts between the two species are occasional and local in the low desert area farther north, as around Picacho and Florence.

In the mountainous area to the north-east, by contrast, *O. fulgida* and *O. spinosior* are generally allopatric but nearly contiguous. Here *O. spinosior* ranges down to 4000 feet or sometimes 3500 feet and is therefore separated by an elevational gap from *O. fulgida* in the foothill zone up to 2700 feet. This elevational gap is accompanied by a microgeographical gap of varying extent. We have found two places—one near Superior and one northeast of Hayden—where populations of the two species occur less than five miles apart. There may well be biotically sympatric contacts between *O. spinosior* and *O. fulgida* in the foothill zone, at least occasional ones, which we have not observed.

Seasonal isolation is the first and perhaps most important mode of reproductive isolation between *O. fulgida* and *O. spinosior* in nature. The two species differ in flowering time. *Opuntia spinosior* blooms in late spring from late April to June, whereas *O. fulgida* blooms in the summer from June to September. Furthermore, the flowers of *O. fulgida* are vespertine and short-lived, while those of *O. spinosior* are diurnal and remain open for days.

In the lowland areas of biotic sympatry between *O. spinosior* and *O. fulgida*, the former species is well past its peak of flowering before the latter begins to bloom, and the seasonal isolation is therefore essentially complete. In the mountains *O. spinosior* blooms somewhat later than it does in the lowlands. Sympatric contacts have not been found in the northeastern mountainous part of our study area, as noted earlier, but could and probably do occur occasionally, and in such instances the seasonal isolation would be expected to break down partially.

It will be recalled that *O. spinosior* is a good seed producer whereas *O. fulgida* produces seeds sparsely and irregularly and then retains them for years in persistent fruit chains on the mother plant. The chances of hybrid seeds forming and germinating where *O. fulgida* is the maternal parent are consequently very slight as compared with the chances of effective hybridization occurring in the opposite direction. Hybrid formation would appear to have the best chance of succeeding where *O. spinosior* serves as the seed parent and *O. fulgida* as the pollen parent.

## THE HYBRIDS

Hybrids between *O. fulgida* and *O. spinosior* are now known from two localities. The oldest known locality is Sacaton, as mentioned previously, while the best stand of hybrid plants is in the vicinity of Kelvin (see Fig. 2). The hybrid plants are intermixed with *O. fulgida* at both Sacaton and Kelvin. *O. spinosior* occurs with both *O. fulgida* and the hybrids in one locality 15 miles southwest of Kelvin; the nearest known plants of *O. spinosior* in the Sacaton area are 13 miles away from the hybrid colony. We have looked for hybrids of this combination in other parts of the study area, including the southern region of extensive biotic sympatry between the parental species, but have not found any indisputable cases, though these may of course turn up with further search.

The morphological features of the hybrids are shown in Table 1 and Figures 3 and 5. From the total array of character differences between *O. fulgida* and *O. spinosior* we selected five vegetative and four fruit characters which could be scored reliably in the field or laboratory, as listed in Table 1. Local populations of *O. fulgida* from Kelvin and *O. spinosior* from Pinal Ranch (between Superior and Globe) will serve as standards of reference for the two parental species. These are compared with hybrids from Kelvin of the type designated *K* and hybrids from Sacaton of type *R* in Table 1. Other hybrid types will be introduced later.

The table shows that hybrid plants of type *K* at Kelvin are intermediate between geographically neighboring races of the parental species in growth habit, joint characteristics, fruit shape, and fruit surface. These hybrids resemble *O. fulgida* in fruit color and *O. spinosior* in fruit attachment. The table also indicates that type *R* hybrids at Sacaton are very similar but not identical to type *K* hybrids at Kelvin. Some of the characters listed in the table are shown in Figure 3.

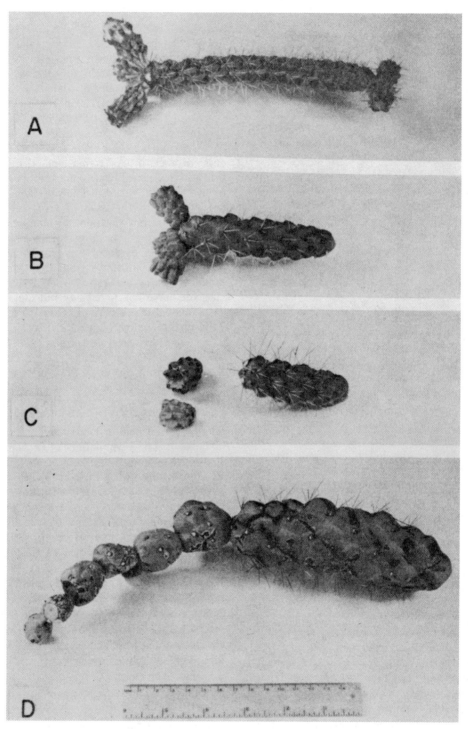

FIG. 3. Terminal stem joints and fruits of *Opuntia spinosior, O. fulgida,* and natural hybrids. (A) *O. spinosior,* Pinal Ranch. (B) Hybrid type *R,* Sacaton. (C) Hybrid type *K,* Kelvin. (D) *O. fulgida,* Superior. All to same scale.

TABLE 1. *Vegetative and fruit characters in* Opuntia fulgida, O. spinosior, *and their hybrids.*

| Character | O. *fulgida*, Kelvin | Hybrid *K*, Kelvin | Hybrid *R*, Sacaton | O. *spinosior*, Pinal Ranch |
|---|---|---|---|---|
| Growth habit | large, tree-like, irregular branching with drooping branches | intermediate | intermediate | med'um-sized, shrubby, branches in whorls |
| Terminal joint diameter | 4.1 cm | 2.6 cm | 2.6 cm | 2.0 cm |
| Terminal joint detachment | drop off freely | detach fairly easily | detach fairly easily | cling on |
| Tubercles, width at base | 1.4 cm | 0.9 cm | 0.8 cm | 0.4 cm |
| Spines, length of longest | 2.4 cm | 2.4 cm | 2.1 cm | 1.4 cm |
| Fruit shape | pear-shaped | olive-shaped | olive-shaped | cone-shaped to oblong |
| Fruit surface | smooth | moderately tuberculate | moderately tuberculate | strongly tuberculate |
| Fruit attachment | forms chains of 3 or more fruits, chains usually long and branched | single | single, or sometimes in short chains of 2 fruits | single or in whorls, but not in chains |
| Fruit color | green | green | green | yellow |

The hybrid populations at Kelvin and Sacaton contain some other variants in addition to the types *K* and *R*. These have been labelled types *C, V* and *W* in the Kelvin area and types *S* and *T* in the Sacaton population. They are very similar morphologically to types *K* and *R,* and consequently are likewise intermediate between *O. fulgida* and *O. spinosior,* but they also differ somewhat from standard types *K* and *R.* The differences will be described briefly here.

In the Sacaton hybrid population, plants of type *S* are larger shrubs than type *R,* they have stouter main trunks, and bear longer chains of bigger fruits. There are four individuals of type *T* in the Sacaton population. These differ from *R* in having a reddish hue on the stem joints and fruits, in flowering early, and in lacking any tendency to form short chains of fruits. Type *S* differs from *R* in the direction of the *O. fulgida* parent and type *T* differs in the direction of *O. spinosior.*

In the Kelvin area, where type *K* plants preponderate, type *C* forms a small uniform colony. Type *C* differs from type *K* in having more irregular and compact branching, smaller flowers, and a tendency to form short chains of two or three fruits. These are variations in the direction of *O. fulgida.* Type *V* in this area is represented by a single known individual growing with type *K* hybrids downstream from Hayden. The *V* plant differs from type *K* in having a more loose open habit of branching, more slender stem joints, strongly tuberculate fruits, and yellow fruits. In these features it approaches *O. spinosior.*

Type *W* is abundant in the desert hills west and southwest of Kelvin. It differs from type *K* in growth habit, being smaller of stature, and in its fruits, which are cone-shaped and quite tuberculate. In these features it approaches the condition found in *O. spinosior.*

Five populations of *O. spinosior* rang-

**111**

ing from mountain to desert habitats were examined for pollen fertility and/or seed fertility. All populations had normal fertility, with 90% or more well-formed pollen grains and abundant sound seeds. Plants of *O. fulgida* from Superior showed 88 to 98% good pollen. Most fruits contain numerous fully plump or half plump seeds. Johnson (1918) has previously shown that some seeds in both size classes have abortive embryos. *O. fulgida* seems to be highly fertile as to pollen and semi-fertile as to seeds.

Four individual plants of hybrid type *K* from Kelvin and two individuals of hybrid type *R* from Sacaton were examined for pollen and seed fertility. The Kelvin plants had 5%, 7%, 12%, and 13% well-formed and well-stained pollen grains; and the Sacaton *R* plants had 8% and 18% well-formed pollen grains. Forty fruits from Kelvin *K* plants all had numerous abortive ovules in the central cavity; most of these fruits had no plump seeds; an occasional fruit had one plump seed. Among six fruits from Sacaton *R* plants, five had no plump seeds, and one fruit had one plump seed. Thus, both the Kelvin *K* plants and the Sacaton *R* plants are highly but not completely sterile.

In general, the other hybrid types are highly sterile but are somewhat more fertile as to pollen or seeds than types *R* and *K*. Thus the *C* plants at Kelvin have 21–22% well-formed pollen and 0–2 plump seeds per fruit. Plants of type *W* exhibited from 28 to 49% good pollen. In the Sacaton colony type *T* has 13–14% well-formed pollen and 0–8 plump seeds per fruit.

The interspecific hybrid constitution of the various types of plants at Kelvin and Sacaton described above is demonstrated by their morphological intermediacy and sterility considered jointly. The morphological intermediacy falls in the middle range and the sterility is greatest in types *K* and *R*, suggesting strongly that these types may well be F₁ hybrids. The plants belonging to types *C*, *V*, *W*, *S*, and *T*

deviate from *K* and *R* in the direction of one parental species or the other with respect to particular morphological characters. And, though fairly sterile, these other hybrid types are slightly more fertile than the putative F₁ types *K* and *R*. This suggests that these other types are later-generation segregation products derived from the original F₁ hybrids.

## THE CLONAL MICROSPECIES

At Sacaton the type *R* hybrids are dispersed throughout an area one-quarter mile square. We counted 24 individual plants of type *R*; there may be a few more which we missed. Type *T* is represented by four identical individuals, standing 15–35 feet apart and forming a small clonal group on one edge of the quarter-mile square area.

Hybrid plants of type *K* are numerous and widespread over a fairly long distance along the Gila River in the vicinity of Kelvin. One subpopulation on a flat ridge near Kelvin proper is three miles long by one-half to one mile wide and includes an estimated 600 individual plants. On one transect through this subpopulation 100 individuals were inspected and observed to be essentially identical. A second subpopulation of *K* plants is strung out in the Gila River valley upstream from Kelvin for a distance of ten miles. A third subpopulation occurs downstream from Kelvin.

The *C* type hybrid plants occur in a disjunct and discrete colony on the Gila River a half-mile downstream from Kelvin. The colony numbers about 100 individuals. These are all alike.

The largest population is that composed of type *W* plants in the Tortilla Mts. west and south of Kelvin. The *W* plants are abundant over an area 20 miles long and 26 miles wide in these desert hills. There must be hundreds of thousands of type *W* individuals in this area.

There is a high degree of individual-to-individual uniformity within a given type of hybrid in both the Sacaton and Kelvin

Fig. 4.  Terminal stem joints and fruits from four sister individuals of hybrid type *R* at Sacaton.

areas (see Fig. 4). This indicates that these hybrid individuals have arisen not by sexual but by asexual means of reproduction. Vegetative propagation from stem joints is the obvious candidate in view of the known reproductive biology of the parental species. The Sacaton hybrids are all old plants and consequently do not furnish direct evidence for this mode of origin. However, direct evidence of vegetative propagation is available for the *K*, *C* and *W* types in the Kelvin area. Young plants of these hybrid types can be seen to be attached laterally at the soil surface to fallen stem joints or the remnants of decayed stem joints.

Some of the expected stages of population growth are well exemplified by the various hybrid types. We have a small clone consisting of four neighboring individuals in the case of hybrid type *T* at Sacaton. The type *R* hybrids at Sacaton and *C* hybrids at Kelvin are medium-sized clones. The type *K* hybrids, on the other hand comprise an endemic clonal micro-

species extending along the Gila River over a linear distance some 20 miles long. And the type *W* plants form a rather widespread clonal microspecies occupying an area of several hundred square miles in the Tortilla Mts.

## DISCUSSION

A clonal complex is a hybrid complex, a taxonomically critical group of basic species and their hybrid derivatives, in which the hybrids reproduce mainly by vegetative means (Grant, 1953). As compared with other types of hybrid complexes—agamic, polyploid, etc.—clonal complexes are rather poorly known. Clonal complexes have been tentatively identified as such in the *O. phaeacantha* and *O. spinosior* groups (Grant, 1953, 1971). In no case, however, has a clonal complex been subjected to a thorough biosystematic and phylogenetic analysis. It is desirable to carry out this task in one or more plant groups in order to round out our picture of the patterns of plant evolution.

The taxonomic studies of Benson (1969a, 1969b, and in press) on western American cacti provide a good foundation on which to build in this direction. Pinkava (unpubl.) is currently investigating the complex O. phaeacantha group from the biosystematic standpoint.

The essential first step in the phylogenetic analysis of a clonal complex is to establish the hybrid constitution and parentage of at least one clonal microspecies. The work reported in the present paper accomplishes this first step. The clonal microspecies designated as Kelvin K consists of sterile, intermediate, vegetatively reproducing hybrids of O. spinosior and O. fulgida.

The K and R hybrids are highly but not completely sterile, retaining the capacity of producing some apparently good pollen and seeds. Evidently some sexual reproduction does take place in the $F_1$ hybrids so as to engender new later-generation segregates.

The plants belonging to types C, V, W, S, and T are probably later-generation segregation products derived from the original $F_1$ hybrids, for they approach one parental species or the other in particular morphological characters, and are slightly more fertile than the putative $F_1$ types K and R. The discrete clonal population C in the Kelvin area provides a concrete case of a later generation of hybrid derivatives which has apparently originated by segregation from the Kelvin K population, and which has multiplied vegetatively in a small local area. Clone T at Sacaton and microspecies W in the Tortilla Mts. have probably arisen independently by parallel processes of hybrid reproduction.

In the southern part of the Tortilla Mts., where O. fulgida and O. spinosior overlap in range, microspecies W is biotically sympatric with both parental species. In one locality in this area we found a hybrid swarm consisting of type W plants and their segregates and apparent backcrosses to O. spinosior. The sexual mode of reproduction is going on here and is engendering a varied array of new hybrid types.

These cholla cacti exhibit an alternation of sexual and asexual processes in hybrid reproduction similar to the alternation which has long been recognized in agamospermous groups and in agamic complexes. This cycle of sexual and asexual reproduction is evidently a successful way of producing and then multiplying new adaptive hybrid types.

The next step in the analysis of a clonal complex will be to refine the analysis by the application of cytotaxonomic and chemotaxonomic methods. The subsequent steps are to extend the investigation to other related parental species and their hybrid products. O. spinosior apparently hybridizes with O. versicolor and O. imbricata (Benson, 1969a). Does this hybridization yield any clonal microspecies? O. fulgida exhibits some seed sterility and reproduces mainly by vegetative propagation. Is it a hybrid derivative of some preexisting species? The answers to these and numerous other specific questions will add up to an understanding of the clonal complex as a whole.

## SUMMARY

This paper describes a case of vegetative multiplication of sterile interspecific hybrids in the cholla cacti. The parental species are O. fulgida and O. spinosior in south-central Arizona. Several types of hybrid products are found at two localities in this area.

Two of these types, which are morphologically intermediate in the middle range between the parental species, and highly but not completely sterile as to pollen and seeds, are identified as probable $F_1$ hybrids. The other hybrid types are identified on morphology and fertility as later-generation segregation products.

Direct evidence was obtained for the vegetative propagation of these hybrid types by means of the fallen stem-joints. One hybrid type has formed a clone of

FIG. 5. *Opuntia kelvinensis*. A plant of type *K* from a low desert ridge above the Gila River near Kelvin, Arizona. Above, plant body; below, terminal stem joint and two fruits.

four identical individuals; another a clone of about 100 identical individuals; still another forms an endemic clonal microspecies; and one other clonal microspecies occurs in large numbers throughout an area of several hundred square miles.

The observed series of clones and microspecies exemplifies the expected but hitherto undocumented stages of growth of clonal hybrid microspecies. Their development involves a cyclical alternation of sexual and asexual processes, as has been found previously in agamospermous microspecies, the only known difference being the method of asexual reproduction.

The results reported here have implications for formal taxonomy, which are taken up in the appendix.

The needs and the categories of formal taxonomy are different from those of population biology. The minor phenomena of population biology do not necessarily warrant formal taxonomic recognition, but microspecies which form a distinctive element in the local flora, on the other hand, may warrant recognition as a taxonomic species. Where to draw the line is of course a matter of taxonomic judgment based on the premise that a taxonomic species is or should be a practically useful category.

The hybrid plants discussed in this paper do form a definite and recognizable element in the cactus flora of Arizona, and therefore we have concluded that it will serve a useful purpose in both southwestern floristics and cactus systematics to describe these plants as a taxonomic species. The description which follows is drawn up so as to include the various clones and clonal microspecies mentioned in this paper in one collective taxonomic entity. It is appropriate to describe the new taxonomic species here in conjunction with the publication of the biological details.

*Opuntia kelvinensis,* sp. nov. (Fig. 5)

Shrubs, 4–6 feet (1.2–1.8 meters) tall, 1–3 stems from base, much branched above. Terminal stem joints 6–10 cm long, 2.5–3.5 cm in diameter. Tubercles on terminal joints rounded and medium high, 1.2–1.8 cm long, 5–9 mm high, 6–10 mm wide at base, 4–5 rows of tubercles visible from one side of stem. Spines 4–8 in a group, with one long spine 1.7–2.5 cm long, one medium-long spine, and several short spines. Spines pale pinkish with deciduous sheaths. Flowers with bright pink or wine-colored perianth, yellow anthers, and yellow stigma. Perianth spreading, 5 cm in diameter. Flowering in May and June. Fruits ovoid to slightly cone-shaped, 2.5–3.5 cm long, green to yellowish-green, moderately tuberculate, with apical cavity present. Fruits borne singly or in whorls of 2 or 3 at tips of joints; not in chains, or occasionally in short chains of 2 or 3 fruits. Seed cavity at maturity containing numerous abortive ovules and occasionally one or a few plump seeds.

Frutex 1.2–1.8 m altus, ramis numerosis; caulis ad ultimum 6–10 cm longus 2.5–3.5 cm diametros, tuberosus; flos color puniceus 5 cm diametros; fructus ovoideus viridis tuberosus 2.5–3.5 cm longus, non in catenis; flos et fructus sterilis vel semisterilis.

*Distinguishing characteristics.*—Closely related to *O. spinosior* (Engelm.) Toumey and *O. fulgida* Engelm. from which derived as hybrid products. Differs from *O. spinosior* as follows: branching

irregular rather than whorled; terminal stem joints short and stout, instead of long and slender; tubercles moderately rounded, rather than sharply ridged; fruits green or greenish, rather than lemon-yellow; fruit wall moderately tuberculate rather than strongly so. Differs from *O. fulgida* as follows: small shrubs, instead of tall arborescent plants with large trunks; terminal stem joints much smaller in diameter than in *O. fulgida*; fruits not forming long chains; fruit wall moderately tuberculate instead of smooth; fruits having a prominent apical cavity instead of a shallow or no cavity.

*Range.*—Flat ridges and hill-tops in desert hills in the saguaro-paloverde zone at 1700–2100 feet elevation. Usually growing with *O. fulgida,* occasionally with *O. spinosior.* From Kelvin southeast to Kearney and beyond; and south and southwest of Kelvin in the desert foothills to the lower elevational limits mentioned above. A disjunct colony occurs in a low desert valley at 1270 feet elevation near Sacaton to the west of Kelvin. Pinal County, south-central Arizona.

*Type.*—V. Grant, collection no. 70–29, just southeast of Kelvin, Pinal County, Arizona, June 15, 1970. Herbarium, University of Texas, Austin.

## LITERATURE CITED

BENSON, L. 1940. The Cacti of Arizona. 1st ed., University of Arizona Press, Tucson, Arizona.

——. 1969a. The Cacti of Arizona. 3rd ed., University of Arizona Press, Tucson, Arizona.

——. 1969b. The Native Cacti of California. Stanford University Press, Stanford, California.

GRANT, V. 1953. The role of hybridization in the evolution of the Leafy-stemmed Gilias. Evolution 7:51–64.

——. 1971. Plant Speciation. Columbia University Press, New York, (in press).

GUSTAFSSON, Å. 1946–1947. Apoximis in Higher Plants. Lunds Univ. Arsskrift, vol. 42–43, Lund, Sweden.

JOHNSON, D. S. 1918. The fruit of *Opuntia fulgida.* A study of perennation and proliferation in the fruits of certain Cactaceae. Carnegie Institution of Washington, Publ. 269.

KEARNEY, T. H. AND R. H. PEEBLES. 1942. Flowering Plants and Ferns of Arizona. U. S. Department of Agriculture, Publ. 423, Washington, D. C.

——. 1964. Arizona Flora. 2nd ed., University of California Press, Berkeley, California.

STEBBINS, G. L. 1950. Variation and Evolution in Plants. Columbia University Press, New York.

——. 1959. The role of hybridization in evolution. Proc. Amer. Phil. Soc. 103:231–251.

# 10

Reprinted from *Genetics* **84**:791–805 (1976)

## THE ORGANIZATION OF GENETIC DIVERSITY IN THE
## PARTHENOGENETIC LIZARD *CNEMIDOPHORUS TESSELATUS*

E. DAVIS PARKER, JR. AND ROBERT K. SELANDER

*Department of Biology, University of Rochester, Rochester, New York 14627*

Manuscript received November 25, 1975

### ABSTRACT

The parthenogenetic lizard species *Cnemidophorus tesselatus* is composed of diploid populations formed by hybridization of the bisexual species *C. tigris* and *C. septemvittatus*, and of triploid populations derived from a cross between diploid *tesselatus* and a third bisexual species, *C. sexlineatus*. An analysis of allozymic variation in proteins encoded by 21 loci revealed that, primarily because of hybrid origin, individual heterozygosity in *tesselatus* is much higher (0.560 in diploids and 0.714 in triploids) than in the parental bisexual species (mean, 0.059). All triploid individuals apparently represent a single clone, but 12 diploid clones were identified on the basis of genotypic diversity occurring at six loci. From one to four clones were recorded in each population sampled. Three possible sources of clonal diversity in the diploid parthenogens were identified: mutation at three loci has produced three clones, each confined to a single locality; genotypic diversity at two loci apparently caused by multiple hybridization of the bisexual species accounts for four clones; and the remaining five clones apparently have arisen through recombination at three loci. The relatively limited clonal diversity of *tesselatus* suggests a recent origin. The evolutionary potential of *tesselatus* and of parthenogenetic forms in general may be less severely limited than has generally been supposed.

THE origin and fate of parthenogenetic organisms are subjects of major concern in theoretical discussions of the evolutionary significance of genetic recombination (MULLER 1932; FISHER 1958; SUOMALAINEN 1962; CROW and KIMURA 1965; BODMER 1970; WHITE 1970; MAYNARD SMITH 1971; SMITH 1971; FELSENSTEIN 1974; STANLEY 1975; WILLIAMS 1975). It has been suggested that parthenogenetic forms evolve more slowly than sexual populations, owing to a reduction or absence of recombination, and consequently are more subject to extinction in changing environments. To account for the development of parthenogenesis in various groups of animals and plants, several immediate selective advantages have been suggested, including, importantly, a lessening or elimination of recombinational load (FISHER 1958; ESHEL and FELDMAN 1970), the avoidance (in dioecious forms) of the "cost of meiosis" (MAYNARD SMITH 1971), and increased reproductive potential (WILLIAMS 1975).

Although knowledge of the amount and organization of genic and genotypic diversity in parthenogenetic organisms is fragmentary, genetic data (KALLMAN 1962; SUOMALAINEN and SAURA 1973; LOKKI *et al.* 1975), together with information derived from cytological and morphological studies (SUOMALAINEN 1962,

1969; ZWEIFEL 1965; UZZELL 1970; ASHER and NACE 1971; WHITE 1973; RÖSSLER and DE BACH 1973), indicate that the genetic structure of parthenogenetic populations frequently is complex, and that the evolutionary potential of parthenogens may not be as severely limited as has sometimes been supposed. Clonal diversity in parthenogenetic species may be generated as a result of polyphyletic origin from bisexual forms, by mutation, and, in some cases, through recombination.

Because the genetic diversity of parthenogens initially, at least, will reflect that present in the bisexual ancestral species, parthenogenetic species formed by hybridization between species should in general exhibit greater diversity than those arising from single individuals or populations. In time, however, diversity in hybrid and nonhybrid parthenogens with similar recombination rates should become equal at levels determined by mutation and selection (WHITE 1970). The extent of recombination in parthenogens depends primarily on the mechanism of egg maturation (ASHER 1970; UZZELL 1970; ASHER and NACE 1971). Apomictic parthenogens have a mitotic egg maturation with no recombination; progeny are genetic replicates of the mother and all loci are expected to become heterozygous with time, as a result of the accumulation of mutations (WHITE 1970). But in the several mechanisms of egg maturation classified as automictic, at least part of normal meiosis is retained (WHITE 1973). Recombination may enforce complete homozygosity in one generation (NUR 1971) or cause a gradual decay of heterozygosity unless countered by heterotic selection (ASHER 1970).

We here analyze the patterns and identify the probable sources of genic and genotypic diversity in the clonal complex *Cnemidophorus tesselatus*, a parthenogenetic lizard of the family Teiidae consisting of diploid populations formed by hybridization between the bisexual species *C. tigris* and *C. septemvittatus*, and of triploid populations derived from a cross between diploid *tesselatus* and a third bisexual species, *C. sexlineatus* (WRIGHT and LOWE 1967; NEAVES 1969; NEAVES and GERALD 1968, 1969). Diploid populations of *tesselatus* occur along the Rio Conchas Valley in northern Mexico, the Rio Grande and Pecos River drainages in western Texas and New Mexico, and the Canadian and Arkansas river drainages in northern New Mexico and in Texas and southern Colorado, respectively. Triploid populations are also distributed along the Arkansas River drainage in Colorado (Figure 1). *C. tesselatus* is one of eight known parthenogenetic species of Cnemidophorus, all of which arose through hybridization of bisexual species (MASLIN 1971).

The maturation of eggs in another parthenogenetic species, *Cnemidophorus uniparens*, is known to involve premeiotic endoduplication of chromosomes, followed by meiosis, with synapsis and crossing-over normally occurring only between identical chromatids (Cuellar 1971). This process presumably occurs in *tesselatus* and other parthenogenetic species of Cnemidophorus, and has also been reported in unisexual salamanders (MACGREGOR and UZZELL 1964), fishes (CIMINO 1972), snails (JACOB 1957), and grasshoppers (WHITE 1966).

Recombination may occur in the course of the maturation of eggs of parthenogenetic Cnemidophorus. If some of the parental chromosomes are homologous but not identical, two different events at metaphase I can lead to segregation of

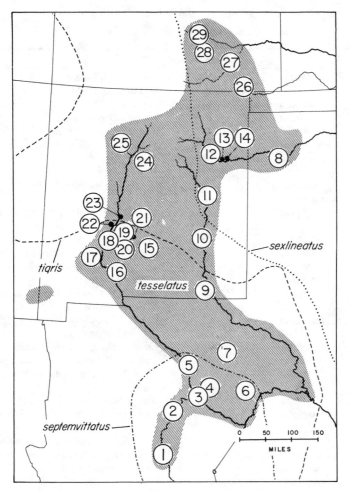

FIGURE 1.—Localities at which samples of Cnemidophorus were collected. Numbers refer to localities listed in Table 1.

the parental genotypes: (1) quadrivalent formation, a process observed in the meiosis of triploid unisexual salamanders (MACGREGOR and UZZELL 1964; ASHER and NACE 1971), and (2) bivalent formation of two of the different parental chromosomes, a condition cytologically indistinguishable from the normal pairing of identical sister chromatids. When crossing-over occurs, the amount of genetic exchange between parental chromatids depends on how the chromatids segregate at anaphase I.

MATERIALS AND METHODS

Samples are listed in Table 1, and collecting localities are shown in Figure 1. Our analysis is based on examination of three populations of *septemvittatus*, two of *tigris*, two of *sexlineatus*, and 27 of *tesselatus*. We have recently examined six additional samples (112 individuals) of *tigris* from Mexico, Texas, and New Mexico. In none of these were there alleles not represented in samples 4 and 7 of *tigris*.

### TABLE 1

*Samples of* Cnemidophorus tesselatus *and parental species*

| Sample number | Species | Sample size | Locality |
|---|---|---|---|
| 1 | *tesselatus* | 19 | Julimes, Chihuahua, Mexico |
| 2 | *tesselatus* | 15 | El Pueblito, Chihuahua, Mexico |
|  | *septemvittatus* | 30 | |
| 3 | *tesselatus* | 18 | Presidio, Presidio Co., Texas |
| 4 | *tigris* | 42 | 5 mi N Presidio, Presidio Co., Texas |
| 5 | *tesselatus* | 13 | Vicinity of Ruidosa, Presidio Co., Texas |
|  | *septemvittatus* | 10 | |
| 6 | *septemvittatus* | 22 | 18 mi S Marathon, Brewster Co., Texas |
| 7 | *tesselatus* | 2 | ½ mi N Balmorhea, Reeves Co., Texas |
|  | *tigris* | 27 | |
| 8 | *tesselatus* | 17 | 9 mi SE Jct. RR 1061 and US Rt. 385 on RR 1061, Potter Co., Texas |
| 9 | *tesselatus* | 19 | White's City, Eddy Co., New Mexico |
| 10 | *tesselatus* | 10 | Roswell city dump, Chaves Co., New Mexico |
| 11 | *tesselatus* | 11 | Ft. Sumner city dump, DeBaca Co., New Mexico |
|  | *sexlineatus* | 19 | |
| 12 | *tesselatus* | 24 | 10½ mi W Conchas Dam, San Miguel Co., New Mexico |
| 13 | *tesselatus* | 5 | Conchas Lake State Park, San Miguel Co., New Mexico |
| 14 | *tesselatus* | 27 | Conchas Dam, San Miguel Co., New Mexico |
| 15 | *tesselatus* | 1 | 2 mi NE Tularosa, Otero Co., New Mexico |
| 16 | *tesselatus* | 16 | 15 mi N Las Cruces, Doña Ana Co., New Mexico |
| 17 | *tesselatus* | 22 | 8 mi E Hillsboro, Sierra Co., New Mexico |
| 18 | *tesselatus* | 9 | S end Elephant Butte Lake, Sierra Co., New Mexico |
| 19 | *tesselatus* | 29 | 5 mi W Engle, Sierra Co., New Mexico |
| 20 | *tesselatus* | 4 | 5 mi S Engle, Sierra Co., New Mexico |
| 21 | *tesselatus* | 16 | Engle, Sierra Co., New Mexico |
| 22 | *tesselatus* | 18 | Springtime Campground Road, Cibola National Forest, Socorro Co., New Mexico |
| 23 | *tesselatus* | 2 | San Antonio, Socorro Co., New Mexico |
| 24 | *tesselatus* | 27 | 3 mi N Placitas, Sandoval Co., New Mexico |
| 25 | *tesselatus* | 4 | 5 mi NW San Ysidro, Sandoval Co., New Mexico |
| 26 | *tesselatus* | 1 | Miser Ranch, 15 mi S and 12 mi W Pritchett, Baca Co., Colorado |
| 27 | *tesselatus** | 20 | Vicinity of Higbee, Otero Co., Colorado |
| 28 | *tesselatus* | 22 | Huerfano Canyon, 23 mi SE Pueblo, Pueblo Co., Colorado |
| 29 | *tesselatus* | 35 | Pueblo city dump, Pueblo Co., Colorado |
|  | *sexlineatus* | 15 | |
| Totals: | *tesselatus* | 406 | (339 diploids and 67 triploids) |
|  | *tigris* | 69 | |
|  | *septemvittatus* | 62 | |
|  | *sexlineatus* | 34 | |

* Sample 27 consisted of 10 diploid and 10 triploid individuals; all individuals of *tesselatus* from localities 28 and 29 were triploid. All other samples of *tesselatus* were composed exclusively of diploids.

**120**

Extracts of heart, liver, kidney, and tail muscle in a tris-EDTA-NADP buffer, pH 7.0, were prepared from lizards that were frozen on dry ice upon capture and stored at —76° in the laboratory. Allozymic variation was studied by means of horizontal starch-gel electrophoresis, using techniques described by SELANDER *et al.* (1971). The following enzymes and other proteins, encoded by 21 loci, were scored: Leucyl-alanine peptidase (*Pep*, demonstrated in extracts of liver and kidney), two malate dehydrogenases (*Mdh-1* and *Mdh-2*, all tissues), malic enzyme (*Me*, liver), 6-phosphogluconate dehydrogenase (*6-Pgd*, liver), two isocitrate dehydrogenases (*Idh-1*, kidney, and *Idh-2*, heart), two lactate dehydrogenases (*Ldh-1*, heart and kidney, and *Ldh-2*, muscle), α-glycerophosphate dehydrogenase (*α-Gpd*, all tissues), phosphoglucose isomerase (*Pgi*, muscle), two phosphoglucomutases (*Pgm-1* and *Pgm-2*, liver), two glutamic oxaloacetic transaminases (*Got-1* and *Got-2*, all tissues), superoxide dismutase (*Sod*, all tissues; see BECKMAN and BECKMAN 1975), two esterases (*Es-1*, liver, and *Es-2*, muscle), albumin (*Alb*, liver), and two muscle proteins (*Gp-1* and *Gp-2*). Alleles were numbered according to the proportional electrophoretic mobilities of their products relative to that of the most common allele (e.g., $Ldh-1^{100}$) in a reference sample of *C. tigris* from Reeves County, Texas.

Average individual heterozygosity (*H*) was estimated from gene frequencies for the bisexual species and from direct counts of heterozygotes for *tesselatus*. Genetic similarity between populations was estimated by the method developed by NEI (1973) and NEI and ROYCHOUDHURY (1974).

RESULTS

## Genotypes in the parthenogens

The hybrid nature of both diploid and triploid forms of *C. tesselatus* is clearly demonstrated by the data presented in Table 2. Individuals were assumed to be triploid when known to carry the *sexlineatus*-specific alleles $Pep^{102}$, $6\text{-}Pgd^{98}$, $Pgm\text{-}1^{99}$, $Idh\text{-}1^{102}$, $Sod^{95}$, $Es\text{-}1^{105}$, and $Alb^{96}$. And for these individuals, a triploid gene dosage also was indicated at the *Es-1*, *Ldh-1*, and *Idh-1* loci by a double intensity of bands (electromorphs) corresponding to the $Es\text{-}1^{100}$, $Ldh\text{-}1^{95}$, and $Idh\text{-}1^{100}$ alleles. (Similar interpretations of triploid allozyme phenotypes were made by NEAVES and GERALD [1969], UZZELL and GOLDBLATT [1967], and VRIJENHOEK [1975].) Triploid dosages were not apparent at the remaining loci. In the cases of *Mdh-1* and *Me*, the *sexlineatus* alleles and those of diploid *tesselatus* encode proteins of such similar mobility that they do not resolve clearly enough to permit an evaluation of relative intensity. Similarly, dosage effects could not be evaluated at the *Got-2* and *Sod* loci because their allozymes do not discretely band in triploids. And in the case of *Pgi*, it was difficult to compare the intensities of bands because they faded soon after they appeared. For other loci (*Pgm-2*, *Got-1*, *Es-2*, and *Gp-2*), the relatively poor resolution of electromorphs may have obscured triploid dosage effects; but, in any event, the allozyme phenotypes of triploids were similar to those of diploids (Table 2).

There are two possible genotypes at the *Pep* locus in triploids: $Pep^{102/98/94}$ or $Pep^{102/94/94}$ (Table 2). In this dimeric enzyme, the mobility of the band representing the *98* allele is intermediate between those of the *102* and *94* alleles, and similar to that of the heterodimeric *102/94* band occurring in heterozygotes. Because the intensity of staining of the *102* and *94* homodimer bands was equal, the genotype of the triploids was assumed to be $Pep^{102/98/94}$.

121

E. D. PARKER, JR. AND R. K. SELANDER

## TABLE 2

*Common alleles (q ≥ .05) in parental species and genotypes in diploid and triploid* Cnemidophorus tesselatus*

| Locus | Species | | | | |
| --- | --- | --- | --- | --- | --- |
| | *tigris* | 2N *tesselatus* | *septemvittatus* | 3N *tesselatus* | *sexlineatus* |
| *Pep* | 100 | 98/94 | 98 | 102/98/94 | 102 |
| | 98 | 98/98 | 94 | | |
| | | 94/94 | | | |
| *Mdh-1* | 100 | 100/98 | 98 | 100/98/(99)† | 99 |
| *Me* | 100 | 100/98 | 99 | 100/98/(99) | 99 |
| | | 100/100 | 98 | | |
| *6-Pgd* | 100 | 100/100 | 100 | 100/(100)/98 | 98 |
| | 102 | | | | |
| *Idh-1* | 100 | 100/100 | 100 | 102/100/100 | 102 |
| | | | | | 104 |
| *Ldh-1* | 100 | 100/95 | 95 | 100/95/95 | 95 |
| | | | 93 | | |
| *Pgi* | −96 | −100/−96 | −100 | −100/(−100)/−96 | −100 |
| | −100 | −99/−96 | −104 | | |
| | | −104/−96 | −99 | | |
| | | −100/0 | | | |
| *Pgm-1* | 100 | 100/97 | 97 | 100/99/97 | 99 |
| *Pgm-2* | 100 | 100/94 | 94 | 100/94/(94) | 94 |
| | 96 | | 92 | | |
| *Got-1* | 100 | 100/95 | 95 | 100/(100)/95 | 95 |
| | | | | | 100 |
| *Got-2* | −100 | −100/−97 | −97 | −100/−97/(−97) | −97 |
| *Sod* | 100 | 105/100 | 105 | 105/(100)/95 | 95 |
| *Es-1* | 100 | 100/100 | 100 | 105/100/100 | 105 |
| | | | 104 | | |
| *Es-2* | 100 | 100/96 | 96 | 100/(100)/96 | 105 |
| | | 96/96 | 99 | | 100 |
| *Alb* | 100 | 100/98 | 98 | 100/98/96 | 96 |
| | | 100/99 | | | 95 |
| *Gp-2* | −100 | −100/−100 | −100 | −100/−100/(−100) | −100 |
| | | −100/−95 | | | |

* All species examined are monomorphic for the same allele at the *Mdh-2, Idh-2, Ldh-2, Gp-1,* and *α-Gpd* loci.

† Alleles in parentheses are probable but not directly demonstrable contributions of *sexlineatus* to the triploid genotype (see text).

## Levels of heterozygosity in parthenogenetic populations

We found no genotypic variation within or among the three samples of triploid *tesselatus* studied (samples 27, 28, and 29). All 67 triploids examined were identically heterozygous at 15 of the 21 loci assayed, with a mean heterozygosity ($H$) of 0.714. Heterozygosity among diploid populations of *tesselatus* ranged from 0.524 (11 of 21 loci in fixed heterozygous state) to 0.571 (12 of 21 loci uniformly heterozygous).

TABLE 3

*Genetic identity* (I) *between populations of bisexual species of* Cnemidophorus

| Species | tigris | | septemvittatus | | | sexlineatus | |
|---|---|---|---|---|---|---|---|
| | 4 | 7 | 2 | 5 | 6 | 11 | 29 |
| 4 | 1.000 | .973 | .473 | .460 | .458 | .320 | .352 |
| 7 | | 1.000 | .492 | .486 | .486 | .357 | .390 |
| 2 | | | 1.000 | .953 | .983 | .568 | .519 |
| 5 | | | | 1.000 | .954 | .574 | .524 |
| 6 | | | | | 1.000 | .581 | .528 |
| 11 | | | | | | 1.000 | .947 |
| 29 | | | | | | | 1.000 |

The high level of heterozygosity in the parthenogens is indicative of marked genic divergence among pairs of the bisexual parental species. Comparing populations of the three species over 21 loci, we obtained the values of NEI's index of identity (I) shown in Table 3. Mean values are as follows: *tigris—septemvittatus*, $\bar{I} = .476$; *tigris—sexlineatus*, $\bar{I} = .355$; and *septemvittatus—sexlineatus*, $\bar{I} = .549$. The uniformly high values of I for comparisons of conspecific populations ($\bar{I} = .962$) suggest that there is little geographic variation in allele representation and frequency in the bisexual species.

### Probable sources of clonal diversity

Tables 4 and 5 present information on variation in the diploid parthenogenetic populations, including locality 27 (Higbee), where we collected both diploid and triploid lizards, corresponding to the two color and pattern types previously described from this locality by ZWEIFEL (1965). Contrary to indications from earlier studies of histocompatibility (MASLIN 1967) and electrophoretic variation (NEAVES 1969) that diploid *tesselatus* are genetically uniform, our analysis has revealed genotypic diversity at six of 21 loci, producing a total of 12 recognizable diploid clones. The number of clones per population ranged from 1 to 4,

TABLE 4

*Genotype frequencies at four variable loci in diploid* Cnemidophorus tesselatus

| Sample number (and size) | Alb | | Gp-2 | | Me | | Es-2 | |
|---|---|---|---|---|---|---|---|---|
| | 100/98 | 100/99 | −100/−100 | −100/−95 | 100/98 | 100/100 | 100/96 | 96/96 |
| 1*(19) | 1.00 | | 1.00 | | 1.00 | | 1.00 | |
| 3 (18) | 1.00 | | .22 | .78 | 1.00 | | 1.00 | |
| 5 (13) | .92 | .08 | 1.00 | | 1.00 | | 1.00 | |
| 7 (2) | 1.00 | | 1.00 | | | 1.00 | 1.00 | |
| 12 (24) | 1.00 | | 1.00 | | 1.00 | | .25 | .75 |
| 14 (27) | 1.00 | | 1.00 | | 1.00 | | .85 | .15 |

* Samples of *tesselatus* from localities 2, 8–11, 13, 15, 16–26, and diploid individuals in sample 27 are genotypically identical to sample 1.

TABLE 5

*Genotype frequencies at* Pgi *and* Pep *loci in diploid* Cnemidophorus tesselatus

| Sample number (and size) | Pgi | | | | Pep | | |
|---|---|---|---|---|---|---|---|
| | −100/−96 | −99/−96 | −104/−96 | −100/0 | 94/94 | 98/94 | 98/98 |
| 1*(19) | 1.00 | | | | | 1.00 | |
| 2 (15) | 1.00 | | | | | .13 | .87 |
| 3 (18) | 1.00 | | | | .78 | | .22 |
| 5 (13) | 1.00 | | | | 1.00 | | |
| 8 (17) | 1.00 | | | | 1.00 | | |
| 11 (11) | | 1.00 | | | | 1.00 | |
| 12 (24) | .83 | .17 | | | | 1.00 | |
| 14 (27) | .78 | .22 | | | | 1.00 | |
| 18 (9) | .78 | | .22 | | | 1.00 | |
| 19 (29) | .52 | | .48 | | | .93 | .07 |
| 20 (4) | .75 | | .25 | | | 1.00 | |
| 21 (16) | .63 | | .37 | | | .19 | .81 |
| 25 (4) | | | | 1.00 | | 1.00 | |

\* Samples of *tesselatus* from localities 7, 9–10, 13, 15–17, 22–24, 26, and diploid individuals in sample 27 are genotypically identical to sample 1.

with a mean of 1.67. The clone characterized by the genotype $Alb^{100/98}$, $Gp\text{-}2^{100/-100}$, $Pgi^{-100/-96}$, $Pep^{98/94}$, $Me^{100/98}$, and $Es\text{-}2^{100/96}$ is by far the most widespread, occurring in all but five (3, 5, 7, 8, and 11) of the 27 diploid populations of *tesselatus* sampled. Each of the other 11 diploid clones apparently is confined in distribution to a small part of the range, and most were recorded at only one locality (Tables 4 and 5).

At each of three loci (*Alb, Gp-2,* and *Pgi*), we found a variant genotype of *tesselatus* involving an allele not represented in our samples of the parental species (*tigris* and *septemvittatus*). Although we cannot exclude the possibility that these alleles formerly were present in the bisexual species or still occur in populations not sampled by us, we have provisionally concluded that they arose by mutation in *tesselatus*. One individual of *tesselatus* from Ruidosa, Texas (sample 5), was heterozygous for $Alb^{100}$ and an allele, $Alb^{99}$, not detected in other populations of *tesselatus* or in the parental species (Table 4). Similarly, 14 of 18 individuals of *tesselatus* from Presidio, Texas (sample 3), were heterozygous at the *Gp-2* locus for the common allele, $Gp\text{-}2^{-100}$, and a unique allele, $Gp\text{-}2^{-95}$. And *Pgi* in lizards from San Ysidro, New Mexico (sample 25), also was heterozygous for a common allele, $Pgi^{-100}$, and a unique allele, $Pgi^0$.

Recombination of hybrid genotypes is the most likely explanation for variation at the *Me* and *Es-2* loci (Table 4). All diploid *tesselatus* were heterozygous $Me^{100/98}$, except for two individuals from Balmorhea, Texas (locality 7), which apparently were homozygous for the $Me^{100}$ allele. At the *Es-2* locus, two samples from the vicinity of Conchas Lake, New Mexico (12 and 14), contained individuals having either the widespread hybrid genotype $Es\text{-}2^{100/96}$ or an apparently segregant genotype $Es\text{-}2^{96/96}$. Because the latter genotype occurred in two clones defined by *Pgi* genotypes ($Pgi^{-100/-96}$ and $Pgi^{-99/-96}$), we presume that two independent recombinations have occurred at the *Es-2* locus.

Variation at the *Pgi* and *Pep* loci is complex with regard to the number of genotypes represented in diploid populations, the combinations of genotypes in clones, and the geographic distribution of these clones. As noted above, the $Pgi^0$ allele occurring as a $Pgi^{-100/0}$ heterozygote at locality 25 apparently is a mutant. $Pgi^{-100/-96}$ is the predominant, widespread genotype, combining the $-100$ allele of *septemvittatus* with the $-96$ allele of *tigris*. The other two *Pgi* genotypes ($Pgi^{-99/-96}$ and $Pgi^{-104/-96}$) also could have been formed by hybridization, since both the $-99$ and the $-104$ alleles occur in *septemvittatus* (Table 2). In *septemvittatus*, the $Pgi^{-99}$ allele was recorded, as a homozygote, in a single individual collected at the Black Gap Wildlife area, 40 miles south of locality 6. (All 22 individuals from locality 6 were homozygous for the $Pgi^{-100}$ allele.) Yet clones of *tesselatus* having the $Pgi^{-99/-96}$ genotype were recorded only at localities 11, 12, and 14, which are far to the north on the upper Pecos River and along the Canadian River; and only the standard genotype ($Pgi^{-100/-96}$) was represented at the intermediate localities 7, 9, and 10. Similarly, the $Pgi^{-104}$ allele was recorded in *septemvittatus* only at Pinto Canyon (locality 5), where it was polymorphic with the $-100$ allele, but clones of *tesselatus* having the $Pgi^{-104/-96}$ genotype apparently are confined to the Elephant Butte Lake region, New Mexico, represented by samples 18, 19, 20, and 21. All *tesselatus* from the southern region (localities 1, 2, 3, 5, and 7) had the standard $Pgi^{-100/-96}$ genotype.

Three genotypes of *Pep* were recorded in diploid *tesselatus* (Table 5). $Pep^{98/94}$ is the predominant genotype, from which $Pep^{94/94}$ presumably was derived by recombination. Perhaps two separate recombinations were involved, one occurring in the southern part of the range (localities 3 and 5) and another at locality 8 along the Canadian River. The $Pep^{98/98}$ genotype, which also has a disjunct distribution (occurring at localities 2 and 3 and at 19 and 21) could have been formed either by hybridization or by recombination of the standard $Pep^{98/94}$ genotype. If hybridization is suggested as the source, we must postulate that at least four hybrid zygotes were involved in order to account for the presence of four clones involving the *Pep* and the *Pgi* loci; all four possible combinations of the $Pep^{98/94}$ and 98/98 genotypes and the $Pgi^{-100/-96}$ and $-104/-96$ genotypes occur in *tesselatus*. And if recombination is invoked, separate recombinational events must be postulated to account for these combinations of *Pep* and *Pgi* genotypes.

## DISCUSSION

The geographical and ecological distributions of the parental bisexual species *tigris* and *septemvittatus* suggest an origin for the diploid form of *tesselatus* in northern Mexico or southern Texas, followed by migration northward in the Rio Grande and Pecos River valleys and thence along the Canadian River in the Texas Panhandle and the drainage basin of the Arkansas River in southeastern Colorado (Figure 1). Subsequently, hybridization between diploid *tesselatus* and the bisexual species *sexlineatus* in Colorado gave rise to the triploid clone of *tesselatus* (WRIGHT and LOWE 1968; SCUDDAY 1971; PARKER and SELANDER, in preparation).

Three of the *septemvittatus* alleles present in *tesselatus* (*Pep⁹⁴*, *Me⁹⁸*, and *Pgi⁻¹⁰⁴*) were found only in the *septemvittatus* population at Pinto Canyon, Texas (locality 5). This suggests that diploid *tesselatus* arose from hybridization of *tigris* and a population of *septemvittatus* more closely related to the Pinto Canyon lizards than to the other two populations sampled, which probably represent two different subspecies of *septemvittatus* (DUELLMAN and ZWEIFEL 1962; J. F. SCUDDAY, personal communication).

Estimates of genic heterozygosity in parthenogenetic and related bisexual animal species are presented in Table 6. For Otiorrhynchus weevils and Solenobia moths, the similarity in heterozygosities of bisexual and parthenogenetic forms is consistent with the hypothesis that hybridization was not involved in the origin of these forms (SUOMALAINEN and SAURA 1973; LOKKI *et al.* 1975). But parthenogenetic forms of Cnemidophorus, which arose by hybridization, show a marked increase in average heterozygosity over the bisexual species. This is apparent both in *tesselatus* and in *C. laredoensis*, which is a diploid parthenogen formed from a cross between *C. sexlineatus* and *C. gularis* (McKINNEY, KAY and

TABLE 6

*Genetic diversity in parthenogenetic and related bisexual animal species*

| Breeding system and species* | Number of populations and individuals | Number of loci | Proportion of loci | |
|---|---|---|---|---|
| | | | Polymorphic per population | Heterozygous per individual |
| **Parthenogenetic** | | | | |
| *Otiorrhynchus scaber* (3N) | 2(162) | 26 | .. | .314 |
| *O. scaber* (4N) | 3(91) | 26 | .. | .317 |
| **Bisexual** | | | | |
| *O. scaber* | 1(68) | 24 | .83 | .309 |
| **Parthenogenetic** | | | | |
| *Solenobia triquetrella* (2N) | 5(59) | 15 | .. | .250 (XY type) |
| | | | | .200 (XO type) |
| *S. triquetrella* (4N) | 2(212) | 15 | .. | .203 |
| **Bisexual** | | | | |
| *S. triquetrella* | 2(44) | 15 | .67 | .230 |
| **Parthenogenetic** | | | | |
| *Cnemidophorus tesselatus* (2N) | 24(339) | 21 | .. | .560 |
| *C. tesselatus* (3N) | 3(67) | 21 | .. | .714 |
| *C. laredoensis* (2N) | 1(72) | 15 | .. | .267 |
| **Bisexual** | | | | |
| *C. tigris* | 2(69) | 21 | .19 | .050 |
| *C. septemvittatus* | 3(62) | 21 | .19 | .058 |
| *C. sexlineatus* | 2(34) | 21 | .24 | .070 |

* References: Otiorrhynchus (SUOMALAINEN and SAURA 1973); Solenobia (LOKKI *et al.* 1975); *Cnemidophorus laredoensis* (McKINNEY, KAY and ANDERSON 1973); other Cnemidophorus, present study.

ANDERSON 1973). Heterozygosity levels in the bisexual species are close to the average reported for vertebrates (SELANDER 1976).

Data presented by UZZELL and DAREVSKY (1975) for parthenogenetic lizards of the genus Lacerta and by VRIJENHOEK, ANGUS and SCHULTZ (1976) for hybridogenetic fishes (*Poeciliopsis*) provide further evidence that high levels of heterozygosity are characteristic of hybrid unisexual organisms.

Among plants, the genetic systems of species having permanent translocation heterozygosity resemble those of hybrid parthenogenetic animals in that the genomes were derived by hybridization of different races or species and recombination is severely limited. Recent studies of allozymic variation in several species of Oenothera and Gaura indicate that levels of heterozygosity and polymorphism are not greater in translocation heterozygotes than in "normal" congeneric species (LEVY and LEVIN 1975; LEVIN 1975). A plausible explanation for this finding is that the ancestral genomes were similar and the structural heterozygotes are of recent origin.

In attempting to analyze genetic diversity in Cnemidophorus, we have provisionally assumed that electromorphs of similar mobility are encoded by isoalleles. Thus, we refer to "alleles" shared by populations of *tesselatus* and those of the bisexual species, although we actually have compared electromorphs that may be allelically heterogeneous (KING and OHTA 1975). It is possible, for example, that what we have called the "$Pgi^{-104}$" allele in *tesselatus* is actually a mutant of the $Pgi^{-100}$ allele encoding an electromorph equivalent in electrophoretic mobility to that of the $Pgi^{-104}$ allele of *septemvittatus*. Another potential source of error stems from our inability to make genetic crosses to determine the genotypes of parthenogenetic individuals presumed to be homozygous at a particular locus on the basis of having a single-banded phenotype. Hence, for example, individuals presumed to have the genotype $Pep^{94/94}$ may in fact be heterozygous for the *94* allele and a mutant "null" or "silent" allele; or, alternatively, they could be $Pep^{98/94}$ heterozygotes, with a regulatory mutant suppressing expression of the *98* allele.

Notwithstanding uncertainties regarding the translation of electrophoretic phenotypes into genotypes, we have attempted to apportion the existing diversity in diploid *tesselatus* among the three potential sources of variation: mutation, hybridization, and recombination. Our objective has been to account for the existence of nine nonstandard genotypes detected at six variable loci. Three genotypes defining three clones (each restricted to a single locality) can be attributed to mutation at three loci, and 19 of the 339 diploid individuals (5.6%) in our collection carry mutant alleles. Hybridization has produced three clones distinguishable by genotypes at the *Pgi* locus. One of the *Pep* genotypes, which in combination with *Pgi* genotypes characterizes two additional clones, may also owe its origin to hybridization. Five of the 12 clones of diploid *tesselatus* apparently have arisen through recombination. These interpretations are made with the understanding that more extensive surveys of the parental species may reveal additional electromorphs, hybrid combinations of which could be responsible for those genotypes of *tesselatus* that we have interpreted as having arisen by recombination or mutation.

ASHER (1970) and ASHER and NACE (1971) have proposed that the maintenance of heterozygosity at equilibrium in parthenogens exhibiting recombination is dependent upon heterosis, since otherwise heterozygosity will decay at a rate determined by the frequency of recombination. In Cnemidophorus, as in all hybrid parthenogens employing premeiotic endoduplication as a mechanism of egg maturation, the frequency of recombination depends on the degree of homology of the two parental chromosomes as this determines the probability of quadrivalent or bivalent formation between the parental homologues. Loci on chromosomes that are insufficiently homologous will remain in fixed heterozygous state, while those on more nearly homologous chromosomes will become partially or completely homozygous, depending on chiasma localization.

There is less genotypic diversity in *tesselatus* than in the parthenogenetic species of insects studied by SUOMALAINEN and SAURA (1973) and LOKKI et al. (1975). As many as 9 clones were observed in one population of the triploid weevil *Otiorrhynchus scaber*, and 15 clones were recognized in the three populations sampled. Tetraploid populations of *O. scaber* consisted of as many as 8 clones, and 27 clones were identified in six populations. Similarly, over 21 clones were found in ten samples of the diploid parthenogenetic moth *Solenobia triquetrella*, and 27 clones were identified in 16 populations of tetraploid Solenobia. In contrast, only four clones were present in the most variable populations of *C. tesselatus* (samples 12, 14, 19, and 21; Tables 4 and 5).

Divergence within and between parthenogenetic populations depends largely on mutation and recombination and may, therefore, be related to age of the parthenogenetic form (WHITE 1970). It has been suggested that the parthenogenetic forms of Otiorrhynchus and Solenobia, which have more genotypic diversity than *Cnemidophorus tesselatus*, arose in the Würm glaciation period in Europe, 12,000 years ago (SUOMALAINEN 1962; SUOMALAINEN and SAURA 1973). Mutation (and possibly some recombination) has contributed to the clonal diversity but has not significantly increased levels of individual heterozygosity over those of the ancestral bisexual populations.

The age of *tesselatus* populations is unknown. The widespread distribution of *tesselatus* (and perhaps also the complexity of geographic patterning of the clones) suggests an early origin, as proposed by AXTELL (1966) for another parthenogen, *C. neomexicanus*, which may have evolved in the Wisconsin period (ca. 12,000 years B.P.). However, the relatively low level of clonal diversity in *tesselatus* may be interpreted as evidence of a more recent origin. It is noteworthy that both the bisexual species and *tesselatus* are associated with man-disturbed habitats over much of their ranges. Possibly *tesselatus* evolved no earlier than 200 years ago, when over grazing of short-grass prairie in the southwestern United States and northern Mexico (LOWE et al. 1970) may have first brought the parental species into contact.

In sum, our analysis of genic variation in *Cnemidophorus tesselatus* adds to the growing body of evidence supporting the view that the evolutionary potential of parthenogenetic organisms may be less severely limited than has been generally supposed. Most of the genetic variability in *tesselatus* is in the form of heterozygosity established at the time of origin through hybridization of its bisexual ancestors, but mutation and recombination make continued adaptive evolution possible, although slow. In conjunction with possible general advantages of heterosis and parthenogenetic reproduction, this potential for generating clonal diversity may explain the success of the parthenogenetic lineages of Cnemidophorus.

Valuable assistance in the field was provided by A. CASTILLO, W. E. MARSHALL, and A. O. STEWART. We are indebted to U. NUR, E. SUOMALAINEN, T. UZZELL and M. J. D. WHITE for critical review of the manuscript. Research supported by NSF Grants BMS73-06856 and GB-41112. E. D. P. is a Predoctoral Trainee on NIH Grant 1-T32-GM-07102.

## LITERATURE CITED

ASHER, J. H., 1970 Parthenogenesis and genetic variability. II. One-locus models for various diploid populations. Genetics **66**: 369–391.

ASHER, J. H. and G. W. NACE, 1971 The genetic structure and evolutionary fate of parthenogenetic amphibian populations as determined by Markovian analysis. Am. Zool. **11**: 381–398.

AXTELL, R. W., 1966 Geographic distribution of the unisexual whiptail *Cnemidophorus neomexicanus* (Sauria: Teiidae)—present and past. Herpetologica **22**: 241–253.

BECKMAN, G. and L. BECKMAN, 1975 Genetics of human superoxide dismutase isozymes. pp. 781–795. In: *Isozymes*, Vol. IV. *Genetics and Evolution*. Edited by C. L. MARKERT. Academic Press, New York.

BODMER, W. F., 1970 The evolutionary significance of recombination in prokaryotes. pp. 279–294. In: *Prokaryotic and Eukaryotic Cells*. Symposia of the Society for General Microbiology No. XX. Edited by H. P. CHARLES and B. C. J. G. KNIGHT. Cambridge Univ. Press, Cambridge.

CIMINO, M. C., 1972 Meiosis in triploid all-female fish (*Poeciliopsis*, Poeciliidae). Science **175**: 1484–1486.

CROW, J. F. and M. KIMURA, 1965 Evolution in sexual and asexual populations. Am. Naturalist **94**: 439–450.

CUELLAR, O., 1971 Reproduction and the mechanism of meiotic restitution in the parthenogenetic lizard *Cnemidophorus uniparens*. J. Morph. **133**: 139–165.

DUELLMAN, W. E. and R. G. ZWEIFEL, 1962 A synopsis of the lizards of the *sexlineatus* group (genus *Cnemidophorus*). Bull. Am. Mus. Natur. History **123**: 155–210.

ESHEL, I. and M. W. FELDMAN, 1970 On the evolutionary effect of recombination. Theoret. Pop. Biol. **1**: 88–100.

FELSENSTEIN, J., 1974 The evolutionary advantage of recombination. Genetics **78**: 737–756.

FISHER, R. A., 1958 *The Genetical Theory of Natural Selection*. 2nd ed. Dover Publications, New York.

JACOB, J., 1957 Cytological studies of Melaniidae (Mollusca) with special reference to parthenogenesis and polyploidy. I. Oögenesis of the parthenogenetic species of *Melanoides* (Prosobranchi–Gastropoda). Trans. Royal Soc. Edinburgh **63**: 341–356.

KALLMAN, K., 1962 Population genetics of the gynogenetic teleost, *Mollienesia formosa* (Girard). Evolution **16**: 487–504.

KING, J. L. and T. OHTA, 1975   Polyallelic mutational equilibria. Genetics **79**: 681–691.

LEVIN, D. A., 1975   Genetic correlates of translocation heterozygosity in plants. Bioscience **25**: 724–728.

LEVY, M. and D. A. LEVIN, 1975   Genic heterozygosity and variation in permanent translocation heterozygotes of the *Oenothera biennis* complex. Genetics **79**: 493–512.

LOKKI, J., E. SUOMALAINEN, A. SAURA and P. LANKINEN, 1975   Genetic polymorphism and evolution in parthenogenetic animals. II. Diploid and polyploid *Solenobia triquetrella* (Lepidoptera: Psychidae). Genetics **79**: 513–525.

LOWE, C. H., J. W. WRIGHT, C. J. COLE and R. L. BEZY, 1970   Natural hybridization between the teiid lizards *Cnemidophorus sonorae* (parthenogenetic) and *C. tigris* (bisexual). Syst. Zool. **19**: 114–127.

MACGREGOR, H. C. and T. M. UZZELL, JR., 1964   Gynogenesis in salamanders related to *Ambystoma jeffersonianum*. Science **143**: 1043–1045.

MASLIN, T. P., 1967   Skin grafting in the bisexual teiid lizard *Cnemidophorus sexlineatus* and in the unisexual *C. tesselatus*. J. Exptl. Zool. **166**: 137–150. ——, 1971   Parthenogenesis in reptiles. Am. Zool. **11**: 361–380.

MAYNARD SMITH, J., 1971   What use is sex? J. Theoret. Biol. **30**: 319–339.

MCKINNEY, C. O., F. R. KAY and R. A. ANDERSON, 1973   A new all-female species of the genus *Cnemidophorus*. Herpetologica **29**: 361–366.

MULLER, H. J., 1932   Some genetic aspects of sex. Am. Naturalist **66**: 118–138.

NEAVES, W. B., 1969   Adenosine deaminase phenotypes among sexual and parthenogenetic lizards in the genus *Cnemidophorus* (Teiidae). J. Exptl. Zool. **171**: 175–184.

NEAVES, W. B. and P. S. GERALD, 1968   Lactate dehydrogenase isozymes in parthenogenetic teiid lizards (*Cnemidophorus*). Science **160**: 1004–1005. ——, 1969   Gene dosage at the lactate dehydrogenase B locus in triploid and diploid teiid lizards. Science **164**: 557–558.

NEI, M., 1973   The theory and estimation of genetic distance. pp. 45–54. In: *Genetic Structure of Populations*. Edited by N. E. MORTON. University of Hawaii Press, Honolulu.

NEI, M. and A. K. ROYCHOUDHURY, 1974   Sampling variances of heterozygosity and genetic distance. Genetics **76**: 379–390.

NUR, U., 1971   Parthenogenesis in coccids (Homoptera). Am. Zool. **11**: 301–308.

RÖSSLER, Y. and P. DeBACH, 1973   Genetic variability in a thelytokous form of *Aphytis mytilaspidis* (LeBaron) (Hymenoptera: Aphelinidae). Hilgardia **42**: 149–176.

SCUDDAY, J., 1971   The biogeography and some ecological aspects of the teiid lizards (*Cnemidophorus*) of Trans-Pecos Texas. Ph.D. Dissertation, Texas A&M University.

SELANDER, R. K., 1976   Genic variation in natural populations. In: *Molecular Evolution*. Edited by FRANCISCO J. AYALA. Sinauer Associates, Sunderland, Massachusetts.

SELANDER, R. K., M. H. SMITH, S. Y. YANG, W. E. JOHNSON and J. B. GENTRY, 1971   Biochemical polymorphism and systematics in the genus *Peromyscus*. I. Variation in the old-field mouse (*Peromyscus polionotus*). Stud. Genet. VI. Univ. Texas Publ. **7103**: 49–90.

SMITH, S. G., 1971   Parthenogenesis and polyploidy in beetles. Am. Zool. **11**: 341–349.

STANLEY, S. M., 1975   Clades versus clones in evolution: why we have sex. Science **190**: 382–384.

SUOMALAINEN, E., 1962   Significance of parthenogenesis in the evolution of insects. Ann. Rev. Entomology **7**: 349–366. ——, 1969   Evolution in parthenogenetic Curculionidae. pp. 261–296. In: *Evolutionary Biology*. Edited by THEODOSIUS DOBZHANSKY, MAX K. HECHT and WILLIAM C. STEERE. Appleton-Century-Crofts, New York.

SUOMALAINEN, E. and A. SAURA, 1973 Genetic polymorphism and evolution in parthenogenetic animals. I. Polyploid Curculionidae. Genetics **74**: 389–508.

UZZELL, T., 1970 Meiotic mechanisms of naturally occurring unisexual vertebrates. Am. Naturalist **104**: 433–445.

UZZELL, T. and I. S. DAREVSKY, 1975 Biochemical evidence for the hybrid origin of the parthenogenetic species of the *Lacerta saxicola* complex (Sauria: Lacertidae), with a discussion of some ecological and evolutionary implications. Copeia **1975**: 204–222.

UZZELL, T. M., JR. and S. M. GOLDBLATT, 1967 Serum proteins of salamanders of the *Ambystoma jeffersonianum* complex, and the origin of the triploid species of this group. Evolution **21**: 345–354.

VRIJENHOEK, R. C., 1975 Gene dosage in diploid and triploid unisexual fishes (*Poeciliopsis*, Poeciliidae). pp. 463–475. In: *Isozymes*. Vol. IV. *Genetics and Evolution*. Edited by C .L. MARKERT. Academic Press, New York.

VRIJENHOEK, R. C., R. ANGUS and R. J. SCHULTZ, 1976 Variation and clonal structure in a unisexual fish. Isozyme Bull. **9**: 60.

WHITE, M. J. D., 1966 Further studies on the cytology and distribution of the Australian parthenogenetic grasshopper, *Moraba virgo*. Rev. Suisse de Zoologie **73**: 383–398. ——, 1970 Heterozygosity and genetic polymorphism in parthenogenetic animals. pp. 237–262. In: *Essays in Evolution and Genetics in Honor of Theodosius Dobzhansky*. Edited by M. K. HECHT and W. C. STEERE. Appleton-Century-Crofts, New York. ——, 1973 *Animal Cytology and Evolution*. 3rd ed. Cambridge University Press, Cambridge.

WILLIAMS, G. C., 1975 *Sex and Evolution*. Princeton University Press, Princeton, New Jersey.

WRIGHT, J. W. and C. H. LOWE, 1967 Evolution of the alloploid parthenospecies *Cnemidophorus tesselatus* (Say). Mammal. Chromo. Newsl. **8**: 95–96. ——, 1968 Weeds, polyploids, parthenogenesis and the geographical and ecological distribution of all-female species of *Cnemidophorus*. Copeia **1968**: 128–138.

ZWEIFEL, R. G., 1965 Variation in and distribution of the unisexual lizard, *Cnemidophorus tesselatus*. Am. Mus. Novitates **2235**: 1–49.

Part IV

# INTROGRESSIVE HYBRIDIZATION

# Editor's Comments
# on Papers 11 Through 15

**11  HUBBS**
*Hybridization Between Fish Species in Nature*

**12  ANDERSON and HUBRICHT**
*Hybridization in* Tradescantia. *III. The Evidence for Intro-gressive Hybridization*

**13  HEISER**
*Hybridization Between the Sunflower Species* Helianthus Annuus *and H. Petiolaris*

**14  ALSTON and TURNER**
*Natural Hybridization among Four Species of* Baptisia (*Lu-guminosae*)

**15  CARSON, NAIR, and SENE**
Drosophila *Hybrids in Nature: Proof of Gene Exchange Be-tween Sympatric Species*

As noted by Raven (1976):

The taxonomist's desire to place all organic diversity into con-venient pigeonholes long concealed the widespread occur-rence of natural hybridization in plants; until no more than 40 years ago, it was commonly assumed that a more critical analysis would eliminate the suspected hybrids by classifying them cor-rectly. Just how remarkable it was that the leading botanists of nineteenth-century Britain, for example, considered hybrids to be extremely rare in nature, if present at all, is demonstrated by recent studies of hybrids in the British flora [Stace, 1975a, b]. In a flora of about 2200 species, at least 850 interspecific hybrid combinations are now known!

Hybridization is much less common in animals than in plants. Mayr (1963) wrote:

In most groups of animals hybridization is sufficiently excep-tional to justify a report in the literature whenever a hybrid is discovered. The suggestion is sometimes made that this rarity

of animal hybrids is more apparent than real, but . . . the frequency of hybrids is as low in groups where hybrids can be discovered easily . . . as where the recognition of hybrids is difficult.

The situation in fish is in contrast to the general rule; it has been
reviewed by Hubbs (Paper 11).

Students of hybridization have faced two problems: demonstration that the intermediate entities growing among two good
species are indeed hybrids; and demonstration that genes of one
species may be incorporated by another species. Anderson and
Hubricht (Paper 12) designated the term *introgression* for the incorporation of genes of one species into the gene pool of another
species as a result of repeated hybridization followed by backcrossing. If introgression were occurring, selected characters should be
more variable in that part of a species range that is sympatric with
a related species, and should vary in the direction of the nonrecurrent parent. Anderson and Hubricht found strong evidence of introgression of *Tradescantia canaliculata* into *T. occidentalis* and
*T. bracteata*. One of the first substantive demonstrations of introgression, this has been followed in the last four decades with numerous others. Heiser (1947, 1949b, 1951a, 1951b, 1954) presented
evidence of extensive introgression in *Helianthus* with *H. annuus*
as the central figure. This species has given rise to introgressive
races varying toward *H. debilis, H. argophyllus, H. petiolaris*, and
perhaps *H. bolanderi*. Experimental hybridizations demonstrated
the plausibility of the introgression hypothesis. Paper 13, the first
in the *Helianthus* series, demonstrates the use of an experimental
breeding program and Anderson's (1936) hybrid index method of
analyzing hybrid swarms.

A similar picture is seen in other genera. For example, the
"plundering" of local gene pools by the wide-ranging grass *Both-
riochloa intermedia* may account for its occupation of divergent
habitats and its broad distribution (Harlan and deWet, 1963). Harlan
and deWet (1963) have proposed the concept of compilospecies
in reference to species that readily assimilate the germplasm of
their allies.

Wide-ranging introgression is not restricted to plants. Perhaps the most intensively investigated analysis of hybridization
associated with habitat disturbance by humans involves *Pipilo erythrophthalmus* and *P. ocai* in Mexico (Sibley, 1950, 1954; Sibley and
West, 1958). The former is widespread in North America and reaches
as far south as Chiapas and Guatemala; the latter ranges from Oaxaca
to Jalisco. In Oaxaca, the species occur together without hybri-

dizing. In Puebla, about 15 percent of the specimens display evidence of hybridization. A series of introgressed hybrid populations extends across the Mexican Plateau from northern Puebla through Naharit and Michoacán to Jalisco. In the west and south, the populations vary toward *P. ocai;* in the east and north, they vary toward *P. erythrophthalmus.* Variable but generally intermediate populations occur along north–south and east–west transects through this region. The extensive interbreeding and introgression between these species suggests that they might be treated as conspecific (Short, 1969). A similar situation to that in *Pipilo* occurs in the bird genera. *Melidectes* (Gillard, 1959), *Criniger* (Rand, 1958), and *Ceyx* (Sims, 1959).

The possibility of widespread introgression in the fruit fly *Dacus* has been studied by Lewontin and Birch (1966). The tropical Queensland species, *D. tryoni,* has extended its range southward in Australia during the past century, presumably as a result of the development of individuals adapted to the more rigorous temperate climate. Evidence from laboratory hybridization and the performance of hybrids over a range of temperature suggests that the genetic variation necessary to allow *D. tryoni* to increase its ecological tolerance came from introgression of genes from *D. neohumeralis.* The introduction of cultivated fruits afforded a food source necessary for survival in the newly occupied territory.

The fate of hybrid swarms through time is poorly understood. One time-lapse analysis of a hybrid swarm involves *Helianthus annuus* and *H. bolanderi,* initially described by Heiser (1949b) and reexamined several times in the 1950s by Stebbins and Daly (1961). The western half of the population maintained a high proportion of *H. bolanderi* types, some like *H. annuus* and some intermediates. In the eastern half, intermediate hybrids initially prevailed, but the variation pattern shifted toward *H. bolanderi* in the early 1950s, and then toward *H. annuus* during the latter part of the decade. Overall, the parental types seemed to become more common.

Three hybrid swarms of the perennial sunflowers, *Helianthus divaricatus,* a species generally of open or semi-open habitats, and *H. microcephalus,* a species generally found in shaded areas, were studied in southern Indiana in 1950 (Smith and Guard, 1958). The same areas were reexamined by Heiser (pers. comm.) in 1961 and 1972. Considerable hybridity was still evident in one population in 1972, probably the result of continued disturbance in the area. Although the habitat has become more closed at the other two sites, *H. divaricatus*-like plants were dominant at one site and the sole representative at the other site in 1972. Heiser (pers. comm.) explains these results in part, by the facts that *H. divaricatus* was the more abundant species in these two localities in 1950 and that it has a greater life span. While it seems possible that the original hybridization was merely transitory with no permanent effects, there is also a possibility that *H. divaricatus* may owe its increased shade tolerance to introgression from *H. microcephalus*.

Jones (1973) followed up an earlier analysis by A. P. Blair (1941) concerning hybridization between *Bufo americanus* and *B. woodhousii* in the Bloomington, Indiana, area. Whereas Blair found that 9.4 percent of the matings were interspecific, Jones found no evidence of contemporary hybridization. Both species, however, have converged in several morphological characters, and are apparently more strongly ethologically isolated. In another study on frogs, Littlejohn and Watson (1976) followed, for seven years, the mating-call structure of a population composed principally of recombination products of *Geocrina laevis* × *G. victoriana*. The population appeared to be a largely self-perpetuating hybrid swarm in which parental phenotypes arise through occasional migration or recombination.

Prior to 1960, the structure of hybrid swarms was analyzed principally by morphological criteria. The use of chromatographically separable secondary products as genomic markers ushered in an era of more critical analysis of hybrid swarms, providing the researcher with numerous codominant, largely independently segregating characters. Phenolic compounds have been the most widely used class of secondary products in the study of natural hybridization. The studies of Alston, Turner, and their associates on *Baptisia* have been most notable. Their analysis (Paper 14) of hybridizing populations involving two, three, and four species demonstrates the application of the technique and the interpretations that it permits. The *Baptisia* work, summarized by Alston (1967), was the prime stimulus for the extensive utilization of phenolic markers as criteria for hybridization in plants, and was the focus of excitement in experimental systematics in the early 1960s. Terpenes also have been used extensively to document hybridization and introgression in plants (Mirov, 1956; Pryor and Bryant, 1958; Critchfield, 1967; Ogilvie and von Rudloff, 1968; Gerhold and Plank, 1970; Irving and Adams, 1973; Hunt and von Rudloff, 1974). Other groups of secondary products have been much less valuable in these regards.

The study of animal hybrids was based almost exclusively on proteins in the early 1960s, and continues as such. Blood proteins

(Fox et al., 1961) and venom constituents (Wittliff, 1964) of two hybridizing species of toads permitted the identification of rare backcross hybrids, and the possibility of unidirectional introgression. Using albumin-like proteins, Crenshaw (1965) reported evidence for gene exchange between conspecific turtle species thought to be reproductively isolated over most of their range. The value of protein markers is well illustrated in the lizard *Cnemidophorus tigris*. A morphological analysis of a hybrid zone between subspecies of *C. tigris* indicated little gene exchange, whereas simple biochemical markers revealed that gene exchange was actually extensive (Dessauer et al., 1962).

The use of chromosomal data to document complex hybridization between species with the same chromosome number has been limited, because many closely related species fail to have chromosome complements that are sufficiently differentiated to identify backcross and advanced-generation hybridization. The genus *Drosophila* is a prime subject for the analysis of hybridization and introgression using chromosomal markers because of species-specific banding patterns. In spite of an extensive literature on the genus extending over fifty years, only two reports suggest the possibility of introgression in nature (Pipkin, 1968; Dobzhansky, 1973). Unequivocal evidence of weak introgression in *Drosophila* has been presented only recently by Carson, Nair, and Sene (Paper 15) for two Hawaiian species. A similiarly low level of introgression has been described subsequently for another pair of Hawaiian *Drosophila* (Kaneshiro and Val, 1977).

Introgression in *Drosophila* involved species with the same chromosome number. However, there are instances in other genera where hybridization and introgression occur between chromosomally distinctive species with different chromosome numbers. One prime example involves the ladybird beetles *Chilocorus hexacyclus* ($2n = 14$) and *C. tricyclus* ($2n = 20$) (Smith, 1966). What is of additional interest is that although the species are readily categorized chromosomally, there are no apparent external differences. Thus the chromosome complement provides proof rather than only corroboration of introgression.

## 11

Reprinted from *Syst. Zool.* **4**:1–20 (1955)

# Hybridization between Fish Species in Nature

CARL L. HUBBS

THE Jordan school of ichthyology, in which I was nurtured, held steadfastly to the view that the lines between fish species are almost never crossed. This mistaken idea was perhaps in part a holdover of the pre-Darwinian concept of the immutability of species, but was, more pointedly, a reaction against the tendency of some European ichthyologists to explain as hybrids specimens that proved difficult to identify. This view that there is little or no natural hybridization between fish species may be further attributed to the circumstance that systematic ichthyology in America from about 1890 to 1920 was still essentially in the exploratory and species-naming stage, and dealt chiefly with marine fishes, which, as I shall point out, hybridize much less often than do freshwater fishes.

Early in my ichthyological career I encountered evidence that fish species do cross in nature, at times in considerable frequency. The analysis of such hybridization has ever since engaged my attention, in so far as other activities would allow. This report lays particular emphasis on the researches that I have conducted during the past thirty-five years, largely with the collaboration of my wife and of several students and colleagues, to all of whom I pay tribute. They might all appear as co-authors of this report. The studies continue.

During the first ten years of my intensive studies of the freshwater fishes of North America, from 1919 to 1929, I gathered strong circumstantial indications that the species lines are rather often crossed in nature, and during the following fifteen years, from 1929 to 1944, I was able, with the constant aid of Mrs. Hubbs, to confirm these indications, not only by the accumulation of much more circumstantial evidence, in part subjected to critical statistical analysis, but also by breeding experiments in the aquarium of the Division of Fishes of the then new Museum of Zoology at the University of Michigan.

## Sunfish Hybrids

The first main project in this aquarium was the experimental verification of extensive hybridization between the well-known species of sunfish. I had already concluded, on the basis of circumstantial evidence, that certain supposed species had been based on interspecific crosses. The commonest of these, known as *Lepomis euryorus*, I had treated as a natural hybrid between the green sunfish and the pumpkinseed. By placing a male of the one species with a female of the other we soon were elated to observe the characteristic nuptial gyration of sunfishes, followed by the laying of eggs, some of which hatched. And from the eggs that hatched we developed a considerable number of the miscegenates into large adults. The characters matched those of the supposed species *euryorus* and were intermediate between those of the parental species. Similarly, we produced fine adults of the hybrid between the green sunfish and the bluegill. This hybrid cross, also with intermediate characters, had been passing under the name of *Lepomis ischyrus*. Other combinations were effected. Most of the interbreeding was quite voluntary; some was induced by pituitary injections.

From our studies of sunfish hybrids we learned many of the basic facts regarding the interspecific hybridization of these fishes. As already mentioned, the hybrids are intermediate in taxonomic characters between the parental species. They differ from either parental type, however, in several respects, all attributable to the still poorly understood phenomenon of heterosis, or hybrid vigor. That they grow faster than either parental species was shown by the analyses of size frequencies at determined ages in nature and by measurements of growth in aquarium and pond. The hybrids dominate in the social hierarchy (known in fishes as the nip order). When a mixed group is fed, the hybrids, larger and more vigorous than the pure species, take the food first. In general activity, as in holding fins erect (an indication of vigor), and in the depth, brightness, and intensity of color, the hybrids excel. The great majority of the hybrids in nature as well as in captivity are males. The hybrid males construct, fan, and guard their gravel nests with unusual vigor over a prolonged period, but to no avail, for they are sterile. Sections of the testes of hybrids, whether bred in tanks or caught in nature, show marked abnormalities. In the hybrids the tubules of the testis are often vacuolated and the spermatozoa, when produced, vary greatly in size. When stripped from the hybrids some of the spermatozoa are relatively so immense that the beating of the tails merely rolls them along. The abnormal spermatogenesis of the natural hybrids confirms their identification as hybrids and helps to explain their infertility.

The frequency of hybridization in nature between sunfish species seems to be determined by several ecological factors. There is a relatively high incidence of hybridization in ponds largely filled with vegetation and with very limited gravel areas, where the spawning fish are unduly crowded. Great scarcity of one species coupled with the abundance of another often leads to hybridization: the individuals of the sparse species seem to have difficulty in finding their proper mates. The introduction of a species into a pond inhabited only by other species is also conducive to interbreeding. This is true even for species that elsewhere have come to live together with little or no crossing. In some ponds the observed incidence of hybridization rises to over 10 percent.

In one stream near Ann Arbor I found about 95 percent of the sunfish to be hybrids between the green sunfish and the pumpkinseed. This finding, made early in the study, baffled me at first. On a visit to the Northville Hatchery near Ann Arbor the superintendent complained that the sunfish he had taken in the stream to serve as breeders had not bred in the pond. He scooped out a net full of what I at once recognized as hybrids. How could sterile hybrids, almost all males, attain a 95 percent dominance? I wracked my brain for an answer, which finally took form. There must be, I hypothesized, a pond downstream inhabited by the parental species, which cross occasionally, and the hybrids because of their superior vigor counter the stream current to reach the vicinity of the hatchery. I called the superintendent and learned that there was indeed a pond five miles downstream. We then seined in the pond just what the "desperation hypothesis" called for—the parental species and a few hybrids!

Ordinarily the sunfish hybrids are definitely intermediate and do not grade into either parental type. In three places, however, it was found that the characters of the especially abundant hybrids graded into those of each parental species. At each place the hybrids were further exceptional in that they were not predominantly males. At these three isolated spots, the genes were probably being exchanged—a point of considerable speciational significance.

### The Case of the "Amazon Molly"

While we were still engaged in the experimental confirmation of interspecific hybridization in the sunfishes, fishes collected in northeastern Mexico by Drs.

Myron Gordon and Edwin P. Creaser posed a problem that cried for experimental attack. *Mollienisia formosa*, a species of the viviparous topminnow family Poeciliidae, was brought back in only one sex: all of the many specimens were female. Since this nominal species is intermediate between the sailfin molly (*Mollienisia latipinna*) and *Mollienisia sphenops*, hybridization was suspected. We secured live material of all three types and in 1932 inaugurated a twelve-year program of intensive breeding experiments with these three species, and many others.

We proceeded at once to test the reproductive behavior of the species of which only females had been collected. We promptly mated some of these females, which we termed "Amazon mollies," with the two suspected parental types. One of these, *Mollienisia latipinna*, has a longer dorsal fin than *formosa*, originating farther forward and with more rays (13 to 15 instead of 10 to 12); and it has a deeper and more strongly striped body. *Mollienisia sphenops*, the other suspected parental form, has in contrast a shorter dorsal fin than *formosa*, originating farther back and with fewer rays (usually 9 in the subspecies used), and it has a slenderer body with the streaks poorly developed. We soon obtained broods of young, all of which, in confirmation of the seemingly impossible hypothesis that I was forced to erect, proved to be females exactly like their mothers, without any trace of the distinctive paternal characters! Crossing *latipinna* and *sphenops* reciprocally, we obtained hybrids almost exactly like the all-female *formosa*, yet these laboratory-bred hybrids were bisexual, and produced intermediate offspring on backcrosses. In our preliminary note in *Science* (Hubbs and Hubbs, 1932b) we concluded:

The conditions demonstrated by this study, so far as we know novel in the biology of the vertebrates are: (1) the abundant occurrence in nature of a form of demonstrated hybrid origin, having nearly all of the characteristics of a natural species; (2) the occurrence of a form as females only, over a wide portion of its range [in fact, we now know, over its entire range]; (3) the consistent and abundant production of wholly female and purely matroclinous young; (4) apparent parthenogenesis in nature.

We continued in our genetic test of the complete production of females and the totally matroclinous inheritance of *Mollienisia formosa*. Year after year and generation after generation, unto the twenty-first generation, our *Mollienisia formosa* females continued to produce all-female broods of fish, many hundreds in all, that were replicas of their mothers, although we had mated the mothers with all the available species of *Mollienisia* and with various simple and multiple hybrids between the species of this genus. We mated them with the black mutant of *Mollienisia*, that always throws black-blotched young when crossed with normally colored specimens of *Mollienisia* of the same or other species. We also had broods from matings of *Mollienisia formosa* with species of the related genus *Limia*, though such broods are not often produced. Twice we obtained two young by mating a *Mollienisia formosa* with a male of the mosquitofish, *Gambusia affinis*, belonging to a distinct subfamily. This amazing result, which we could hardly believe, suspecting an experimental error, led us to repeat trial matings of this heterogeneous combination. After several failures we planned and executed a forced mating. We injected a *Gambusia* male and an Amazon molly with pituitary extract to induce rapid and simultaneous maturity. We then injected some sperm stripped from the *Gambusia* into the tiny oviduct of the molly. In due term she gave birth to twelve healthy young, which all grew up into adult females—again just like the mother, with no trace of *Gambusia* characters. But the best characters in these fishes lie in the gonopodium, which is the anal fin of the male highly modified to serve as the intromittent organ. Being females, the offspring of the outcrossing showed no male characters—until we bathed some of them in the dilution of testosterone that we had found to be effective and thus forced them to develop the secondary male characters. The gonopodium that developed, normally, showed no trace of *Gambusia* characters, but was strictly that of a *Mollienisia*. Furthermore, the colors of the nuptial male of *Mollienisia* became evident.

Always the result was the same—all female young, fully typical of *Mollienisia formosa* in every detail of character, not only in the first generation, but also in subsequent backcrosses and outcrossings. Thus the male used may have had 6, 8, 10, 13, or 17 dorsal rays, but the progeny, modally, had 11 like their mother. Since we have found no trace of such genetic behavior in systematic characters of any other fishes, in this or other groups, and since *formosa* is very fertile, giving birth to full complements of healthy young that often all develop to maturity, we consider the genetic evidence as next to conclusive, that *formosa* transmits to all its offspring only its own maternal chromatin.

*Mollienisia formosa* seems to have arisen as a hybrid, but now has almost all the characteristics of a species. It is completely differentiated from its supposed parental species *latipinna* and *sphenops*, and certainly breeds true, much truer in all probability than most species. It justifies systematic consideration on ecological and historical grounds, which in this case as in general I regard as being quite as significant as the genetic potential: it has increased and spread, for it now occupies a considerable and definite range, including areas where only one of the parental species exists or has likely ever existed. In some places it outnumbers the parental species. The circumstances that its fecundity is high and that 100 percent of its population instead of little more than 50 percent are female breeders would seem to give it a definite selective advantage within the range of the genus, since it mates freely with the males of all species of *Mollienisia* and since the males of

poeciliid fishes promiscuously mate with many females. We think it best to treat the Amazon molly as a species, presumably of hybrid origin. This is a special but particularly interesting and significant crossing of the species line.

## Intermediacy of Hybrids

In all our work with hybrid fishes in this or any other family, the only taxonomic character that we have been able to attribute to the action of a single gene is the dorsal and anal ray numbers in the subspecies of the mosquitofish, *Gambusia affinis*, (Hubbs and Walker, ms.). The Atlantic Coast form, *G. a. holbrooki*, has 7 dorsal and 10 anal rays, the Mississippi Valley form, *G. a. affinis*, one fewer in each fin (6 dorsal and 9 anal). Intergrades, obviously produced through hybridization near the Gulf of Mexico in Mississippi and Alabama, have either 6 or 7 dorsal rays and either 9 or 10 anal rays. The sum of the ray counts is therefore 17 in the Atlantic Coast form, 15 in the Mississippi Valley form, and 15, 16, or 17 in the intergrades. The intergradation though narrow in area is gradual. First-generation hybrids have 7 dorsal and 10 anal rays, showing that the Atlantic Coast type is completely dominant in this regard. In the $F_2$ generation the condition shown by the intergrades is at once attained, for the dorsal rays are either 6 or 7, the anal rays either 9 or 10. Other characters, however, such as those of the gonopodium and of the sensory pores on the head, remain intermediate and give evidence of being due, like nearly all taxonomic characters, to multiple genes.

It has proved to be an almost universally valid rule that natural interspecific hybrids are intermediate between their parental species in all characters in which those species differ, whether they be external or internal, of shape, color, form, structure, or numbers of parts (vertebrae, gillrakers, fin rays, teeth)—except for some features that reflect hybrid vigor.

This intermediacy has been most extensively analyzed for the numerous hybrids we have described (Hubbs, Hubbs, and Johnson, 1943) in the family of suckers, Catostomidae, from many streams in the western states. Intermediacy is seen in such external characters as the coloration, the general body form, the size of the head, the length and protrusion of the snout, the size of the scales. The lips are definitely transitional in structure. When statistically analyzed the proportions and counts of the hybrids are seen to be not only definitely intermediate, but also of the same order of variability as shown by the parental species—strong evidence that the hybrids are of the sterile $F_1$ generation. The intermediacy may involve the type of relative growth. Internal characters, such as those of the skeleton, may also be transitional. When the values for the hybrids are computed as an index, on a percentage scale grading from the value for one parent, set at 0, to the value for the other, set at 100, the indices form a frequency distribution with the mode very close to 50, which represents ideal interjacency.

The most extensive data that we have analyzed for any one hybrid combination pertain to mass hybridization and species replacement in the Mojave River system. In our published study (Hubbs and Miller, 1942) we reported on 3350 individuals of the species *Gila orcutti*, that we now feel sure was introduced, on 1812 of the native chub, *Siphateles mohavensis*, and on 442 hybrids, constituting 7.9 percent of the entire random sample (or 9 percent if we exclude the fish collected from one spring, into which the introduced species has no access). The hybrid proved on analysis to be intermediate in nearly all characters of color, form, and number of parts. The variable intermediacy extends to such fine characters as are exhibited by the scales. Particularly significant is the intermediacy of the hybrids, with very little overlap on the frequency distribution of either parental species, in the number of gillrakers. The native chub, though confined in Recent times to trick-

ling streams, has very numerous rakers, which were obviously adapted for the straining of plankton in the large lakes that occupied the depressions of the Mojave Desert in Pluvial times. The minnow introduced (as bait) from the coastal streams of southern California has very few rakers, since its diet is essentially the bottom invertebrates, for the consumption of which it has no need for a fine straining apparatus. Data from successive years showed a marked increase in the relative numbers of the introduced species, which seem much better adapted than the relict lake chub to the present-day desert trickles. Collections made in 1953 indicate that the better-adapted introduced species has now almost completely replaced the doomed native form. Hybridization has doubtless co-acted with competition in this rapid replacement of a less well-adapted native species by a better-fitted introduced form.

Although it has long been recognized that interspecific hybridization is a common and important phenomenon in the Plant Kingdom, it has generally been held that in the Animal Kingdom species lines are only rarely, sporadically, and insignificantly crossed. In order to appraise the significance of interspecific hybridization among fishes, we will need to consider for different groups and in different habitats, the extent of such crossing, in terms of the number or proportion of species that cross and, in given combinations, in terms of the number or proportion of hybrid individuals. We also need data on the fertility of the hybrids when intermated, when backcrossed to each of the parental species, and when outcrossed with still others. We also need to know the viability of the hybrids and their chances of survival under different conditions. Some of these numerical relations have been determined; other determinations await much more observational, statistical, and experimental evidence.

### Marine Fishes

An extended search by us for natural hybrids, involving the careful examination of about one million specimens of freshwater fishes and about an equal multitude of marine species, has yielded, along with the results of research by others, some idea of the frequency of hybridization. Two conclusions stand out vividly, (1) that frequency of hybridization, like speciation in general, is to a large degree a function of the environment, and (2) that among marine fishes in general, as contrasted with northern freshwater fishes, hybrids are excessively rare.

The one group of marine fishes in which hybridization is both fairly common and well authenticated is the flatfishes. In Europe there have been discovered hybrids of one interspecific (and intergeneric) combination among the turbots and of four combinations among the right-eyed flounders. One combination, between flounder and plaice, is rather common and has been experimentally produced. Stocks of hybrids of this combination have been planted to test migrations and survivals. Two combinations in the Pleuronectidae have been identified from western North America, and we have analyzed in detail the rather common hybridization of two genera in Japan (Hubbs and Kuronuma, 1942). Hybridization among the flatfishes has been attributed to their life on very uniform sand and mud bottom and to promiscuous mass spawning. Years ago I discovered that two species of silversides (family Atherinidae) living on the sand bottom of the South Atlantic coast hybridize rather frequently, and this case has been analyzed by Gosline (1948).

With these exceptions, hybridization between marine species has been detected or definitely suspected to occur in only a few scattered combinations and with extreme rarity in these few combinations. This is true despite the fact that in cool and more especially in warm seas many closely related species live together. True, much of the study of marine fishes has been less extensive and less intensive than the researches that have led to the recognition of so much crossing of the

species in fresh waters, but along the West Coast, from Alaska to Mexico, and to a considerable degree in Japan and on the Atlantic coast of North America, we have done systematic work comparable to that which we have done on the freshwater fishes of North America. The researches we are now undertaking on pelagic and bathypelagic fishes are disclosing no hybrids. Before we attempt an appraisal of this difference in the incidence of hybridization, let us briefly review the evidence on interspecific hybridization in different families of fishes in the fresh waters of North America.

### Hybrids in Freshwater Fishes

Among the lampreys (Petromyzontidae) we have moderately conclusive evidence of occasional hybridization between two freshwater species of *Ichthyomyzon* (Hubbs and Trautman, 1937).

Among the trout we find occasional intergeneric hybrids in nature as well as in culture. Most of these are between the native brook trout and the introduced brown trout. In the West the cutthroat and rainbow trouts, both belonging to the genus *Salmo*, live side by side with little or no crossing in many coastal streams, but in the interior, where the cutthroat alone was native, the introduction of rainbows has repeatedly led to very extensive hybridization, and frequently to the elimination of the cutthroat, through a combination of hybridization and superior competition. The rainbow and golden trouts hybridize similarly. In the hatcheries many crosses have been produced between good species of chars, trouts, and salmons. Some of these produce fertile offspring, and by mating one hybrid with one of another cross, four species of the Pacific salmons have been combined in one individual (among the poeciliids we have thus combined as many as five species and twelve subspecies or races). One combination, between the brook trout and the lake trout, is being propagated in Canada with promise of use in fish management. The parental species were formerly classed as distinct genera, but in the current lumping spree both are put in *Salvelinus* —which action does not lessen the difference between the species.

Hybrids between the whitefish and the lake herring or cisco of the Great Lakes, of the related family Coregonidae, have been found in nature (rarely) and have been experimentally produced (Koelz, 1929). They also represented an intergeneric cross a few years ago, but only an interspecific cross now, owing to the revised generic delimitations—a subjective matter. I have interpreted other specimens as interspecific hybrids within the subgenus *Leucichthys*.

Attention has already been given to hybridization in the suckers (Catostomidae), with reference to the characters of the hybrids, but we need to indicate the frequency of hybridization and its relation to the phylogeny of the group. In our first and most comprehensive study (Hubbs, Hubbs, and Johnson, 1943), we identified from the Western states 182 interspecific hybrids representing nine interspecific and in part intergeneric combinations. The 182 hybrids were 11.5 percent as numerous as one parental species taken in the same collections and 7.2 percent as numerous as the other parental species. In terms of all specimens examined from the same river system, the hybrids were 1.8 percent as numerous as one parental species, 1.5 percent as many as the other. In some crosses the hybrid ratio was higher; in one intergeneric cross in the Columbia River system the ratios of hybrids to the two parental species were 15.1 and 21.6 percent for collections containing hybrids, and 6.4 and 4.2 percent for all specimens known from the entire river system. Owing to the mass collecting technique of modern times there has been little selection of hybrid specimens in the field. It seems safe to conclude now, as it was when we published the main paper, that among the suckers in all the lakes and streams of the

Western states at least one in a hundred is an interspecific hybrid. Conditions conducive to hybridization include the association of a few individuals of one species with an abundance of another, and the introduction of one species into the natural range of another. When thus thrown together, types that avoid hybridizing, where they occur together naturally, will cross. Subsequent studies (Hubbs and Hubbs, 1947; Hubbs and Miller, 1953; and ms.) have confirmed and expanded the conclusions reached in the paper cited.

For the suckers, and for several other freshwater fish families of North America, charts (Figs. 1 to 8) have been prepared to show by white rectangles the combinations of which no species occur in the same region, and hence cannot be expected to hybridize in nature; by gray rectangles, those that do occur together, but are not yet known to include hybrids; and by black rectangles, the combinations that include, to the best of our knowledge or judgment, at least one interspecific hybrid. Each such entry may represent several different interspecific combinations; of each combination we may have studied many series of specimens, and the total number of specimens may run into dozens or even hundreds. The recognized subfamilies are separated by heavy lines and the tribes by medium-weight lines. The genera are in full capitals; the subgenera, bracketed together, in lower case. The sequence is intended to be phylogenetic (in so far as may be indicated by a single column), so that combinations among most closely related generic groups should cluster along the diagonal, along which lie the combinations within each of the charted units. Note that for the Catostomidae (Fig. 1) the known hybrids are confined to tribal limits. All of the many sucker hybrids already discussed belong in the one tribe Catostomini. The other hybrid suckers that have been observed are between two buffalofishes (subfamily Ictiobinae).

The largest number of hybrid combina-

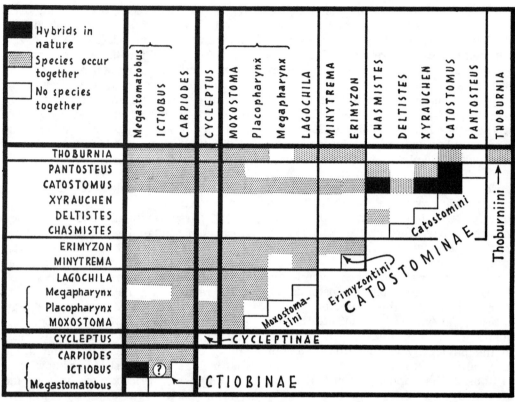

Fig. 1. Classification of North American Catostomidae, with indication of recognized hybrids between subgenera (printed in lower case) and genera.

tions occurs in the Cyprinidae, which is by far the most speciose family of North American freshwater fishes. The subgenera and genera of minnows are so numerous that we are forced to set up separate diagrams for the Pacific and Atlantic slopes. Taking the chart for the western slope (Fig. 2), we note first (at the lower left) hybridization between the carp and the goldfish in the subfamily Cyprininae. This cross has been produced in captivity and occurs very commonly in the East, especially about Lake Erie, where goldfish that escaped in a storm from a rearing pond got into Lake Erie, multiplied, and started interbreeding with the carp, previously established. The hybrids have become rather abundant locally. In one small creek in Ohio we found that the physiologically active hybrids had sorted themselves out, as sunfish hybrids some-

times do, so that, locally, they greatly outnumbered the parental species combined. In their native home, in eastern Asia, the two genera live together with little or no miscegenation, though they have been crossed there in captivity. Seemingly because of separation for a few decades or centuries, the two types have lost the art of avoiding the wasteful practice of producing sterile offspring (the uniformly intermediate characters of the hybrids indicate sterility). The carp-goldfish hybrids are uniformly intermediate, with about the same variation as each parental species. They are being reared for sale as bait, as they grow rapidly, are very hardy, and, being sterile, will not become established when bait is released.

The other hybrids among western cyprinid fishes involve crosses between rather diverse genera. In certain waters

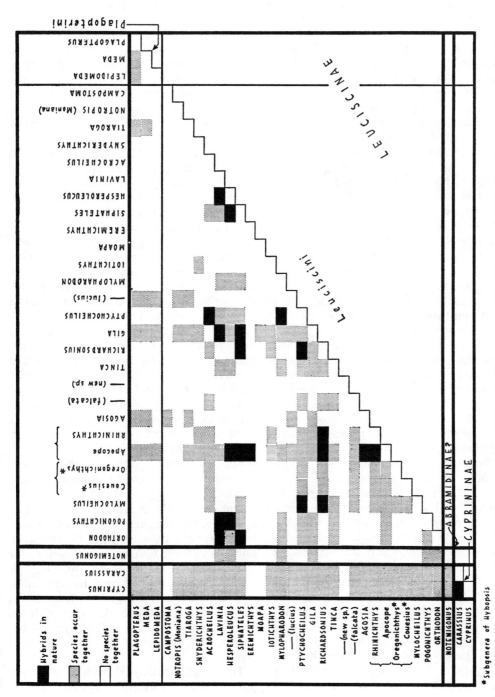

Fig. 2. Classification of the Cyprinidae of the Pacific slope north of Mexico, with indication of recognized hybrids between subgenera (printed in lower case) and genera (omitting *Tinca*, an introduced genus not known to hybridize with other genera).

most of the genera hybridize with one another in different combinations, sometimes rather commonly so. In one rather isolated stretch of the San Juan River of the Salinas system the genera *Hesperoleucus* and *Lavinia*, like certain plant species, have crossed to produce a hybrid swarm, involving almost the entire population and, locally, breaking down completely the bars between the species. Because of the sharp differentiation of the Western freshwater fish faunas, a large proportion of the genera and subgenera do not cohabit. Of the 120 pairs that do occur together, 17 percent have already been found to produce hybrids in nature. Three more of the combinations produce hybrids on the Atlantic slope.

In the chart for the Atlantic drainage (Fig. 3) we have omitted the Cyprininae, containing the carp and the goldfish and their numerous hybrids, already discussed, and three other exotic genera that are not known to hybridize in any way. For some unknown reason, the golden shiner, *Notemigonus*, seems to be immune to crossing on either slope. In the East a much higher percentage of the generic groups live together and hybridization is scattered throughout the series of native fishes. Formerly these minnows were placed in several subfamilies, but I am now inclined to class them in a single subfamily, even in a single tribe (Leuciscini). The apparent freedom of crossing is one reason for this changed evaluation. The record would seem to indicate that the cyprinids did not populate America prior to the Miocene and that at most a few immigrants have through radiative adaptation erupted into the very rich Recent fauna. Hybridization has remained, or has become very widespread. The very highly aberrant genus *Campostoma*, with the intestine wound spirally around the gasbladder, crosses with seven other genera. The subgenus *Luxilus* of the very speciose genus *Notropis*, containing the common shiner *Notropis cornuta*, hybridizes with fourteen other subgenera and genera. Of the 608 pairs of subgenera

and genera that live together, 68 (11 percent) have hybridized in nature, according to our best evidence. The interspecific combinations are more frequent than the subgeneric in number, of course, but not proportionwise. The chart of hybridization has grown with the years, until most of the major areas of possible hybrids have received at least one entry. It has been like filling in the periodic system of the elements.

Hybridization in the minnows has been attributed largely to the chance meeting of sperm and egg. As many as six or more species may be seen breeding violently on a single small gravel patch, commonly the nest of a large species. Very frequently *Notropis rubella* spawns in midwater a few inches above *Notropis cornuta*, which spawns on the bottom. The hybrids between these two species are perhaps the most commonly taken of any combination. The actual nuptial embrace of one species by another has not been observed, but is unnecessary to explain the frequent hybridization. Since the hybrids between *Notropis rubella* and *N. cornuta* appear to bridge the gap between the two species and have a normal sex ratio, it is assumed that they are at least partially fertile. The numerous hybrids between *Chrosomus eos* and *Chrosomus neogaeus* (formerly *Pfrille neogaea*) are invariably or almost invariably female.

Among the North American freshwater catfishes, the Ameiuridae, the few known hybrids are between species of *Ameiurus* and between species of *Schilbeodes* (Trautman, 1948; W. R. Taylor, ms.). A close connection between hybridization and classification is again indicated. In the South American catfish family Pimelodidae I have found a few hybrids between two species of *Rhamdia* in Guatemala.

In the pike family Esocidae (with one genus *Esox*) hybridization is rather common in nature and has been checked experimentally, but only the species of adjacent size seem to cross: the 10-inch

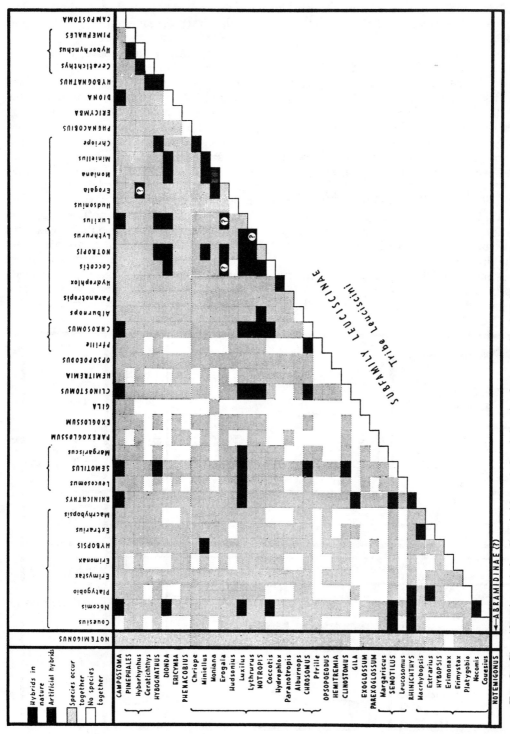

FIG. 3. Classification of the Cyprinidae of the Atlantic slope north of Mexico, with indication of recognized hybrids between subgenera (printed in lower case) and genera (omitting the Cyprininae, which are included on Fig. 2, and three genera, *Tinca*, *Scardinius*, and *Rhodeus*, that are not known to hybridize in North America).

bulldog pickerel (*americanus*) with the 18-inch chain pickerel (*niger*); the chain pickerel with the 3-foot northern pike (*lucius*); and the pike with the 5-foot muskellunge (*masquinongy*). The last-named cross has received some attention in fish culture (Eddy, 1944; Black and Williamson, 1947, with other citations).

Among the killifishes, or oviparous cyprinodonts (Fig. 4), a few of which range into brackish or even salt water, three interspecific combinations are known, all within the primitive tribe Fundulini (Hubbs, Walker, and Johnson, 1943). Hybridization in this group seems to be conditioned by cohabitation of a few individuals of one species with an abundance of the other.

Among the viviparous cyprinodonts two species of the Mexican family Goodeidae, regarded as generically distinct though closely related, have been found to hybridize rather commonly (Hubbs and Turner, 1939).

In the larger and more widespread, exclusively American family Poeciliidae very few natural hybrids are known, though members of this family have been more extensively hybridized than any other in home aquaria and in genetic laboratories. One reason seems to be that this group has a very wide range, from central United States to Argentina, so that many of the genera are of such limited range that they occur with few others, some with no other. Another reason is that copulation in this group follows complex nuptial performances, often characteristic of genera. In the large subfamily Gambusiinae (Fig. 5) very few hybrids have been identified, representing two subgeneric crosses in the mosquitofish

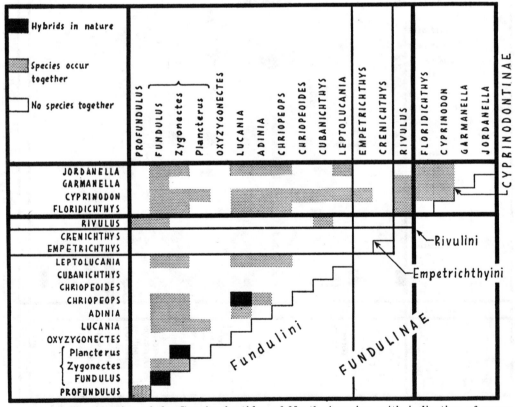

Fig. 4. Classification of the Cyprinodontidae of North America, with indication of recognized hybrids between subgenera and genera.

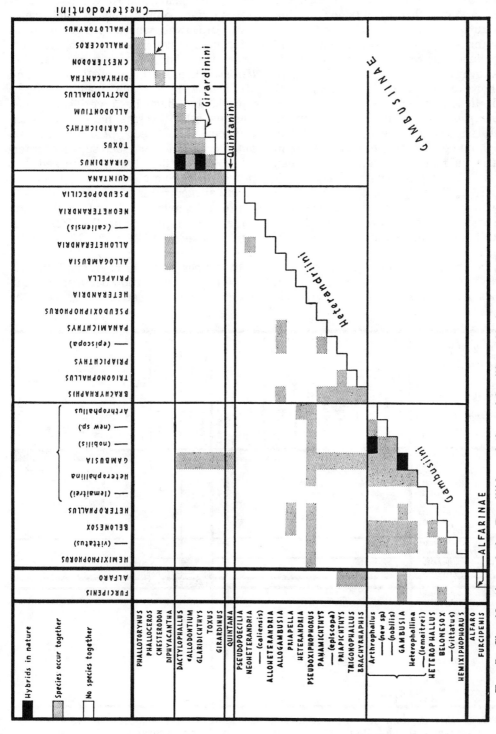

Fig. 5. Classification of the Poecilliidae of the subfamilies Alfarinae and Gambusiinae, with indication of recognized hybrids between subgenera and genera.

genus *Gambusia* and two intergeneric combinations in the Cuban tribe Girardinini (Hubbs and Rivas, ms.).

In the subfamilies Poeciliopsinae and Poeciliinae, also, very few natural hybrids are known (Fig. 6). Dr. Robert Rush Miller, however, has found considerable hybridization between certain species of *Poeciliopsis*. In the tribe Xiphophorini no natural hybrid has yet been identified, though all species seem to be crossable in aquaria and most combinations have been effected. The very extensive genetic work by Gordon, Kosswig, and others has been done in this tribe, with crosses between the various species of swordtails and platies (all now classified in the genus *Xiphophorus*).

In the tribe Poeciliini we have found in nature a very few natural hybrids between species of *Mollienisia,* in Yucatan (Hubbs,

1936), Guatemala, and Hawaii (a different interspecific combination in each country). In Hawaii, where both genera were introduced, I have also found hybrids between the related genera *Limia* and *Mollienisia*. Crosses between *Lebistes* and *Mollienisia* have been made in home aquaria and have been verified by me. Various other crosses have often been claimed, as between *Lebistes* and *Xiphophorus*, but every such supposed intertribal cross that I have checked has proved to be referable to one or the other of the alleged parental species.

Probable hybrids of one combination have been reported in the freshwater Atherinidae (Jordan and Hubbs, 1919, p. 77).

In the perch family (Percidae) we find again a close correlation between hybridization and classification (Fig. 7). The

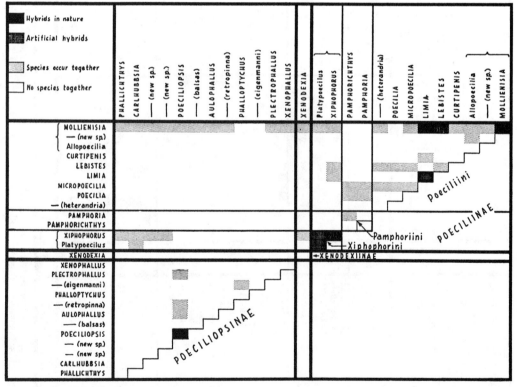

FIG. 6. Classification of the Poeciliidae of the subfamilies Poeciliopsinae, Xenodexiinae, and Poeciliinae, with indication of known experimental and natural hybrids between subgenera and genera.

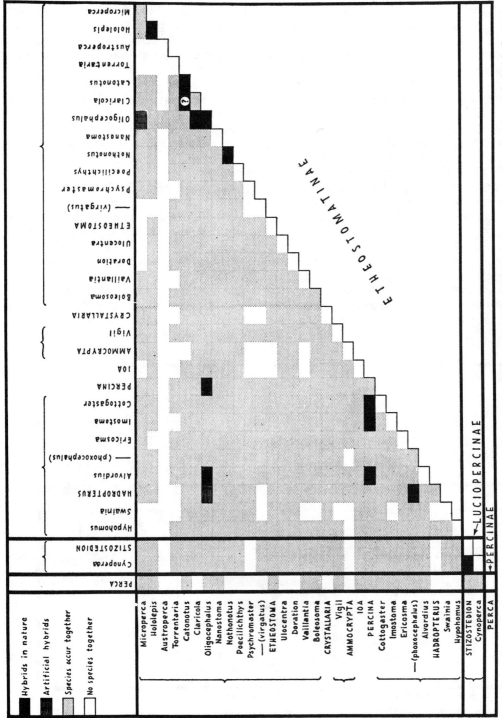

Fig. 7. Classification of the North American Percidae, with indication of known experimental and natural hybrids between subgenera and genera.

yellow perch, representing the Percinae, does not hybridize with any other fish. The two main species of the Luciopercinae hybridize occassionally, and the yellow and the blue walleye appear to cross so often that they are now treated as subspecies (the decision is in part arbitrary). No member of the Luciopercinae hybridizes, so far as is known, with any genus in either of the other subfamilies.

In the subfamily Etheostomatinae, the diminutive perches known as "darters," hybridization is moderately common but in nature is restricted largely to the less reduced and more primitive genera and subgenera. A few hybrids are known to occur between several species among the smaller and more brilliant, specialized darters, within a subgenus or between closely related subgenera. Among the thousands of specimens examined by me, only two represent natural crosses between the more primitive and the more specialized genera. Several laboratory

crosses are being made by Clark Hubbs and by Allan Linder, with rather ready success among the highly specialized darters, but in general with poor success in crosses between the primitive and the specialized genera. Here again is evidence of a close correlation between hybridization and relationship. Crossing of the species line in this family as in the minnows may well result from the chance meeting of eggs and sperm when two species breed in close proximity, as they often do.

In the sunfish family Centrarchidae (Fig. 8), as well, hybridization is confined to the limits of single tribes. We have found several hybrids between species of blackbass, but only in the subgenus *Micropterus* (Hubbs and Bailey, 1940). Within the sunfish tribe Lepomini, comprising nine genera as recently classified but only two at present, by reason of lumping, 18 (about one-half) of the 38 possible subgeneric crossings have already

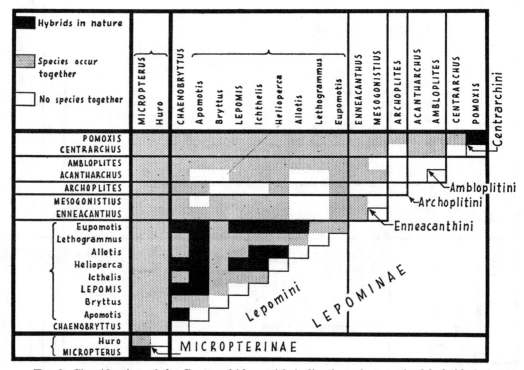

FIG. 8. Classification of the Centrarchidae, with indication of recognized hybrids between subgenera and genera.

been realized. The main features of the abundant hybridization in this tribe have already been outlined. The other interspecific hybrids in the family are between the two species of *Pomoxis,* in another tribe.

The only other natural hybrid recognized in the fresh waters of North America is between two species of sculpins, of the genus *Cottus,* and this cross needs further study.

The frequent crossing of the species line in the freshwater fish fauna of eastern North America is duplicated in Europe in the minnow and trout families (where it would be expected). A few minnow hybrids have been reported from Asia and some of the Indian cyprinids have hybridized in aquaria. In Africa and South America, at least some interspecific hybridization can be predicted, though it has seldom been indicated. It is not at all improbable that hybrids in those continents have been studied and named as distinct species, just as the sunfish hybrids were in North America before they came under close scrutiny.

It would be of particular interest to determine whether and to what degree the species lines are crossed in the large lakes, especially the rift lakes of Africa, where radiative speciation has dramatically exploded, to produce rapidly, with very little geographic isolation, a great number of species. Under such conditions I would expect much crossing of the line between the recently differentiated species. Dr. Ethelwynn Trewavas of the British Museum has told me that she has reason to suspect that some of the variants and some of the nominal species encountered in the study of the extraordinarily numerous cichlid species of the African lakes may be interspecific hybrids. In critically examining Dr. Tchernavin's report on the multitudinous species of the cyprinodont genus *Orestias,* endemic in Lake Titicaca, Perú, I found evidence of some interspecific hybridization, and in treating the atherine genus *Chirostoma,* which has proliferated extensively in the

lakes of the Río Lerma system of México, Jordan and Hubbs (1919, p. 31) discussed three probable interspecific hybrids of one combination.

## The Role of Environmental Factors

Even if no further evidence of interspecific crossing among freshwater fishes should be discovered, the conclusion would stand out boldly that such hybridization is vastly more common in fresh waters than in the sea. This contrast cannot be attributed solely to a difference in the genetic potential of families respectively more or less confined to fresh and ocean waters, for the hybridization occurs in all of the larger families in North American fresh waters, including some families that have close relatives in the sea.

It is evident that the hybridization is conditioned by environmental factors. Some of these factors have already been indicated. Species that rarely or never hybridize in nature can often be crossed, sometimes very freely, in captivity. Merely placing a male of one species with a female of another in an aquarium may lead to crossing. Thousands of such matings between the swordtails and the platies have been consummated, yet not a single hybrid between the two has ever been taken in nature, despite the fact that they often occur together and have been very extensively collected and very critically studied. Species of *Mollienisia* that hybridize in nature with extreme infrequency cross very freely in aquaria, in some combinations with no decrease in fecundity in the first or even in the succeeding generations. Many combinations that seldom or never cross voluntarily, even in planned matings within the confines of an aquarium, prove their basic genetic compatibility when the breeders have been injected with hormones, or when artificial impregnation is practiced. The genetic potential often remains unrealized.

Similarly, where the spawning areas in

nature are greatly limited and the breeding fish are thus forced into close proximity, the incidence of hybridization is increased. Spawning in proximity may lead to the chance meeting of eggs and sperm, but may also lead to the mating of species that breed with complex pairing behavior. The cohabitation of a few individuals of one species with a multitude of a related species is another circumstance definitely conducive to crossing, especially when the scarcer species has been introduced. When ready to spawn, the species that is locally rare may outcross if the proper mates are not at hand. As in plants, hybridization is often a function of the intergradation of the habitat: species that are segregated in breeding by differential responses to any features of the physical, chemical, or organic environment tend to breed together where the environment is rendered intermediate, either through natural causes or through modifications by man. And species that normally breed at different seasons may cross where conditions are such as to cause an overlap in the breeding seasons.

In the light of these thoughts, how can we explain the high incidence of hybridization in the fresh waters of North America (and of Europe), in contrast to the infrequent breakdown of species lines in the sea? We attribute the frequent hybridization in the fresh waters in part to the greater opportunity for chance meeting of egg and sperm, but, more especially, to the ephemeral and changing nature of the environment. Lakes are temporary, often suddenly formed but all doomed to eventual extinction. Most lakes, of limited area and depth, are very transient. As they change in character and approach extinction the habitats are altered or destroyed. Streams also change in character in glaciated areas. In North America and Europe the almost catastrophic changes in climate through the Ice Age, modifying conditions and forcing dispersals, vastly decreased the long-term stability of the ecological niches, and it is in the regions affected sharply by Pleisto-

cene stresses that hybridization has been encountered most often. In North American fresh waters, where hybridization has probably been most frequent, most of the existing fish fauna, furthermore, appears to date only from invasions during the Miocene, Pliocene, and Pleistocene, leaving no very great time for full adjustments. In addition, most freshwater fishes are so limited in possible movements that they cannot reach the full assortment of habitats that might otherwise be available.

In the sea, in contrast, and to a large degree perhaps in more nearly equatorial fresh waters, speciation has had a longer play. Local changes in both time and space may be rapid and profound, but the various ecological niches to which related species have become differentially adjusted remain available, if not in the immediate vicinity, at least within reach along the much freer highways of dispersal. The organic environment has apparently also remained more stable. Much greater opportunities have existed for the development and operation of the multitudinous fine adjustments involved in the location, with precise timing, of the proper breeding grounds and the proper mates.

Contrast in the regional incidence of hybridization may probably be drawn not only between fresh waters and the sea, but also between the temperate regions and the tropical. The environmental and particularly the organismal diversity of tropical regions seems to have been accompanied by very little interspecific crossing (though, as already noted, more critical and complete systematic studies may well disclose some hybridization in tropical fresh waters, and possibly even in tropical seas). The seeming lack of hybridization in the swarming reef fauna of the Indo-Pacific region is amazing. The fact that most of the few marine fish hybrids thus far recognized have come from temperate northern waters may be due not only to more complete studies there, but also to faunal disruptions induced to some extent in the littoral waters, as well as in fresh waters, by the vast shifts in Pleistocene climate. I venture to predict that the incidence of natural hybridization between species will show a clinal gradation from north temperate to tropical zones, as well as from fresh waters to the sea, with the highest frequency in Holarctic freshwater regions strongly affected by Pleistocene events and the lowest incidence in tropical marine waters. North temperate marine and tropical freshwater fishes may show an intermediate and rather low incidence. Similar relations may well hold for other groups of animals.

Selective advantage as well as conditions imposed more directly by the environment may help to explain not only the general avoidance of hybridization that I have attributed to environmental stability, but also the relatively frequent crossing of the species that seems to be a result of environmental instability. The general avoidance of crossing, where environmental stability permits, seems to be attributable not only to reduced fertility of the hybrids, but also to the selective advantage of the fine adjustments, between species and environmental niches, that can be attained only with relatively uniform genetic constitution. But where the environment is in a state of flux, increased premium may be inherent in the genetic variability that hybridization produces. Material is then needed for new adjustments to the rapidly changing environment. Under these conditions the advantages of genetic variability may outweigh the advantages of uniform constitution. Under more stable environments these opposing advantages may be reversed.

Frequency of hybridization appears to be inversely correlated with the number of species in given areas. It is certainly true in the Catostomidae (the suckers) and I believe also in the Cyprinidae (the minnows) that hybridization is more frequent in the fresh waters of western North America than in the more speciose waters of eastern North America. In the abounding fauna of the tropics many more species come in contact with one another, yet hybridization seems least common. One reason may be that the diversity lies largely in the organismal environment, under which conditions species recognition and specialized reproductive responses and behavior are presumably enhanced (the highly diverse species-specific coloration of tropical reef fishes comes to mind in this connection).

The absence or scarcity of hybridization where species are most numerous is one of several lines of evidence that belittle interspecific hybridization as an important factor in the speciation of fishes. The frequent if not usual sterility or partial sterility of the hybrids is another of the lines of evidence. There is a considerable body of circumstantial evidence, however, to indicate that in at least some groups of fishes in certain regions the crossing of the species line has been a factor of some significance in speciation, particularly in the way of introgressive hybridization.

## REFERENCES

BAILEY, R. M., and LAGLER, K. F. 1938. An analysis of hybridization in a population of stunted sunfishes in New York. *Pap. Mich. Acad. Sci., Arts, Letters, 23* (1937):577–604, figs. 1–5.

BLACK, J. D., and WILLIAMSON, L. O. 1947. Artificial hybrids between muskellunge and northern pike. *Trans. Wisc. Acad. Sci., Arts, Letters, 38*:299–314, figs. 1–19. (With bibliography on esocid hybrids.)

EDDY, S. 1944. Hybridization between northern pike (*Esox lucius*) and muskellunge (*Esox masquinongy*). *Proc. Minn. Acad. Sci., 12*:38–43.

GOSLINE, W. A. 1948. Speciation in the fishes of the genus *Menidia*. *Evolution, 2*:306–313.

HUBBS, CARL L. 1920. Notes on hybrid sunfishes. *Aquatic Life, 5*:101–103.

——— 1933. Species and hybrids of *Mollienisia*. *Aquarium, 1*:263–268, 277, figs. 1–3, 5–9. (Reprinted in same journal, 1942, *10*: 162–168, figs. 1–3, 5–9.)

——— 1934. Double-crossing Molly. *Home Aqu. Bull., 4*:5–11, 3 figs.

——— 1936. Fishes of the Yucatan Peninsula. *Carnegie Inst. Wash. Publ.*, No. 457, 157–287, fig. 1, pls. 1–15.

—— 1940. Speciation of fishes. *Amer. Nat.,* 74:198–211. (Reprinted in *Biol. Symp.,* 1941, 2:7–20.)

—— 1942. Sexual dimorphism in the cyprinid fishes, *Margariscus* and *Couesius,* and alleged hybridization between these genera. *Occ. Pap. Mus. Zool. Univ. Mich.,* No. 468, 1–6.

—— 1951. The American cyprinid fish *Notropis germanus* Hay interpreted as an intergeneric hybrid. *Amer. Midl. Nat., 45:* 446–454.

Hubbs, Carl L., and Bailey, R. M. 1940. A revision of the black basses (*Micropterus* and *Huro*), with descriptions of four new forms. *Misc. Publ. Mus. Zool. Univ. Mich.,* No. 48. 1–51, fig. 1, pls. 1–6, maps 1–2.

—— 1952. Identification of *Oxygeneum pulverulentum* Forbes, from Illinois, as a hybrid cyprinid fish. *Pap. Mich. Acad. Sci., Arts, Letters,* 37 (1951):143–152, pl. 1.

Hubbs, Carl L., and Brown, D. E. S. 1929. Materials for a distributional study of Ontario fishes. *Trans. Roy. Can. Inst.,* 17:1–56.

Hubbs, Carl L., and Hubbs, Laura C. 1931. Increased growth in hybrid sunfishes. *Pap. Mich. Acad. Sci., Arts, Letters,* 13 (1930): 291–301, figs. 45–46.

—— 1932a. Experimental verification of natural hybridization between distinct genera of sunfishes. *Pap. Mich. Acad. Sci., Arts, Letters,* 15 (1931):427–437.

—— 1932b. Apparent parthenogenesis in nature, in a form of fish of hybrid origin. *Science,* 76:628–630.

—— 1933. The increased growth, predominant maleness, and apparent infertility of hybrid sunfishes. *Pap. Mich. Acad. Sci., Arts, Letters,* 17 (1932):613–641, figs. 69–71, pls. 64–65. (With references.)

—— 1946. Experimental breeding of the Amazon molly. *Aqu. Jour.,* 17:4–6, pl. 1.

—— 1947. Natural hybrids between two species of catostomid fishes. *Pap. Mich. Acad. Sci., Arts, Letters, 31* (1945):147–167, figs. 1–2.

Hubbs, Carl L., Hubbs, Laura C., and Johnson, R. E. 1943. Hybridization in nature between species of catostomid fishes. *Contr. Lab. Vert. Biol. Univ. Mich.,* No. 22, 1–76, figs. 1–8, pls. 1–7. (With references to key papers on cyprinid hybrids.)

Hubbs, Carl L., and Kuronuma, K. 1942. Hybridization in nature between two genera of flounders in Japan. *Pap. Mich. Acad. Sci., Arts, Letters, 27* (1941):267–306, figs 1–5, pls. 1–4. (With references to literature on hybrid flatfishes.)

Hubbs, Carl L., and Miller, R. R. 1942. Mass hybridization between two genera of cyprinid fishes in the Mohave Desert, California. *Pap. Mich. Acad. Sci., Arts, Letters,* 28 (1942):343–378, figs. 1–2, pls. 1–4.

—— 1953. Hybridization in nature between the fish genera *Catostomus* and *Xyrauchen.* *Pap. Mich. Acad. Sci., Arts, Letters, 38* (1952):207–233, figs. 1–3, pls. 1–4.

Hubbs, Carl L., and Schultz, L. P. 1931. The scientific name of the Columbia River chub. *Occ. Pap. Mus. Zool. Univ. Mich.,* No. 232, 1–6, pl. 1.

Hubbs, Carl L., and Trautman, M. B. 1937. A revision of the lamprey genus *Ichthyomyzon. Misc. Publ. Mus. Zool. Univ. Mich.,* No. 35, 1–109, figs. 1–5, map 1, pls. 1–2.

Hubbs, Carl L., and Turner, C. L. 1939. Studies of the fishes of the order Cyprinodontes. XVI. A revision of the Goodeidae. *Misc. Publ. Mus. Zool. Univ. Mich.,* No. 42, 1–80, pls. 1–5.

Hubbs, Carl L., Walker, B. W., and Johnson, R. E. 1943. Hybridization in nature between species of American cyprinodont fishes. *Contr. Lab. Vert. Biol. Univ. Mich.,* No. 23, 1–21, pls. 1–6.

Jordan, D. S., and Hubbs, Carl L. 1919. Studies in ichthyology. A monographic review of the Atherinidae or silversides. *Leland Stanford Jr. Univ. Publ. (Univ. Ser.).* 87 pp., pls. 1–12.

Koelz, W. 1929. Coregonid fishes of the Great Lakes. *Bull. U. S. Bur. Fish., 43* (1927):297–643, figs. 1–31.

Meyer, H. 1938. Investigations concerning the reproductive behaviour of *Mollienisia "formosa." Jour. Genet.,* 36:329–366, figs. 1–13, pl. 10.

Trautman, M. B. 1948. A natural hybrid catfish, *Schilbeodes miurus* × *Schilbeodes mollis. Copeia,* 1948: 166–174, pl. 1.

**CARL L. HUBBS,** Professor of Zoology at the Scripps Institution of Oceanography, University of California, is a Past President of the Society of Systematic Zoology. This article has been extracted from the 1954 Faculty Research Lecture of the University of California (Southern Section), "Crossing the Species Line," repeated under the auspices of the Society of Systematic Zoology and the American Society of Ichthyologists and Herpetologists at the Berkeley meeting of the American Association for the Advancement of Science. Contributions from the Scripps Institution of Oceanography, New Series, No. 759.

# 12

Reprinted from *Am. J. Bot.* 25:396–402 (1938)

## HYBRIDIZATION IN TRADESCANTIA. III.
## THE EVIDENCE FOR INTROGRESSIVE HYBRIDIZATION [1]

Edgar Anderson and Leslie Hubricht

PREVIOUS STUDIES of the American species of *Tradescantia* have shown that interspecific hybridization is comparatively frequent between the eighteen or more species closely related to *Tradescantia virginiana* (Anderson and Sax, 1936; Anderson and Woodson, 1935). Hybridizations which were inferred from herbarium and field work have been produced experimentally in the breeding plot (Anderson, 1936a). Detailed morphological analyses of hybrid populations have shown that the ultimate effects of interspecific hybridization are various (Anderson, 1936c). Apparently the commonest result is that through repeated back-crossing of the hybrids to the parental species there is an infiltration of the germplasm of one species into that of another. If, for instance, two species, 'A' and 'B' come into effective contact, they usually do so under conditions which greatly favor either 'A' or 'B.' If 'A' and 'B' differ in habitat preferences, seldom or never is there a habitat equally acceptable to both; they usually meet, if at all, in a situation quite favorable to one of the species but just fairly so to the other. Therefore, if hybrids are produced, they tend to cross back to the more abundant species. The progeny of these secondary hybrids are likewise crossed back again, and so on. The final result will depend upon the balance between the deleterious effects of the foreign germplasm and its advantageous effects in the areas where the hybridization has taken place or to which the hybrids may spread.

Preliminary analyses of a number of genera of the flowering plants (Anderson, 1936b; Riley, 1936, 1937; Goodwin, 1937; Delisle, 1937) have shown that while such is not the only effect of hybridization between species, it is certainly one of the commonest. We have therefore given it a distinctive name, introgressive hybridization. In discussing the effects of introgressive hybridization, we shall speak of the hybridization of one species into another rather than hybridization with another. This terminology is chosen as a matter of convenience in discussing particular cases and avoids needless repetition of explanatory phrases.

The application of the terms "hybrid" and "species" becomes a difficult problem in dealing with successive back-crosses between an original first generation hybrid and one of the parental species. The F₁ is clearly entitled to the term hybrid, but among the progeny of its first cross back to the parent there will be a number of individuals which resemble that species very closely indeed, and each successive back-cross will increase the percentage of these indistinguishable or almost indistinguishable mongrels. After a few back-crosses most of the individuals cannot be

distinguished by morphological means from the pure species, though even then a study of the group as a whole would indicate by its departure from the average of the species something of what had taken place. Further back-crossing would weaken even the effect upon the group average.

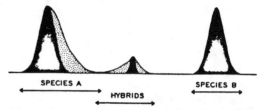

Fig. 1. Diagram illustrating the application of the terms "species" and "hybrids" to a case of introgressive hybridization. Solid black, original species and first generation hybrids. Dotted, later hybrid generations and back-crosses. Further explanation in the text.

Since it is therefore a practical impossibility to distinguish between individuals of partially hybrid ancestry and those of uncontaminated pedigree, we shall in the following discussion restrict the term hybrid to the more or less obvious intermediates between recognized species. The term species we shall use in a broad enough sense to include the barely perceptible variants, which may sometimes be of partially hybrid ancestry. Figure 1 illustrates the use of these terms in a case which, though hypothetical, is not unlike the relation between *T. canaliculata* and *T. occidentalis*. The solid black indicates the two species 'A' and 'B' and their first generation hybrid. The stippled area shows the secondary hybrids (second generation and back cross to species 'A'). The arrows at the base of the figure designate the approximate limits of the terms "species" and "hybrids" as we shall use them. This use of terms is loose and somewhat illogical, but it avoids the difficulties which would attend an attempt at greater precision.

Anyone who has undertaken monographic work recognizes that there are slight regional differences within many species and varieties. Most taxonomists have been of the opinion that if such differences are too tenuous for cataloguing purposes, they should be ignored in taxonomic work, though they might very well be of considerable biological significance. A few botanists, like E. L. Greene or P. A. Rydberg, have advocated nomenclatorial recognition for such variants. These rather vague local or regional differences are common in some genera and rare or lacking in others. In the genus *Uvularia*, which the senior author has studied intensively (Anderson and Whit-

[1] Received for publication February 28, 1938.

aker, 1934), no such differences are perceptible, not even by the use of detailed biometrical methods. In the genus *Solidago*, on the other hand, the species in any one locality are usually comparatively stable, though fifty or one hundred miles away these same species may be significantly different. The causes of such differentiation are undoubtedly various and include such factors as the direct and selective effect of the environment, isolation, mutation, in addition to introgressive hybridization.

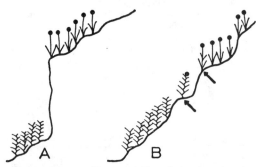

Fig. 2. Diagram illustrating the relation between the habitats occupied by *Tradescantia canaliculata* (above) and *T. subaspera* var. *typica* (below). A, above and below a cliff; B, in ravine at side of cliff, arrows point to back-cross hybrids. Further explanation in the text.

In the case of wide ranging and common species which have been extensively collected, it should be possible to obtain a rough estimate of the importance of such hybridization from a study of herbarium material. If, for instance, introgressive hybridization of *T. canaliculata* into *T. occidentalis* has been an important factor in producing regional differences in the latter species, then we should find that *T. occi-*

be objectively scored, and (3) if there is enough herbarium material for the calculation of significant averages. Among the American Tradescantias, these conditions are met in three cases: (1) *T. subaspera typica* into *T. canaliculata;* (2) *T. canaliculata* into *T. occidentalis;* (3) *T. canaliculata* into *T. bracteata.*

TRADESCANTIA SUBASPERA VAR. TYPICA INTO T. CANALICULATA. — Throughout its range, which is entirely within that of *T. canaliculata, T. subaspera typica* tends to grow in strikingly different habitats from those occupied by the former species. In addition to this barrier there are differences in flowering season, and internal differences which cause the hybrids, once they have occurred, to be partially sterile. In spite of these various barriers, hybrids between the two species are to be found wherever the habitats come into contact without being sharply discontinuous. In much of the Ozark region, *Tradescantia canaliculata* is to be found growing in full sun at the top of cliffs, while *T. subaspera typica* is in deep shade in rich soil at the foot of the same cliffs. In this case the two habitats come into close contact, but with no intermediate zone, and consequently no hybrids. However, at those points where erosion has make a break in the face of the cliff, there are intermediate zones in which hybrids can nearly always be found. Figure 2 diagrams such a situation at Bat Cave, near Stanton, Franklin County, Missouri, which is typical of numerous other localities in the Ozarks.

When transplanted to experimental plots, the suspected hybrids have maintained their intermediate characteristics. Since they are partially sterile, and since they resemble very closely the artificial hybrids which have been raised between the two species (Anderson, 1936a), their hybrid nature can be taken as proved beyond a reasonable doubt.

TABLE 1. *Comparisons of herbarium material of* TRADESCANTIA SUBASPERA *var.* TYPICA *marked "sub.",* T. CANALICULATA *outside the range of* T. SUBASPERA *marked " can.", and* T. CANALICULATA *within the range of* T. SUBASPERA *marked " can. (sub.) ".*

| | Node number | | | | | | | | | | | | Leaf number | | | | | | | | | | | | | Internode | |
| | 2 | 3 | 4 | 5 | 6 | 7 | 8 | 9 | 10 | 11 | 12 | 13 | 6 | 7 | 8 | 9 | 10 | 11 | 12 | 13 | 14 | 15 | 16 | 17 | 18 | Increase | Decrease |
|---|---|---|---|---|---|---|---|---|---|---|---|---|---|---|---|---|---|---|---|---|---|---|---|---|---|---|---|
| Can. ...... | 2 | 12 | 15 | 15 | 12 | 5 | 3 | 1 | | | | | 3 | 10 | 22 | 12 | 10 | 2 | | 2 | 3 | | | | | 50 | 15 |
| Can. (sub.) | | | 3 | 4 | 2 | | | | | | | | | | 3 | 4 | 2 | | | | | | | | | 6 | 3 |
| Sub. ...... | | | | | | 1 | 1 | 2 | 5 | 2 | | 1 | | | | 2 | | 3 | 4 | 2 | | | | 1 | | | 12 |

*dentalis* within and adjacent to the known range of *T. canaliculata* should be significantly different, on the average, from *T. occidentalis* outside that range. Furthermore the difference, however slight, should be in the direction of *T. canaliculata.*

Such a test can be applied only (1) if the distributions of the two species overlap only in part, (2) if there are a number of specific differences which can

The hybrids vary among themselves, some of them being obviously intermediate between the two species, others resembling one or the other parent more or less closely. Primary and secondary hybrids are therefore occurring at a large number of localities and presumably have been occurring for a considerable time. Is there any evidence that *T. subaspera* var. *typica* has introgressed into *T. canaliculata?*

The evidence on this point is assembled in table 1. It will be seen that there are three characters, readily scored on herbarium specimens, by which the two species differ. These are: (1) The number of evident nodes on the stem. *T. canaliculata* averages around five, while *subaspera* has about twice as many. (2) The number of leaves. Due to the fact that internodes may be so condensed at the apex and base of the stem, the number of leaves is higher than the number of apparent nodes. (3) In *T. subaspera* the upper internodes are shorter than the lower ones, usually decreasing in a harmonious sequence. In *T. canaliculata* there is no such sequence, and the upper internodes are often the longest.

When the specimens of *T. canaliculata* are assembled according to whether they are inside or outside the known range of *T. subaspera typica*, it will be seen (table 1) that there is little evidence for the introgression of *T. subaspera*. In spite of decisive differences between the two species, in all three cases the average difference between the two sets of *T. canaliculata* is very slight and of no statistical significance. Furthermore, in one case the direction of the differences is actually away from the value of *T. subaspera typica*.

specific races and strains, too vaguely defined for nomenclatorial recognition, but of great biological interest. It now becomes our purpose to examine the variation within *T. occidentalis* in detail and to learn if any portion of that variation may reasonably be assigned to the introgressive hybridization of *T. canaliculata*.

*Tradescantia occidentalis* var. *typica*, as defined by Anderson and Woodson (1935), is possessed of one strongly marked geographic race which is perhaps worthy of nomenclatorial recognition. It occupies the northwest corner of the range of that variety and differs from its fellows to the south and east by being less glaucous and more pubescent with wider bracts and larger sepals. It was to such plants that the names *T. laramiensis* and *T. universitatis* were given, but the differences between them and *T. occidentalis* from Texas are slight, and they seem to intergrade completely. It has not been possible to separate them satisfactorily on the basis of herbarium work alone, though possibly extensive field work would yield critical evidence.

In so far as the relationship of *T. occidentalis* to *T. canaliculata* is concerned, these *laramiensis* variants are of little or no importance. They occur well

TABLE 2. *Comparisons of herbarium material of* T. CANALICULATA *marked* "can.", T. OCCIDENTALIS *var.* TYPICA *outside the range of* T. CANALICULATA *marked* "occ.", T. OCCIDENTALIS *within the range of* T. CANALICULATA *marked* "occ. (can.)".

| | Node number | | | | | | | | Leaf number | | | | | | | | | Internode | | Tuft | | |
| | 2 | 3 | 4 | 5 | 6 | 7 | 8 | 9 | 6 | 7 | 8 | 9 | 10 | 11 | 12 | 13 | 14 | Increase | Decrease | None | Weak | Strong |
|---|---|---|---|---|---|---|---|---|---|---|---|---|---|---|---|---|---|---|---|---|---|---|
| Occ. ....... | | 3 | 5 | | | | | | 1 | 3 | 3 | 1 | | | | | | 4 | 4 | 8 | 4 | 4 |
| Occ. (can.) . | | 6 | 11 | 2 | | | | | 3 | 5 | 5 | 4 | 1 | 1 | | | | 14 | 5 | 6 | 5 | 18 |
| Can. ....... | 2 | 12 | 18 | 19 | 14 | 5 | 3 | 1 | 3 | 11 | 24 | 14 | 10 | 2 | | 2 | 3 | 56 | 18 | | 26 | 52 |

If there is introgressive hybridization of *T. subaspera* into *T. canaliculata*, it therefore is either so recent or so slight that it cannot be detected. The reverse introgression (*T. canaliculata* into *T. subaspera*) cannot be studied by this method, since the range of *T. subaspera* is entirely within that of *T. canaliculata*. This is unfortunate, since from various kinds of evidence we are led to suspect a strong introgression in that direction.

TRADESCANTIA CANALICULATA INTO T. OCCIDENTALIS VAR. TYPICA.—In more than one sense these two species are the counterparts of each other, the former in the middle west, the latter on the Great Plains. Each has a range of approximately 1,200,000 square miles. Each is clearly made up of two different elements: (1) truly indigenous strains and (2) ubiquitous weeds and plants of waste places, which, while they may be in part anciently established, are mostly post-Columbian. Each species is predominately tetraploid, though each has diploid races in certain restricted areas. Each species has numerous intra-

outside the area in which *T. occidentalis* and *T. canaliculata* commingle. Hence, their removal will not interfere with a study of the relationships of these two species.

Table 2, therefore, is a test of the introgressive effect of *T. canaliculata* in the southern and eastern parts of the range of *T. occidentalis* var. *typica*. While the numbers of specimens are too small to be particularly significant, it will be seen that the difference between the two sets of *T. occidentalis* in each of the four cases is in the direction of *T. canaliculata*. For the last two characters the differences are much more significant than mere numbers might indicate, since the difference is qualitative as well as quantitative. There is conceivably an almost endless series of forms and pubescence patterns which might occur in the genus *Tradescantia*. To find that within the range of *T. canaliculata* the only perceptible variation in the sepals of *T. occidentalis* is in the exact direction of *T. canaliculata* is most significant. These facts become even more illuminating when studied in connection with the distribution of the two species (fig. 3). It will be seen that there is a steady increase in the percentage of *canaliculata* characteristics as one approaches the range of that species. If, for instance, we divide the range of *T. occidentalis* by the 90th, 95th, 100th, and 105th meridians, the percentages of specimens with tufts from west to east are 0, 25, 50, and 100 per cent.

To summarize, a study of herbarium specimens of *T. occidentalis* var. *typica* arranged with reference to the known range of *T. canaliculata* demonstrates the following points: (1) There is a slight difference between the specimens within the range of *T. canaliculata* and those from outside the range of *T. canaliculata*. (2) This difference, though slight, is in the direction of *T. canaliculata*. (3) There is some evidence that the intensity of the difference increases with the comparative frequency of *T. canaliculata*.

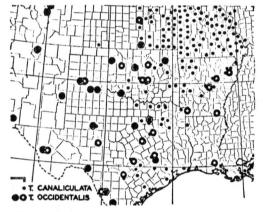

Fig. 3. Illustrating the relation between sepal pubescence in *T. occidentalis* var. *typica* and the distribution of *T. canaliculata*. The white centered dots represent specimens of *T. occidentalis* var. *typica* with barbate eglandular hairs on the sepals.

From these three points and from the known fact that *T. occidentalis* and *T. canaliculata* cross readily, we conclude that there is a strong introgression of *T. canaliculata* into *T. occidentalis*. This conclusion receives confirmation from such field studies as it has been possible to make of the variation within *T. occidentalis*.

We may anticipate the evidence given in detail below by saying that variation in natural population of *T. occidentalis* is proportional to (1) the weediness of the environment, (2) the number of other species of *Tradescantia* in the neighborhood, and (3) the comparative frequency of other species of *Tradescantia*.

If introgressive hybridization takes place, we should expect to find the strongest evidence for it in those regions and at those points where there is greatest opportunity for hybridization. The greater the number of closely related species in the vicinity, the greater the chance that at least one of them will come into effective contact with the species being studied. Similarly, the greater the frequency of any one of these foreign species the greater is the chance that crossing may take place. While the connection between weediness and introgressive hybridization may not be so transparent, it is even more important (Wiegand, 1935). In the genus *Tradescantia* one of the chief barriers between closely related species is differences in habitat preference. When a railroad is constructed, or a roadway, or ditches, or when in various other ways man disturbs the ecological complexion of the countryside, these barriers are broken down. Species which before were kept apart now grow side by side, and what is even more important a whole set of various new intermediate environments is provided in which hybrids may establish themselves. We should therefore predict that the more that natural ecological conditions have been upset the greater would be the opportunity for introgressive hybridization. The analysis of variation in the following natural populations proves this to be the case.

*Mirando.*—These plants were growing near Mirando, Webb County, Texas in typical mesquite shrubland. The land had been fenced for livestock and had been grazed, but natural conditions had not been greatly altered. Although some of the plants were growing between the fence and the highway, there was no evidence that they had been brought in by road construction. No other species of *Tradescantia* are known from the immediate neighborhood. Twenty plants were found in bloom and collected.

Detailed scoring of certain features is presented for ten of these in table 3, and summaries for the entire group are shown in table 4. As the table indicates, there is little variation from plant to plant, and in such characters as have been used for specific delimitation, there is none at all.

*Rainbow Bridge.*—For this collection we are indebted to Hugh C. Cutler who collected it near Rainbow Bridge, San Juan County, Utah. It grows there in fairly stable sands throughout Bridge Canyon. The spot is remote from highways, and grazing animals are few. It may be seen from table 4 that there is little variation between the twenty-one plants which were studied and that they resemble the plants from Mirando.

*Inscription House.*—This collection was also made by Mr. Cutler. The population of which this is a sample was growing in deep sand on a slowly moving dune, eleven miles north of Inscription House Post, Navajo County, Arizona. The spot was 300–400 yards west of the road leading to Rainbow Lodge. Although the region is practically a wilderness area, the roadway is an ancient one, having originally been an Indian trail. The plants are, for the most part, fairly uniform and resemble those from Rainbow Bridge very closely. Unlike that collection, however, about a third of the plants show eglandular tufts on the sepals and have glabrous pedicels. These characters, in our opinion, can be explained only as due to introgression from *T. canaliculata*. If this be true, it demonstrates how extremely difficult, in such a species as *T. occidentalis*, is the determination of the exact extent to which introgressive hybridization is due to man's interference.

*Norman.* — This collection consists of thirty-two plants collected along the dirt roadway at Indian Springs, four miles south of Norman, Cleveland County, Oklahoma. Some of the plants were growing in the actual roadway, and the whole population was evidently spreading along the road. There are no other species of *Tradescantia* in the immediate neighborhood, although *Tradescantia canaliculata* is known to occur a few miles away. All the plants possessed the typical glandular, sepal hairs of *T. occidentalis*, although a quarter of them were glabrous on the pedicels, or nearly so. Furthermore, three plants had eglandular tufts on the sepals. Both of these are characteristics of *T. canaliculata*, and the eglandular tufts are unique in that species.

*College Station.*—These plants were collected along a railroad right-of-way south of College Station, Brazos County, Texas, and in the door-yards of negro cabins adjacent to the railroad. A much larger collection was made but was unfortunately lost before detailed records were made. However, we know from a study of it that these few plants are typical of the lot as a whole. These eleven were cultivated in the greenhouses of the Bussey Institution for several years and retained their distinctive characteristics throughout that period. All the plants exhibit the typical glandular hairs of *T. occidentalis*, although a majority of them show an eglandular tuft at the apex, like *T. canaliculata*. This latter species has been collected at various nearby points. Particularly significant is the fact that it has been found several miles farther north, along the same railroad right-of-way.

*Scott–Dusan.*—These forty-eight plants are from a large colony which extends more or less continuously along the railroad right-of-way between Scott and Dusan, Lafayette County, Louisiana. At the eastern end of the colony there are a number of plants of *T. hirsutiflora*, a variable weed along the gulf coastal plain. As its name indicates, the calyx is more or less covered with long, scattered, eglandular hairs. It hybridizes more or less extensively with *T. canaliculata* and with various other species of *Tradescantia*. Of the plants collected at Scott–Dusan the majority were either typical of *T. occidentalis* or differed only by a slight eglandular tuft at the apex. There were a few clear-cut plants of *T. hirsutiflora* and a complete set of intergrades between *T. hirsutiflora* and *T. occidentalis*.

*Comal County.* — This collection of thirty-seven plants was made on the edge of the Edwards escarpment along the highway running west from New Braunfels, Texas. In table 3 the detailed scoring of ten plants is presented in contrast to a similar ten from Mirando, Texas. The Comal County population had almost the maximum chance for contamination.

In the first place, Comal County is in the very center of diversity for the American species of *Tradescantia*. In the second place, the collection was made along a roadside ditch where there had been repeated grading, gravelling, and other alterations of the original terrain.

TABLE 3. *Comparative variation between 10 plants of* T. occidentalis *var.* typica *in "weedy" (Comal Co.) and "non-weedy" (Mirando) environments.*

| Locality | Glandular hairs on pedicels | Length of pedicel hairs | Length of sepals (in flower) | Glandular hairs on sepals | Eglandular tuft on tip of sepals | Eglandular hairs scattered over sepals |
|---|---|---|---|---|---|---|
| Mirando, Texas | X | .4 | 6 | X | | |
| | X | .4 | 6 | X | | |
| | X | .3 | 6 | X | | |
| | X | .5 | 6 | X | | |
| | X | .4 | 6 | X | | |
| | X | .3 | 6 | X | | |
| | X | .3 | 5 | X | | |
| | X | .4 | 6 | X | | |
| | X | .3 | 7 | X | | |
| | X | .3 | 5 | X | | |
| Comal Co., Texas | | .0 | 11 | X | X | X |
| | X | .4 | 8 | X | X | |
| | X | .5 | 8 | X | X | X |
| | X | .4 | 9 | X | | X |
| | | .0 | 9 | X | | X |
| | | .0 | 9 | X | X | |
| | X | .4 | 9 | X | X | |
| | X | .2 | 11 | X | X | X |
| | X | .4 | 12 | X | X | X |
| | X | .3 | 10 | X | X | X |

Two miles eastward, on the edge of the escarpment, *T. gigantea* spreads down from the cliffs to the side of the same road. Under *Opuntia* in neighboring pastures there were extensive colonies of *T. humilis*, a low-growing, hairy species. The plants along the roadside, on the whole, were more like *T. occidentalis* than like any other species. The majority of them, encountered singly in the herbarium, would have been classified as belonging to that species or to have resulted from hybridization with it. One or two plants were morphologically almost like *T. canaliculata*; others showed extensive contamination with *T. humilis*, and for the most part these particular intermediates were growing at the edge of the roadway, on the side towards the adjoining pastures.

Of particular significance was the fact that the Comal County plants when examined cytologically proved to be tetraploids, though *T. occidentalis*, as well as all other native species of *Tradescantia*, is usually diploid in central Texas (Anderson and Sax, 1936; Anderson, unpublished). In *Tradescantia*, as in many of the higher plants, tetraploids tend to be more ubiquitous and to have more extensive distributions than the diploids from which they arose (Müntzing, 1936).

**TABLE 4.** *Summaries of seven populations of T. OCCIDENTALIS var. TYPICA arranged according to the magnitude of introgression.*

| | Total number of specimens | Glandular hairs on pedicels | Glandular hairs on sepals | Eglandular tuft on tip of sepals | Eglandular hairs scattered over sepals |
|---|---|---|---|---|---|
| Mirando, Texas ...... | 20 | 100% | 100% | 0% | 0% |
| Rainbow Bridge, Utah . | 21 | 100 | 100 | 0 | 0 |
| Inscription House, Ariz. | 30 | 73 | 100 | 63 | 0 |
| Norman, Okla. ....... | 32 | 25 | 100 | 12 | 0 |
| College Station, Texas . | 11 | 100 | 100 | 82 | 0 |
| Scott-Dusan, La. ..... | 48 | 81 | 92 | 94 | 27 |
| Comal Co., Texas .... | 37 | 73 | 97 | 78 | 57 |

From the mere fact of tetraploidy we would have predicted that the Comal County collection was not a part of the indigenous Tradescantias of the region, but was in some way derivative. It was the only tetraploid found during two weeks of collecting in central Texas. It was also the "weediest" population in that area.

way in the middle west. Although *T. bracteata* is a diploid and *T. canaliculata* prevailingly tetraploid (Anderson and Sax, 1936), and although *T. bracteata*, like most diploids, has a short blooming period, hybridization between the two species does occasionally take place, and from the variation in such populations it is evident that the hybrids must be partially fertile.

Table 5 shows that there is apparently an introgression of *T. canaliculata* into *T. bracteata*. In every case the average of the difference is in the direction to indicate introgression and the numbers are large enough to be fairly significant. The evidence is somewhat weakened, however, by the peculiar status of *T. bracteata*. This species, quite aside from any possible introgression of *T. canaliculata*, presents the most complicated pattern of intra-specific variability which has yet been encountered among the American Tradescantias. Until this phenomenon has been studied exhaustively, the evidence for an introgression by *T. canaliculata* will be less conclusive.

CONCLUSIONS. — *Taxonomic.* — The morphological effect of introgression in these species is too slight to merit nomenclatorial recognition. While it may be of interest to taxonomists, introgression is here of slight taxonomic significance.

*The evolutionary importance of introgressive hybridization.*—In genera such as *Tradescantia* the presence of introgressive hybridization greatly complicates the study of the evolutionary dynamics of populations. Were such genera as *Tradescantia* rare, evidence like that presented above could be dismissed as rather a special case. There are, however, a very large number of genera of the flowering plants which resemble *Tradescantia* in that closely related species

**TABLE 5.** *Comparisons of herbarium material of T. CANALICULATA marked "can.", T. BRACTEATA outside the range of T. CANALICULATA marked "bract.", and T. BRACTEATA within the range of T. CANALICULATA marked "bract. (can.)".*

| | Node number | | | | | | | | Leaf number | | | | | | | | | | Internode | | Tuft | | |
|---|---|---|---|---|---|---|---|---|---|---|---|---|---|---|---|---|---|---|---|---|---|---|---|
| | 2 | 3 | 4 | 5 | 6 | 7 | 8 | 9 | 5 | 6 | 7 | 8 | 9 | 10 | 11 | 12 | 13 | 14 | Increase | Decrease | None | Weak | Strong |
| Bract. ......... | | 11 | 1 | | | | | | | 1 | 7 | 3 | 1 | | | | | | 6 | 6 | 4 | 7 | 1 |
| Bract. (Can.) .. | 2 | 9 | 9 | 2 | | | | | | 1 | 8 | 8 | 4 | 1 | | | | | 12 | 10 | 1 | 3 | 19 |
| Can. .......... | | 2 | 12 | 18 | 19 | 14 | 5 | 3 | 1 | | 3 | 10 | 22 | 12 | 10 | 2 | | 2 | 3 | 56 | 18 | | 26 | 42 |

TRADESCANTIA CANALICULATA INTO T. BRACTEATA.— *Tradescantia bracteata* is confined to the middlewestern prairies and to the northeastern edge of the great plains. Under natural conditions it is typically found in rich soil at the edge of prairie swales. Under cultivation it persists along the edge of drainage ditches and may become very abundant along roadways or railroads if they pass through an area with rich soil and a high water table. In such situations it comes into close contact with *T. canaliculata*, which is almost universally present along railroad rights-of-

are separated largely by ecological barriers (Wiegand, 1935). In such genera it will require exceedingly critical data to demonstrate (1) the relative importance of hybridization and mutation and (2) the relative importance of introgressive hybridization in weed populations and under strictly natural conditions.

In most of the populations described above, the effect of introgressive hybridization was so overwhelming that the effect of such presumably more basic evolutionary factors as gene mutation was completely obscured. Much of the introgression in these populations is post-Columbian. To estimate the role which introgression may have played before the ecological complexion of the Great Plains was catastrophically altered by the Caucasian invasion will require detailed analyses of many strictly indigenous populations. It is to be hoped that such populations may be found.

In the last decade a number of techniques, chiefly cytological, have made possible the precise exploration of the germplasm. Such pioneer work as that of Dobzhansky and Sturtevant (1938) is demonstrating the precision which these discoveries bring to the study of phylogeny. Although, in the higher plants, the microstructure of the germplasm cannot yet be determined by direct exploration, as in the diptera studied by Dobzhansky and Sturtevant, much pertinent information can be obtained by inference (Darlington, 1937; Upcott, 1937). These methods are of great promise for the determination of the evolutionary importance of such factors as segmental inversion. Unfortunately, in genera like *Tradescantia*, the importance of inversion cannot be estimated, much less demonstrated, without an understanding of the nature and degree of introgressive hybridization.

In such genera the cytological complexion of the population is a reflection of the ecological complexion of the environment. Ecological data are therefore fundamental to the interpretation of cytological data. Refined cytological analyses of the germplasm will require equally refined ecological analyses of the environment before the potentialities of the data can be realized.

## SUMMARY

Introgressive hybridization is described and defined. Analyses of intraspecific variation in herbarium material indicate a strong introgression of *Tradescantia canaliculata* into *T. occidentalis* and of *T. canaliculata* into *T. bracteata*. There is no evidence for an introgression of *T. subaspera* into *T. canaliculata*, although these species are known to hybridize.

The above conclusions are supported by an analysis of variation in natural populations of *T. occidentalis*. This analysis further demonstrates that introgression is roughly proportional to the frequency of the introgressive species and that it is greater when plants are growing as weeds than when they occupy more natural habitats.

MISSOURI BOTANICAL GARDEN,
WASHINGTON UNIVERSITY, ST. LOUIS

## LITERATURE CITED

ANDERSON, E. 1936a. A morphological comparison of triploid and tetraploid interspecific hybrids in *Tradescantia*. Genetics 21: 61–65.

———. 1936b. An experimental study of hybridization in the genus *Apocynum*. Annals Missouri Bot. Gard. 23: 159–168.

———. 1936c. Hybridization in American Tradescantias. I. A method for measuring species hybrids. II. Hybridization between *T. virginiana* and *T. canaliculata*. Annals Missouri Bot. Gard. 23: 511–525.

———. 1937. Cytology in its relation to taxonomy. Bot. Rev. 3: 335–350.

———, AND K. SAX. 1936. A cytological monograph of the American species of *Tradescantia*. Bot. Gaz. 97: 433–476.

———, AND T. W. WHITAKER. 1934. Speciation in *Uvularia*. Jour. Arnold Arboretum 15: 28–42.

———, AND R. E. WOODSON. 1935. The species of *Tradescantia* indigenous to the United States. Contrib. Arnold Arboretum 9: 1–132, 12 pl.

DARLINGTON, C. D. 1937. Chromosome behaviour and structural hybridity in the *Tradescantiae*. II. Jour. Genetics 35: 259–280.

DELISLE, A. L. 1937. Cytogenetical studies on the polymorphy of two species of *Aster*. Gen. Prog. A. A. A. S. 101 meet. p. 121.

DOBZHANSKY, TH., AND A. H. STURTEVANT. 1938. Inversions in the chromosomes of *Drosophila pseudoobscura*. Genetics 23: 28–64.

GOODWIN, R. H. 1937. The cyto-genetics of two species of *Solidago* and its bearing on their polymorphy in nature. Amer. Jour. Bot. 24: 425–432.

MÜNTZING, A. 1936. The evolutionary significance of autopolyploidy. Hereditas 21: 264–378.

RILEY, H. P. 1936. A character analysis of a colony of *Iris* hybrids. Gen. Prog. A. A. A. S. 99 meet. p. 68.

———. 1937. Hybridization in a colony of *Tradescantia*. (Abstract) Genetics 22: 206–207.

UPCOTT, MARGARET. 1937. The genetic structure of *Tulipa*. II. Structural hybridity. Jour. Genetics 34: 339–399.

WIEGAND, K. M. 1935. A taxonomist's experience with hybrids in the wild. Science 81: 161–166.

## 13

Reprinted from *Evolution* 1:249–262 (1947)

# HYBRIDIZATION BETWEEN THE SUNFLOWER SPECIES *HELIANTHUS ANNUUS* AND *H. PETIOLARIS*

Charles B. Heiser, Jr.

*Department of Botany, Indiana University, Bloomington, Indiana*

Received July 20, 1947

## INTRODUCTION

A study of *Helianthus annuus* was begun by the writer in 1943. The first hybrid swarm of *H. annuus* and *H. petiolaris* was encountered in the summer of 1945, and subsequently two other hybrid swarms were discovered. Although the studies are still being carried on, it seems desirable at the present time to present a preliminary survey of the two species.

This work has included both the collection of specimens in the field and the growing of plants in the greenhouse and in the field. The artificial hybrid between the two species has been produced and analyzed. In addition, cytological studies have been made of the two species and their hybrid.

Herbarium specimens have been consulted from a number of institutions. To the curators of these herbaria the writer wishes to express his thanks. The writer also would like to make acknowledgments to the following: Dr. Edgar Anderson, who first suggested the study of *Helianthus;* the Missouri Botanical Garden for a grant to travel to Arizona; Dr. Harold S. Colton for placing the facilities of the Museum of Northern Arizona at the writer's disposal; and Dorothy Heiser for help with the statistics.

## TAXONOMY

Fortunately there are few nomenclatorial difficulties to be dealt with in the treatment of the two species. *Helianthus annuus* was described by Linnaeus in 1753. Of the many synonyms of *H. annuus* only one, *H. aridus* of Rydberg, requires discussion here. Cockerell (1915) thought that *H. aridus* might possibly be a hybrid of *H. annuus* × *H. petiolaris,* but later (1918) after making the artificial cross he concluded that *H. aridus* was a variety of *H. annuus* subsp. *lenticularis.* The New York Botanical Garden was unable to locate the type of *H. aridus* for examination by this writer, but from the description of the species the possibility that it may represent a hybrid derivative of *H. annuus* × *H. petiolaris* should not be ruled out. Six specimens from the New York Botanical Garden annotated by Rydberg as *H. aridus* have been examined and it was found that five of them should be referred to *H. annuus* and one to *H. petiolaris.* From the examination of other herbarium specimens it appears that the name *H. aridus* has served as a pigeon hole for the filing of misfit specimens of *H. annuus* and, to a lesser extent, of *H. petiolaris.* These misfits are either depauperate plants or hybrid derivatives. Hence the writer is unable to give taxonomic recognition to *H. aridus* and suggests that it be placed in synonymy under *H. annuus* as has been done by Watson (1929) and Blake (1942).

*Helianthus petiolaris* was first described by Nuttall, and its synonymy has been discussed by both Cockerell (1918) and Watson (1929). The variety *canescens* of A. Gray requires mention here. This variety which occurs in the southwestern United States extending into Mexico was raised to specific rank by Wooton and Standley under the name *H. canus* (*H. petiolaris* var. *canus* Britton) and has been retained as a species by Watson. The chief characteristic separating it from *H. petiolaris* lies in its dense cancescent pubescence. In view of the facts that the pubescence appears to be a variable character and that the two appear to intergrade, judging from the examination of herbarium specimens, it seems best to regard it as merely a variety as has been done by Blake (1942) until experimental studies can be undertaken.

Cockerell's *H. petiolaris* var. *phenax* is a form with yellow disk which, for reasons to be given below, this writer regards as an introgressive type.

No new names have been provided for the hybrids. The $F_1$ is simply designated as *H. annuus* × *H. petiolaris*, and no attempt has been made to give variety or form names to the many introgressive types which have been encountered in this study.

The literature citations of the species and combinations discussed are given below.

H. annuus L., Sp. Pl. 904. 1753.
  *H. lenticularis* Dougl. in Lindl., Edwards' Bot. Reg., **15**: pl. 1265. 1829.
  *H. annuus* ssp. *lenticularis* Ckll., Science, n. s., **40**: 284. 1914.
  *H. aridus* Rydb., Bull. Torrey Bot. Club, **32**: 127. 1905.
H. petiolaris Nutt., Jour. Acad. Nat. Sci. Phila., **2**: 115. 1821.
  *H. petiolaris* var. *phenax* Ckll., Nature (London), **66**: 174. 1902.
H. petiolaris var. canescens A. Gray, Smith. Contr. Knowl. (Pl. Wright., **1**: 108), **3**: 108. 1852.
  *H. petiolaris* var. *canus* Britton, Mem. Torrey Bot. Club, **5**: 334. 1894.
  *H. canus* Wooton and Standley, Contr. U. S. Nat. Herb., **6**: 190. 1913.

GEOGRAPHICAL DISTRIBUTION

Both *H. annuus* and *H. petiolaris* are widespread in North America. Both are common in the western United States and occasional eastward. *Helianthus annuus* is the more widely distributed of the two. The two species are native to North America but any effort to delimit their natural boundaries before the appearance of man and their subsequent spread as weeds meets with little success. Most authors have held that both species are indigenous to the Great Plains region west of the Mississippi River and have since been introduced elsewhere. Both are now rather common in the southwest.

Although both species grow together in a variety of localities there do seem to be some slight ecological preferences. In general, *H. annuus* seems to be more restricted to heavy soils and *H. petiolaris* to sandy soils. Deam (1940) writes concerning *H. petiolaris* in Indiana: "This species has just begun to invade the state. It was first reported in 1905. . . . It grows in very sandy soil and within the area of its distribution in the state where the sand has been disturbed it has become an abundant weed in cities and along roads and railroads." This writer has observed that *H. petiolaris* is more frequent as a weed of railroads and that *H. annuus* is more common in waste places about cities and along roadsides.

The observations of Dr. G. L. Stebbins, Jr. are of interest here. He writes (in litt.) after a trip cross-country: "The most impressive thing to me was the continuous, dense population of both *H. petiolaris* and *H. annuus* extending along the roadside as well as the railroad tracks all of the way from the beginning of the short-grass plains near Cheyenne, Wyoming, to their eastern border in Nebraska. *Helianthus petiolaris* was in full bloom, *H. annuus* not yet out, so the seasonal difference between them was marked, although there must be a good deal of overlap later on. *Helianthus annuus* was apparently only in the more fertile and better watered areas."

The seasonal difference which Stebbins observed in time of blooming is entirely in accord with the writer's observations.

*H. petiolaris* in most areas comes into bloom in June and *H. annuus* generally not until July. However, both continue to bloom for a rather long period of time, *H. petiolaris* through July into August and *H. annuus* on into September.

Dr. Stebbins and the writer are in agreement that the distributions of the two species were originally allopatric and that they have been brought together through man's interference. Dr. Stebbins goes on to write: ". . . before man came wild *H. annuus* was probably a plant of disturbed soil-bluffs, mud bars, etc.; *H. petiolaris,* on the other hand, may have originated in a dryer climate with a shorter growing season—probably on the high plains of Colorado or Wyoming, along creek bottoms or sandy washes of

'bad land' areas." In view of our present knowledge this hypothesis appears plausible.

## MORPHOLOGY

A detailed discussion of the morphology of the two species is not necessary here because descriptions of both species are given in Watson (1929, p. 351 and p. 357).

The usual many and well defined differences between the species are occasionally bridged by hybrid swarms. However, a high degree of sterility among the intermediates is indicated by a study of seed and pollen fertilities. The introgressants or introgressive forms, those showing the influence of genes from the other species, generally show little or no reduction in

TABLE 1.   *Comparison of artificial F₁ of* Helianthus annuus × H. petiolaris *with parent species from greenhouse plants*

|  | H. petiolaris (602) (5 plants) | Hybrid (601) (6 plants) | H. annuus (603) (5 plants) |
|---|---|---|---|
| Height | .35–.45 m. | .55–.70 m. | .80–.90 m. |
| Pubescence of stem | glabrate | hispidulous | hispid |
| Branching | much branched | much branched | little branched |
| Lower leaves: (fig. 4) | | | |
|   shape | lanceolate | lanceolate to ovate-lanceolate | lanceolate to ovate-lanceolate |
|   blade length | 6.5–8.4 cm. | 7.5–10.5 cm. | 7.5–9.6 cm. |
|   width | 2.0–2.6 cm. | 3.3–4.7 cm. | 3.2–5.8 cm. |
|   serration | entire or obscurely serrulate | irregularly to regularly serrate | regularly serrate |
|   pubescence | soft | soft | rough |
| Length of peduncles | 5–15 cm. | mostly about 5 cm. | 4 cm. or less |
| Disk diameter | 2.3 cm. or less | 2.5–2.8 cm. | 3.5 cm. or more |
| Bracts of the involucre: | | | |
|   (fig. 4) | gradually attenuate | intermediate | abruptly attenuate |
|   width | 3 mm. | 4 mm. | 7 mm. |
|   pubescence of margin | hispidulous in 4 plants, somewhat hispid in one | approaching that of *H. annuus* | hispid |
| Chaff: (fig. 4) | | | |
|   pubescence of middle awns | hairs long | hairs intermediate in length | hairs short |
|   color in age | straw colored | purple | purple |
| Rays: | | | |
|   number | 13–16 | 15–21 | 21–30 |
|   length | 2.3–2.7 cm. | 2.7–3.4 cm. | 4.0–4.5 cm. |
|   width | .8– .9 cm. | .9–1.2 cm. | 1 3–1.6 cm. |
| Length of corolla of disk flowers | 4.0–5.0 mm. | 5.0–6.0 mm. | 6.0–6.5 mm. |
| Achenes: | | | |
|   length | 5 mm. | 5–6 mm. | 6–7 mm. |
|   width | 1.5–2.0 mm. | 2.5–3.0 mm. | 3 mm. |

fertility. In spite of introgression of certain characters from the other species there remains no doubt as to which of the two species these forms should be referred.

It is difficult to define the limits of a normal plant of *H. annuus* and *H. petiolaris*. In both species the phenotype may be greatly modified by the environment. Since both are annuals and weeds this is not surprising. Secondly, both species consist of many genetic races. Table 1 gives a comparison of a single population of each species and the artificial hybrid grown under greenhouse conditions. The plants of both species, particularly *H. annuus*, when grown in pots are much smaller in many respects than those encountered in nature.

The height of the two species does require brief discussion. *Helianthus annuus* is normally from 1.0 to 3.0 meters tall (excluding cultivated forms); *H. petiolaris* is normally about 0.9 meters tall. Watson gives the height of the latter species as 0.4 to 4.0 meters. He mentions finding one individual of *H. petiolaris* four meters tall growing in a colony of *H. annuus*. He writes that such tall plants are occasionally found where *H. petiolaris* grows abundantly and that it would be interesting to know the explanation of the gigantic size "whether it be a mutation or the result of hybridism with *H. annuus*."

The color of the corolla lobes of the disk-flowers of the two species also requires brief explanation. In *H. annuus* the color is generally red or purple although in some populations yellow lobes are quite common. The color of the lobes in *H. petiolaris* is red or purple, although a form with yellow lobes has been described by Cockerell (1902). The writer has found a single plant of *H. petiolaris* in a hybrid swarm which possessed yellow lobes. Except for the color of lobes this latter plant appeared to have all of the characteristics of *H. petiolaris* and was fully fertile.

### CYTOLOGY

Both *H. annuus* and *H. petiolaris* have the haploid count of seventeen chromosomes. *Helianthus annuus* has been reported with this number by several work-

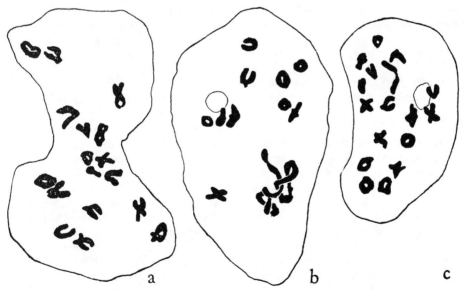

Fig. 1. Camera lucida drawings of chromosomes of *Helianthus annuus, H. petiolaris,* and artificial hybrid: (a) *H. annuus* (752), seventeen pairs, (b) *H. annuus* × *H. petiolaris* (781), ten pairs, two chains, and one ring, (c) *H. petiolaris* (755), seventeen pairs. All × 1370.

ers (see Geisler, 1931, for bibliography). This writer has verified the count of seventeen for *H. annuus* from both cultivated and wild material from a number of collections (figure 1, a). *Helianthus petiolaris* is reported here for the first time with seventeen pairs (figure 1, c).

The hybrid between the two species was grown in the greenhouse in Berkeley during the winter of 1945–46. Although no difficulty was experienced in making suitable aceto-carmine smears of the parent species from greenhouse material, great difficulty was encountered in making smears of the hybrids suitable for detailed study. However, it could be ascertained that meiosis did not proceed normally in the hybrid material. Rings and chains of chromosomes were observed at metaphase I and chromatid bridges with acentric fragments at anaphase I. This would tend to indicate that the parent species differ for both translocations and inversions. Meiosis was later studied in fifty cells of a hybrid between cultivated *H. annuus* ("Mammoth Russian") and *H. petiolaris*. In the latter hybrid it was found that there were from 7 to 11 bivalents at metaphase I, an occasional univalent, and that the remainder of the chromosomes were associated in two or three chains or rings of several chromosomes each (figure 1, b).

### Analysis and Discussion of Hybrid Swarms and Artificial Hybrids

#### First hybrid swarm

In the early fall of 1944 a hybrid swarm of *H. annuus* and *H. petiolaris* was found in St. Louis, Mo., near the intersection of Des Peres Avenue and the University street car tracks. The plants which were few in number were mostly depauperate individuals, but several of the plants were obviously morphological intermediates between the two species. Moreover, those plants intermediate in appearance set few if any good seeds. Since these plants were mostly depauperate and do not fit in well with the methods of scoring the

other populations no attempt is made in presenting the scoring of this population here. The good seeds from the putative hybrids were planted the following winter but only two germinated. The two plants secured, however, did show segregation for *H. annuus* and *H. petiolaris*-like characters. The first plant (*622–1*) had deeply serrate leaves (figure 4, 1), disk-flowers with yellow corolla lobes, and involucral bracts which in shape and pubescence resembled those of *H. petiolaris*. The second plant (*622–2*) had entire leaves (figure 4, k), disk-flowers with purple corolla lobes, and bracts similar to those of *H. annuus*. Both plants possessed less than 20 per cent good pollen and set no seed on crossing.

The next summer this locality was visited again, but unfortunately most of the plants had been mowed down. Of the few plants remaining one was of exceptional interest. This plant, previously mentioned, had over 90 per cent good pollen and fitted the description of *H. petiolaris* except for the yellow lobes of the disk-corollas. One other such plant has been described by Cockerell as *H. petiolaris* var. *phenax*. It is the opinion of the writer that plants of *H. petiolaris* with yellow disk-corolla lobes have probably come about through the introgression of a gene or genes from *H. annuus*.

#### Second hybrid swarm

After finding the first hybrids between *H. annuus* and *H. petiolaris* the writer reexamined his collections made in the summer of 1944 near Flagstaff, Arizona, where the two species had been found growing in close proximity. One plant (*893*) was found to be more or less intermediate morphologically between the two species and was characterized by low pollen fertility.

A second trip was made to Arizona in the late summer of 1945 with the purpose of examining the sunflowers more critically. The trip was rewarded by the finding of a large hybrid swarm by the roadside near the Museum of Northern

Arizona, north of Flagstaff. Sixty-six plants were scored in the field for leaf width and margin, width of the involucral bracts, disk diameter, and pubescence of the chaff. Since it was late in the season the rays were not in suitable condition for scoring. The characters scored were assigned values following the method of Anderson (1936). Characters which were *H. annuus*-like were assigned a value of 2, those which were like *H. petiolaris* were assigned a value of 0, and any intermediate condition was given a value of 1. The resulting histogram showed a distinctly bimodal distribution (figure 2).

Thirty-six plants were collected from the hybrid swarm for a study of pollen and seed fertilities. Pollen counts were made by smearing anthers in lactophenol and cotton blue and those grains which took a dark blue stain were counted as

good. The seed fertility was measured by counting the number of filled and unfilled achenes to the head. Those plants with low percentages of good pollen (68 per cent or less) and low seed set (less than 50 per cent) were scored as hybrids or hybrid derivatives. The remainder of the plants were placed with the non-hybrid group. A few plants with as low as 55 per cent good pollen were placed with the latter group since they showed no reduction from the normal seed set. This group of non-hybrids was easily subdivided into two groups depending upon their morphological resemblance to either *H. annuus* (*1642b*) or *H. petiolaris* (*1642a*). Some of the hybrid group (*1642c*) approached either *H. annuus* or *H. petiolaris* rather closely in appearance, but for the most part they were more or less intermediate.

For most of the characters measured the variability of the group classed as hybrids was greater than that of either parent. The measurements of the disk diameter are presented in table 2. The coefficient of variability of the disk diameter for the plants of *H. annuus* was 14.5, for *H. petiolaris* 9.0, and for the hybrids 16.5.

The great variability of the hybrids can be accounted for if it is assumed that the hybrid swarm is one of several years standing and the hybrid group consists of backcrosses, $F_2$'s, and more complicated crosses, in addition to $F_1$'s. The pollen fertilities tend to bear out this conclusion. The amount of good pollen of the hybrid group ranged from 4 to 68 per cent with a mean of 18 per cent. The plants of both *H. annuus* and *H. petiolaris* also showed rather great ranges of pollen fertility. The percentage of good pollen of *H. annuus* ranged from 55 to 100 per cent with the mean at 81 per cent, that of *H. petiolaris* from 61 to 100 per cent with the mean at 87 per cent.

Seeds were collected from this population but germination was very poor and no attempt will be made here to discuss the few greenhouse plants obtained.

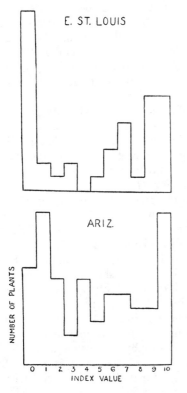

FIG. 2. Histograms of two hybrid swarms of *Helianthus annuus* and *H. petiolaris*. (See text for explanation.)

TABLE 2. *Measurements of disk diameter of* Helianthus annuus, H. petiolaris, *and hybrids from Arizona and Illinois*

|  | Number of plants | Mean | Standard deviation | Coefficient of variability |
|---|---|---|---|---|
| Flagstaff, Arizona | | | | |
| H. annuus (*1642b*) | 10 | 3.3 | 0.48 | 14.5 |
| H. petiolaris (*1642a*) | 13 | 2.1 | 0.19 | 9.0 |
| hybrids (*1642c*) | 13 | 2.3 | 0.38 | 16.5 |
| East St. Louis, Illinois | | | | |
| H. annuus (*1807b*) | 17 | 3.3 | 0.48 | 14.5 |
| H. petiolaris (*1807a*) | 17 | 2.3 | 0.33 | 14.3 |
| hybrids (*1807c*) | 9 | 2.9 | 0.29 | 10.0 |
| H. annuus (*1807d*) | 19 | 3.4 | 0.29 | 8.5 |

*Third hybrid swarm*

In the summer of 1946 another hybrid swarm was found in the railroad yards along the Mississippi River near Eads Bridge in East St. Louis, Ill. *Helianthus annuus* becomes quite conspicuous in the loam soil along the roadside after one crosses the bridge from St. Louis. The plants of *H. annuus* extend down onto the cindery railroad beds where *H. petiolaris* occurs. Forty-three specimens were taken where the two species came together, and these were tentatively identified as *H. annuus* (*1807b*), *H. petiolaris* (*1807a*), and hybrids (*1807c*). All of these identifications with a single exception were verified upon later examination. The plants identified as *H. annuus* and *H. petiolaris* proved to have high percentages of good pollen (over 90 per cent) and, with the one exception mentioned, the plants determined as hybrids had low percentages of good pollen (less than 30 per cent). Subsequent examination of the one plant which had nearly perfect pollen proved it to be a depauperate plant of *H. annuus*. Watson (1929) has previously commented upon the difficulty of identifying depauperate specimens of annual sunflowers.

In addition to the plants collected from the hybrid swarm a number of plants of *H. annuus* (*1807d*) which were growing nearby along the roadside in loam soil were collected.

The plants of both *H. annuus* and *H. petiolaris* as well as the hybrids differed considerably from the Arizona specimens. These differences are not unexpected inasmuch as both species differ genetically throughout their ranges. The Arizona plants of *H. petiolaris* seemed to show some influence of the southwestern variety, *H. petiolaris* var. *canescens*, in the pubescence of their stems which was lacking in the St. Louis plants. Widely different ecological conditions, of course, would be expected to account for some of the other differences.

The pattern of variability was also different as suggested by the measurements of the disk diameter (table 2). For example, the coefficient of variability of the disk diameter for *H. annuus* (*1807b*) was 14.5, for *H. petiolaris* 14.3, and for the hybrids 10.0. The plants of *H. annuus* from along the roadside showed still less variability; their coefficient of variability for disk diameter was 8.5. The ray size, ray number, width of involucral bracts, and leaf index showed the same general trend of variability as that encountered in the disk diameter.

The possible causes for less variation among the hybrids than among the parents is worthy of consideration. The low variability of the hybrids could be accounted for if all of the hybrids were $F_1$'s. The pollen fertilities of this hybrid group showed the same range of good pollen (0–30 per cent) as did the artificial $F_1$

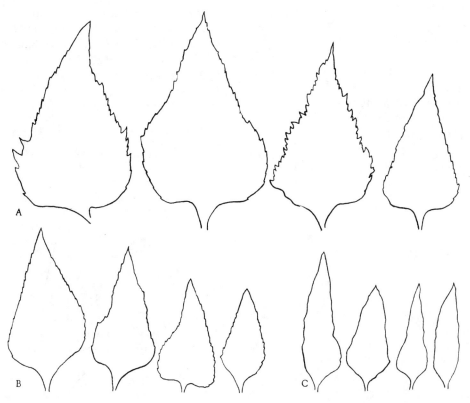

FIG. 3. Leaf tracings from plants of East St. Louis hybrid swarm: (A) *Helianthus annuus* (top), (B) hybrids (lower left), (C) *H. petiolaris* (lower right). × approx. ⅖.

(discussed below) although the mean (20 per cent) for the natural hybrids was slightly higher. However, the morphological variation in the natural hybrids is rather great for $F_1$'s even though the parent stocks may have been highly heterozygous. A second possible explanation is that the hybrids consisted merely of $F_1$'s, possibly $F_2$'s, and first generation back crosses without the more complicated types of backcrosses which were probably present in the Arizona hybrid swarm. This hypothesis is supported by the fact that in the Arizona population it was very difficult to separate the hybrids from the species on the basis of pollen fertility, whereas no such difficulty was encountered with the East St. Louis plants.

The great difference in the variability of the plants of *H. annuus* from along the roadside (*1807d*) and those from the hybrid swarm (*1807b*) is probably due in part to differences in soil conditions as well as to the greater amount of introgression of *H. petiolaris* into the latter (*1807b*).

A histogram for this hybrid swarm utilizing the same characters employed for the Arizona population is presented in figure 2. A number of other histograms were made using characters of the rays, pubescence, and disk-flowers in addition to those previously mentioned. All of the histograms showed a bimodal distribution for the population.

Later in the summer of 1946 achenes were collected from this population, and greenhouse plantings were made the next winter at Davis, California. Progeny of the hybrid plants again showed some segregation for *H. annuus-* and *H. petiolaris-*like characters.

*Artificial F$_1$*

The hybrid between *H. annuus* and *H. petiolaris* was made in the summer of 1945, employing essentially the same technique as that of Putt (1941).[1] The re-

ciprocal cross was made, but the plant of *H. annuus* was injured and did not mature its seeds. The *H. annuus* plant selected was fairly typical of the wild sunflowers from St. Louis except that it possessed some red in its rays. The *H. petiolaris* plant used was from seed of St.

[1] Since that time it has been found that most of the wild annual species of *Helianthus* are rather highly self-sterile, and heads are rubbed in making the crosses instead of employing the more time consuming emasculating and hand pollinating of the individual flowers.

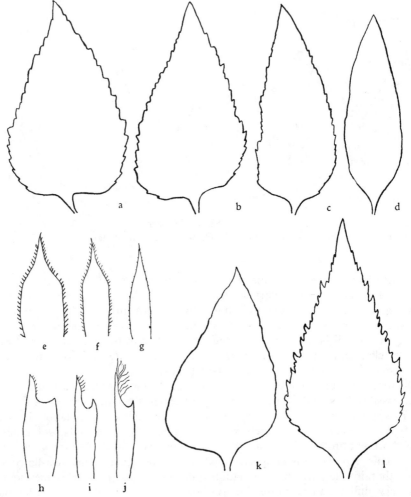

FIG. 4.  a–d, Leaf tracings of artificial hybrid and parents, × ⅔: (a) *Helianthus annuus* (*603*), (b, c) F$_1$ (*601*), (d) *H. petiolaris* (*602*). e–g, Involucral bracts of artificial hybrid and parents, × 2: (e) *H. annuus* (*603*), (f) F$_1$ (*601*), (g) *H. petiolaris* (*602*). h–j, Chaffy bracts from center of heads of artificial hybrid and parents, × 3⅓: (h) *H. annuus* (*603*), (i) F$_1$ (*601*), (j) *H. petiolaris* (*602*). k–l, Leaf tracings of progeny of a natural hybrid, × ⅔: (k) no. *622–2*, (l) no. *622–1*.

Louis *H. petiolaris*. A large amount of good seed was obtained from the cross.

The $F_1$ was grown in the greenhouse at Berkeley, California, during the winter of 1945–46. All of the plants were very vigorous and bloomed much longer than either parent. A comparison of the $F_1$ with the parent species which were grown at the same time is given in table 1. Drawings of leaves, involucral bracts, and chaff are given in figure 4. Direct size comparisons cannot be made between greenhouse and field populations, but the appearance of the synthetic $F_1$ lends additional proof to the hypothesis that the putative hybrids are truly of hybrid origin. The $F_1$ agrees rather closely with that previously described by Cockerell (1918).

The pollen fertilities in the ten $F_1$ plants were very low, ranging from 0 to 30 per cent good pollen with a mean of 14 per cent. Sister plants were crossed by rubbing heads together, and the seed set was less than 1 per cent.

*Artificial $F_2$*

The $F_2$ consisted of five plants which were grown in the greenhouse at Davis during the winter of 1946–47. No two plants were alike but in general they resembled *H. annuus* more nearly than *H. petiolaris*. However, none of the plants was like *H. annuus* in all particulars. Plant no. *768–3* showed the nearest approach, agreeing with *H. annuus* in seven morphological characters, intermediate in four, and approaching *H. petiolaris* in the pubescence of the chaff. The pollen fertilities varied greatly. The percentages of good pollen for the five plants were 13, 39, 48, 50, and 97 per cent. It appears that it is possible for the sterility barrier to be overcome to some extent in the $F_2$. A plant similar to the $F_2$ plant with 97 per cent good pollen if found growing wild would be called an extreme introgressant.

Another interesting feature among the recombinations was the yellow disk-corolla lobes of plant no. *768–2*. Neither of the parent species had yellow disks,

although the factor may have been carried in the recessive state in one or the·other, probably in *H. annuus*. No attempt will be made to go into the full genetic implications here, but it should be pointed out again that two plants of *H. petiolaris* with yellow disk-corolla lobes have turned up in nature. Thus the evidence from the $F_2$ offers further support of the hypothesis that the occurrence of the yellow lobes of the disk-corollas in *H. petiolaris* has come about as a result of hybridization with *H. annuus*.

No weakness was observed in the $F_2$ although progeny of putative natural hybrids growing under the same conditions did show some weakness. Many of the latter plants were dwarfs, and furthermore they were subject to attack by red spider to which the parent species and artificial $F_2$'s were immune.

Backcrosses of the $F_1$ to the two parents have not yet been grown.

*Herbarium studies*

A number of herbarium specimens have been consulted in the present study. Both species show a great deal of variability throughout their range, and it is entirely possible that some of this variability could have come about through the introduction of the germ plasm from one of the species to the other. The following characteristics which show up in some herbarium specimens of *H. petiolaris* might have come about through introgression: broad leaves, serrate leaves, hispid involucral bracts, and a general increase in size of all parts of the plant. Similarly, plants of *H. annuus* have been seen which embody certain characteristics which might have been derived from *H. petiolaris*. They are principally as follows: narrow leaves, narrow involucral bracts, smaller disks, densely pubescent chaff, and a reduced number of rays. It is, of course, almost impossible in most cases to judge from herbarium specimens whether the size differences are genetic or merely represent an ecological modification of the phenotype. The writer fully realizes that

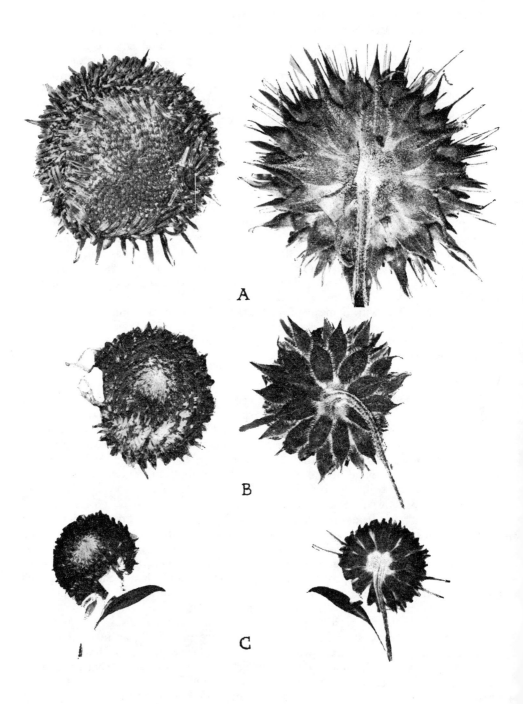

FIG. 5. Heads of *Helianthus annuus,* natural hybrid, and *H. petiolaris.* (A) *H. annuus* from St. Louis, (B) a natural hybrid from East St. Louis hybrid swarm, (C) *H. petiolaris* from St. Louis. On the left, face views showing disk flowers and chaff. On the right, rear views showing the involucral bracts. Rays not shown. Slightly enlarged.

mutation may equally well be responsible for some of these differences, but in view of the fact that the two species do hybridize it would appear that introgression is, at least, partially responsible. The full extent of the introgression is a problem that will require a great deal of additional field work in the West.

## Conclusions

The term "introgressive hybridization" was first used by Anderson and Hubricht (1938) to explain the infiltration of the germ plasm of one species of *Tradescantia* into another as the result of hybridization. Since that time further evidence for introgression has been obtained from many plants. Goodwin (1937), Riley (1938, 1939) and Wetmore and Delisle (1939) have made rather detailed studies utilizing experimental methods. A review of the literature on the work dealing with introgression and a discussion of the mechanisms and consequences of introgression will be published elsewhere.

The result of hybridization between *H. annuus* and *H. petiolaris* certainly falls into the category of introgressive hybridization. Although the hybrid is rather highly sterile some good pollen is produced which would allow for backcrossing to the parent species. That such backcrossing of the hybrid occurs is evident from the examination of hybrid swarms.

Dobzhansky (1941) is inclined to be rather skeptical of the great number of hybrids that have been reported. He points out that the putative hybrids may be remnants of the ancestral population from which the two supposed species have differentiated, and further that mutations arising within one species may resemble certain of the characters found in another species. The writer feels, however, that in the present case there can be little doubt that the putative hybrids are actual hybrids. The natural hybrids morphologically agree very closely with the synthetic hybrids. Furthermore, when seeds from a putative hybrid are planted they give rise to diverse types approaching the parent species in appearance. In addition, the high degree of sterility in the putative hybrids indicates that these plants are of hybrid origin.

Of course, as Huskins (1929) has pointed out, what constitutes a hybrid depends upon what one calls a species. It is perhaps well to state here the writer's reasons for considering *H. annuus* and *H. petiolaris* species. Morphologically there are many differences between the two, implying that they differ for numerous genes. The hybrid between the two shows abnormalities at meiosis, indicating that there are chromosomal differences. There are also slight differences in ecological preference and time of blooming; thus the species show reproductive isolation from each other. Although the two are now sympatric they may have at one time been allopatric.

It is possible that the two species may have descended from the same prototype. With the present knowledge of the genus *Helianthus* that can be neither proved nor disproved. Introgressive hybridization is known to take place between *Helianthus annuus* and *H. Bolanderi* (Heiser, 1947), and is suspected to occur between *H. annuus* and *H. cucumerifolius*. It may be found that all of the annual sunflowers are capable of exchanging genes. However, it is the opinion of the writer that all of the annual sunflowers, defined in the narrow sense, have existed as independent species for long periods of time and that hybridization between them may not have occurred before the coming of man.

The two species under consideration here would fall into the category of "ecospecies" following the definition of Clausen, Keck, and Hiesey, and the annual sunflowers as a whole would then belong to a single "cenospecies."

Dobzhansky stated in 1941 that it was impossible to appraise the evolutionary significance of introgressive hybridization at the present time. However, he does suggest three possible roles of introgression. (1) Physiological barriers may develop in response to the challenge of hybridization and therefore introgression may be a passing stage in species formation. (2) Introgression may lead to an obliteration of differences between incipient species and to their fusion into a single highly variable one. (3) If the environment in which the incipient species was formed should cease to exist, some of the recombination products might be more fitted to the new environment than were the parental species; introgressive hybridization might then result in the emergence of superior genotypes.

It is difficult to evaluate fully these possible roles for the case of *H. annuus* and *H. petiolaris*. Although complete physiological barriers to hybridization may appear, none seems to have done so as yet. There is no evidence that introgression is merely a passing stage in the formation of these two species, nor is there any evidence that the hybridization is resulting in the formation of a single highly variable species. Although the introgressants may show characteristics of the other species they can usually be regarded quite definitely as one species or another. Marsden-Jones and Turrill (1929) found that hybridization between *Silene maritima* and *S. vulgaris* had not resulted in their amalgamation into one polymorphic species.

In the case of *Helianthus annuus* and *H. petiolaris* introgression apparently has resulted in the appearance of some superior genotypes among the many recombinations. The spread of both species into new habitats as weeds may have come about through aggressive recombination products. These successful recombination products, representing introgressants, have been able to inhabit new areas. It may prove that many weeds have evolved through a similar process and that they owe much of their aggressiveness and weedy tendencies to introgression.

Although introgression has increased the polymorphy of the two species, it is not the sole cause of the great variability. Gene mutation may have played an important role, and in the case of *H. annuus* hybridization with other annual species and crossing with the cultivated sunflower (Heiser, 1947) have been important factors. Introgression is not the only answer but certainly it should help contribute to the explanation of many problems that have long puzzled taxonomists.

## SUMMARY

The sunflower species *Helianthus annuus* and *H. petiolaris* are widespread weeds in North America. Most botanists have held that the two species are indigenous to the Great Plains region. Although the two species are now sympatric in their distribution, they may have been allopatric at one time. There are slight differences in ecological preference and in time of blooming between the two species. Both species possess seventeen pairs of chromosomes. The $F_1$ shows some abnormalities at meiosis and consequently is rather highly sterile.

Hybrid swarms between the two species have been found in Arizona, Illinois, and Missouri. Two of these hybrid swarms are analyzed on the basis of pollen and seed fertility and the variability of the disk diameter. Comparison of the artificial with the putative natural hybrids tends to verify that the latter are truly of hybrid origin.

In nature it appears that introgression has occurred through backcrossing of the hybrids to their parents. The main consequence of the introgression has been to increase the variability of the two species without any tendency for an amalgamation of the two into a single highly variable species. It is suggested that introgression has been of importance in the development of superior genotypes which may have increased the ability of the two species to spread as weeds.

## LITERATURE CITED

ANDERSON, E. 1936. Hybridization in American Tradescantias. Ann. Mo. Bot. Gard., 23: 511–525.

—— AND L. HUBRICHT. 1938. Hybridization in *Tradescantia*. The evidence for introgressive hybridization. Am. Jour. Bot., **25**: 396–402.

BLAKE, S. F. 1942. Compositae, in Kearney and Peebles, Flowering plants and ferns of Arizona, U. S. Dept. Agr. misc. publ. 423.

CLAUSEN, J., D. D. KECK, AND W. M. HIESEY. 1939. The concept of species based on experiment. Am. Jour. Bot., **26**: 103–106.

COCKERELL, T. D. A. 1902. De Vriesian species. Nature (London), **66**: 174.

——. 1915. Specific and varietal characters in annual sunflowers. Am. Nat., **49**: 609–622.

——. 1918. A new hybrid sunflower. Torreya, **18**: 11–14.

DEAM, C. C. 1940. Flora of Indiana. Dept. of Conservation. Indianapolis, Ind.

DOBZHANSKY, TH. 1941. Genetics and the origin of species. Revised Edition. Columbia University Press.

GEISLER, F. 1931. Chromosome numbers in certain species of *Helianthus*. Butler Univ. Bot. Studies, **11**: 53–62.

GOODWIN, R. H. 1937. The cytogenetics of two species of *Solidago* and its bearing on their polymorphy in nature. Am. Jour. Bot., **24**: 425–432.

HEISER, C. B. 1947. Variability and hybridization in the sunflower species *Helianthus annuus* and *H. Bolanderi* in California. Ph.D. Thesis (Unpub.), Univ. of Calif. Library, Berkeley.

HUSKINS, C. L. 1929. Criteria of hybridity. Science (n. s.), **69**: 399–400.

MARSDEN-JONES, E. M., AND W. B. TURRILL. 1929. Researches on *Silene maritima* and *S. vulgaris* II. Kew Bull. of Misc. Information, **2**: 33–38.

PUTT, E. D. 1941. Investigations of breeding technique for the sunflower (*Helianthus annuus* L.). Scientific Agriculture, **21**: 689–702.

RILEY, H. P. 1938. A character analysis of colonies of *Iris fulva*, *I. hexagona* var. *giganticaerulea* and natural hybrids. Am. Jour. Bot., **25**: 727–738.

—— 1939. Introgressive hybridization in a natural population of *Tradescantia*. Genetics, **24**: 753–769.

WATSON, E. E. 1929. Contributions to a monograph of the genus *Helianthus*. Papers Mich. Acad. of Science, Arts, and Letters, **9**: 305–475.

WETMORE, R. H., AND A. L. DELISLE. 1939. Studies in the genetics and cytology of two species in the genus *Aster* and their polymorphy in nature. Am. Jour. Bot., **26**: 1–12.

# 14

Reprinted from *Am. J. Bot.* **50**:159–173 (1963)

## NATURAL HYBRIDIZATION AMONG FOUR SPECIES OF BAPTISIA (LEGUMINOSAE)[1]

R. E. ALSTON AND B. L. TURNER

The Plant Research Institute and Department of Botany, The University of Texas, Austin, Texas

### ABSTRACT

ALSTON, R. E., and B. L. TURNER. (U. Texas, Austin.) Natural hybridization among four species of Baptisia (Leguminosae). Amer. Jour. Bot. 50(2): 159–173. Illus. 1963.—Interspecific hybridization involving 4 species of *Baptisia* (*B. leucophaea*, *B. sphaerocarpa*, *B. nuttalliana*, and *B. leucantha*) has been studied by means of extensive field work and subsequent morphological and chromatographic analyses. As a result of these studies, numerous hybridizing populations involving any 2, 3 and, in 1 instance, 4 species have been located. Near Dayton, Texas, all 4 species and all 6 of the possible 2-way hybrid combinations have been found in a single field. Approximately 125 different chemical compounds have now been detected in the 4 species. Many of these compounds serve as species specific markers useful in the validation of specific hybrid types. Hybrids between *B. leucophaea* and *B. sphaerocarpa* and between *B. leucophaea* and *B. nuttalliana* are numerous, and in these large hybrid swarms a chromatographic and morphological analysis of population structure is possible. The former combination provides an excellent opportunity for the utilization of chemical markers as criteria for introgressive hybridization. The hybrid *B. leucantha* × *B. sphaerocarpa* is frequently encountered and contains a large number of compounds species-specific for one or the other parental species. The other 3 hybrid types have been found infrequently. Certain hybrid types are generally similar morphologically (e.g., *B. leucantha* × *B. sphaerocarpa* as opposed to *B. leucantha* × *B. nuttalliana*), and chromatographic techniques are of great value in the absolute identification of such plants, especially in complex populations where backcrossing further complicates the interpretation of the background of a plant from exomorphic features alone.

THE 5 SPECIES of *Baptisia* native to Texas are *B. minor*, *B. leucophaea* (including *B. laevicaulis*), *B. nuttalliana*, *B. leucantha*, and *B. sphaerocarpa* (including *B. viridis*) (Turner, 1959).[2] Although *B. minor* apparently hybridizes with *B. leucophaea* and possibly with others, hybrids involving *B. minor* have not been studied by the present workers, and the species is omitted from consideration at this time. *Baptisia leucophaea* and *B. leucantha* are widely distributed in the central United States; *B. sphaerocarpa* and *B. nuttalliana*, more restricted in their distribution, are found in only 4 or 5 states in the south central United States (Fig. 1, 2). These species are morphologically unlike and may be readily distinguished even at great distances, yet they hybridize without apparent sterility or other incompatibility, as do many other species of *Baptisia*.

[1] Received for publication July 12, 1962.

Supported by National Science Foundation Grant 15890.

The authors wish to express their appreciation for the technical assistance of Mrs. Virginia Findeisen.

[2] *Baptisia laevicaulis* and *B. viridis*, treated as species in the author's earlier papers, are now believed to be better treated as infraspecific categories under the older names *B. leucophaea* and *B. sphaerocarpa*.

TABLE 1. *Locations of specific hybrid combinations*

A.  *B. leucophaea* × *B. nuttalliana*

1. Sebastian Co., Ark., U.S. 71, 3 miles N.W. of Greenwood
2. Evangeline Pa., La., Dirt road between Mamou and Chantaignier
3. Choctaw Co., Okla., U.S. 70, 5 miles east of Hugo
4. Freestone Co., Tex., Farm Rd. 488, 1.6 miles south int. Farm Rd. 2548
5. Grimes Co., Tex., State Hwy. 30, 4.5 miles east of Brazos Co. line
6. Grimes Co., Tex., State Hwy. 158, 2 miles west of Roan's Prairie
7. Hardin Co., Tex., State Hwy. 326, 0.5 miles north of Pine Island Bayou
8. Hardin Co., Tex., Farm Rd. 105, 200 yds. north int. Farm Rd. 770
9. Harris Co., Tex., Farm Rd. 2100, 2 miles south of Huffman
10. Hopkins Co., Tex., U.S. 30, ¼ mile west int. State Hwy. 19
11. Leon Co., Tex., U.S. 79 just east of Jewett
12. Liberty Co., Tex., U.S. 90, 2 miles west of Dayton
13. Madison Co., Tex., U.S. 75, 3.2 miles north of Madisonville
14. Newton Co., Tex., State Hwy. 63, 4 miles east of Burkeville
15. Van Zandt Co., Tex., State Hwy. 19, 3.3 miles north int. State Hwy. 64
16. Van Zandt Co., Tex., State Hwy. 19, between farm Rds. 858 and 1256
17. Walker Co., Tex., U.S. 190, 0.6 miles east int. Farm Rd. 405
18. Walker Co., Tex., U.S. 190, just east int. Farm Rd. 2296
19. Walker Co., Tex., State Hwy. 30, 6.9 miles east of Shiro
20. Walker Co., Tex., U.S. 75, 3.0 miles north of Crabb's Prairie
21. Walker Co., Tex., U.S. 19, 6 miles north of Huntsville, scattered to Trinity River

B.  *B. leucantha* × *B. sphaerocarpa*

1. Jefferson Davis Pa., La., int. State Hwys. 99 and 380
2. Hardin Co., Tex., State Hwy. 105, 1.3 miles east of Sour Lake
3. Harris Co., Tex., Farm Rd. 2100 north of Crosby
4. Harris Co., Tex., int. U.S. 90 and 90A, Houston
5. Jefferson Co., Tex., State Hwy. 326 just south of Pine Island Bayou
6. Jefferson Co., Tex., U.S. 90, 200 yds. west int. Farm Rd. 364, Beaumont
7. Jefferson Co., Tex., U.S. 90, 4.7 miles east of China
8. Liberty Co., Tex., Farm Rd. 1960, 0.2 miles west of int. Farm Rd. 686
9. Liberty Co., Tex., U.S. 90, 2.7 miles west of Nome
10. Liberty Co., Tex., U.S. 90, 2 miles west of Dayton

C.  *B. leucophaea*[b] × *B. sphaerocarpa*

1. Sebastian Co., Ark., U.S. 71, 10.4 miles N.W. of Greenwood
2. Acadia Pa., La., U.S. 90 just east of Mementau
3. Bryan Co., Okla., U.S. 69, 7 miles N.E. of Colbert
4. Brazoria Co., Tex., Farm Rd. 1301, just north of West Columbia
5. Fannin Co., Tex., State Hwy. 121, 2 miles south of Bonham
6. Galveston Co., Tex., State Hwy. 146, 3 miles N.W. Texas City near Moses Lake
7. Hardin Co., Tex., State Hwy. 105, 1.3 miles east of Sour Lake
8. Hardin Co., Tex., State Hwy. 105 just north of int. Farm Rd. 770
9. Harris Co., Tex., int. U.S. 90 and 90-A, Houston
10. Harris Co., Tex., Farm Rd. 2100 north of Crosby
11. Jackson Co., Tex., State Hwy. 111, 2.3 miles east of Edna
12. Liberty Co., Tex., Farm Rd. 1960, 0.2 miles west of Farm Rd. 686
13. Liberty Co., Tex., State Hwy. 61, 2 miles east of Devers
14. Liberty Co., Tex., 2 miles west of Dayton

D.  *B. nuttalliana* × *B. sphaerocarpa*[b]

1. Houston Co., Tex., State Hwy. 21, 1 mile north of Farm Rd. 304
2. Liberty Co., Tex., U.S. 90, 2 miles west of Dayton
3. San Jacinto Co., Tex., State Hwy. 150, 0.4 miles east of Evergreen
4. Walker Co., Tex., U.S. 75, 2 miles north of Huntsville

E.  *B. leucantha* × *B. nuttalliana*

1. Jefferson Davis Pa., La., U.S. 196, 6 miles south of Kinder
2. St. Landry Pa., La., U.S. 190, just west of Lawtell
3. Hardin Co., Tex., State Hwy. 326, north of Pine Island Bayou
4. Jefferson Co., Tex., State Hwy. 326, south of Pine Island Bayou
5. Liberty Co., Tex., U.S. 90, 2 miles west of Dayton

TABLE 1. *Continued*

F.  *B. leucantha* × *B. leucophaea*

    1.  Allen Pa., La., U.S. 190, 4 miles west of Reeves
    2.  Hardin Co., Tex., State Hwy. 162, 1 mile west of Batson
    3.  Harris Co., Tex., Farm Rd. 2100 between Huffman and Crosby
    4.  Liberty Co., Tex., U.S. 90, 2 miles west of Dayton

[a] int. = intersection.
[b] Numerous hybridizing populations may be found to the west and southwest of Houston, in the coastal prairie, to Jackson Co.

Despite the fact that all 4 species are geographically broadly sympatric in Texas, they are adapted to quite different ecological conditions, and consequently one rarely encounters the 4 species together in a single population. Two species commonly occur together and less frequently 3 species may be found together, generally in disturbed areas (Fig. 3). The authors have been engaged in extensive field work in Texas and parts of Louisiana, Arkansas and Oklahoma. One objective has been to determine the extent and characteristics of natural hybridizing populations involving *Baptisia* in these areas. Figures 4–9 show the distribution of various hybrid individuals or hybridizing populations which have been observed. Exact locations of the hybrid sites are listed in Table 1. Prior to this study, only 2 of the 6 possible hybrid combinations were certainly known. Larisey (1940a) reported the hybrid combination, *B. leucantha* × *B. sphaerocarpa* (treated as *B. viridis*), and Turner and Alston (1959) reported details of the combination *B. leucophaea* × *B. sphaerocarpa*. Larisey (1940b), in her mono-

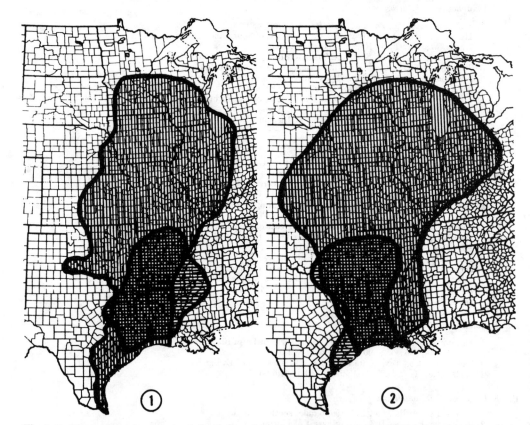

Fig. 1, 2.—Fig. 1. Approximate natural distribution of *Baptisia leucophaea* (vertical lines) and *B. nuttalliana* (horizontal lines); region of sympatry cross-hatched.—Fig. 2. Approximate natural distribution of *B. leucantha* (vertical lines) and *B. sphaerocarpa* (horizontal lines); region of sympatry cross-hatched. Areas of recent local introductions are not included in the maps.

graphic treatment, described apparent hybrids
between *B. leucophaea* and *B. nuttalliana* or *B.
leucantha* and *B. leucophaea* as new species. Her
inability to recognize these hybrids from herbarium specimens underscores the wide morphological separation of the species (i.e., the hybrid
is distinctive itself as a "good species"), and is
also indicative of the difficulty in deducing
hybridity from morphological evidence without
familiarity with the natural populations.

Now, all 6 of the hybrid combinations have
been validated by both morphological and chromatographic data. Details of each combination
are given in a later section. The frequency with
which hybrids of different origins occur varies
greatly. An actual count of parental species and
hybrids has been obtained in only 1 instance: in
a population within Houston, Texas (intersection
of U.S. 90 and U. S. 90A), 2 acres of plants contained: 162 *B. sphaerocarpa*, 64 *B. leucophaea*,
49 *B. leucantha*, and 33 hybrid types. Estimates
of a larger population at Sour Lake, Texas,
including the same 3 species, indicate a lower
percentage of hybrids. The frequency of hybrids
in a particular population is dependent upon a
number of variables which affect, collectively,
the population structure.

In some areas it is impossible to maintain a
simple concept of discrete populations of *Baptisia*.
The pattern may be very complex. For example,
except where interrupted by towns and especially
by the Trinity River bottomland, roughly 75
miles of U. S. 90 between Houston and Beaumont
is a linear mosaic of *Baptisia* populations consisting of 1, 2, 3 and, in 1 instance, 4 species and
assorted hybrids (Fig. 3). Along this highway 2
miles west of Dayton, Texas, all 6 hybrid combinations plus the 4 parental species have been
identified with certainty in a single pasture.

OBSERVATIONS AND RESULTS—*Morphological
attributes of the individual species*—Although the
4 species considered here differ greatly in their
general form and gross appearance, one must
rely, in addition, upon a group of specific key
characters in the detection and validation of
hybrids. In general, hybrids are intermediate in
their morphological characters, and *Baptisia*
hybrids are not exceptional. Table 2 presents a
summary of the distinguishing features of the 4
species. Figures 10, 11 illustrate the 4 species and
the 6 hybrids.

*Chromatographic attributes of the individual
species*—We have analyzed the constituents of
these and other *Baptisia* species intensively by
1- and 2-dimensional chromatography utilizing
many solvent combinations and extraction techniques. Many plant parts have been investigated, including stems, leaves, petals, fruits,
sepals and even the stamens. Compounds such
as free amino acids, the lupine alkaloids and

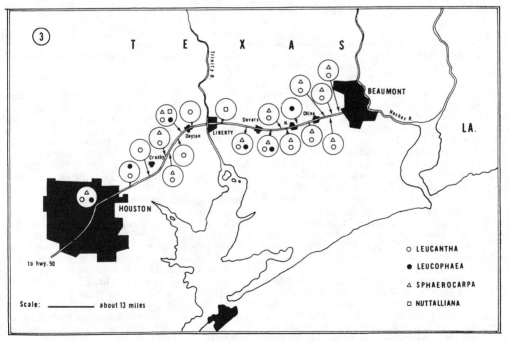

Fig. 3. Mosaic of *Baptisia* populations as they occur along U.S. Highway 90 between Houston and Beaumont, Texas.
Sympatry at any one site is shown by the symbols included in the circles.

TABLE 2. *Morphological characters which distinguish the species of Baptisia being considered*[a]

| Character | Species | | | | List of all 6 possible 2-way hybrids between the species |
| | B. sphaerocarpa | B. leucophaea | B. nuttalliana | B. leucantha | |
| --- | --- | --- | --- | --- | --- |
| **Ovary** | | | | | |
| Pubescence | glabrous | densely villous | densely villous | glabrous | B. leucophaea × sphaerocarpa |
| Ovary shape | globose | ovoid-elliptic | ovoid to sub-globose | obovoid | B. nuttalliana × sphaerocarpa |
| Ovary wall | thick, indurate | thin, fragile | thin, fragile | thin, firm | B. leucantha × sphaerocarpa |
| Ovules per ovary | 2–5 | 11–15 | 6–10 | 16–24 | B. leucophaea × nuttalliana |
| Flower color | bright yellow | pale yellow | bright yellow | white | B. leucantha × leucophaea |
| Stipules | absent | conspicuous persistent (20–40 mm long) | absent or caducous | conspicuous persistent (10–15 mm long) | B. leucantha × nuttalliana |
| Pedicel length | 1–5 mm | 25–40 mm | 2–4 mm | 10–15 mm | (So far as known, when the characters are combined in the $F_1$ the resulting characters are intermediate or nearly so.) |
| Flower arrangement | in racemes, evenly disposed | in racemes, secund | single in leaf axil | in racemes, whorled at nodes | |
| Petiole length | 2–7 mm | 3–6 mm | 0.5–1 mm | 8–15 mm | |
| Raceme condition | stiffly erect | abruptly reflexed | racemes absent (flowers single) | erect, elongate, lax | |

[a] Ten selected characters which serve, singly or in combination, to identify the 4 *Baptisia* species in this study. At least 20 additional characters could be added to the list.

anthocyanins, although useful in other ways, have not proven to be of much practical value in the documentation of hybrids. We have concentrated upon miscellaneous substances extractable in 0.5% HCl in methanol. These extracts were chromatographed in 2 solvents: (1) 22 hr in t-butanol: acetic acid: water (3:1:1); and (2) 4 hr in 15% acetic acid. The substances visible in ultraviolet light with and without ammonia and after spraying with dinitroaniline are recorded in these investigations. Many of these substances are adjudged to be phenolic or polyphenolic, but the group may be quite heterogeneous. For example, spot #23 (Table 3), one of the few substances identified thus far, is the coumarin, scopoletin.

The chromatographic results are summarized in tabular form (Table 3) and graphically (Fig. 14). There is no need to discuss further the nature of the biochemical criteria. The chromatographic "profiles" are essentially a composite. Rare indeed is the single plant which exhibits every one of the major and minor spots. Yet, the general pattern of basic components is so

reliable that one can identify a species from these patterns as rapidly and with as much accuracy as could be done from a living specimen in full flower. The work of Brehm (1962) may be consulted for documentation of the nature and extent of variation in chromatographic constituents in one of the species, *B. leucophaea*. Although the data of Table 3 and Fig. 14 represent, as accurately as possible, our present knowledge of the chromatographic patterns in these species, it is obvious that some further modifications, mostly in the nature of additions, can be expected. For example, Brehm (1962) has recently added 2 components to the profile of *B. leucophaea* which were detected in a chemical race growing along the Gulf Coast. Had these substances appeared first in a hybrid they might have been considered tentatively to be "hybrid substances."

The following sections contain brief descriptions of the 6 hybrid combinations with notes on the frequency of hybridization and methods of validating the hybrids.

*Baptisia leucophaea × B. sphaerocarpa*—Extensive hybridization of these taxa occurs in

TABLE 3. *Characteristics of components illustrated in Fig. 14*[a]

| Number in Fig. 14 | Appearance in various treatments | | | | Occurrence and characteristics | | | |
|---|---|---|---|---|---|---|---|---|
| | Daylight (NH₃) | U.V. | U.V. (NH₃) | Dinitro-aniline | *B. leucophaea* | *B. nutt.* | *B. sphaer.* | *B. leucantha* |
| 1 | Y | D | bY | Y → lBr | 3-A | 3-A | 3-A | |
| 2 | Y | D | bY | Y → lBr | 3-D | 3-D | | |
| 3 | Y | D | Y | pY → lBr | 3-A | 3-A | 1-F | |
| 4 | — | D | pY | lY | 3-E | 3-E | | |
| 5 | — | D | pCr | slGyGr | 3-B | 3-B | | |
| 6 | — | D | YGr | slBr | 3-B | 3-B | 3-B | |
| 7 | — | D | D | slBr | 3-B | 3-B | 3-B | {1-B {2-A |
| 8 | — | lB | bBGr | | 3-E | 3-E | 2-A | |
| 9 | — | W | W | — | 3-F | | | |
| 10 | — | lB | BGr | — | 3-A | 3-A | 3-A | 3-A |
| 11 | — | lB | BGr | — | 1-F | 1-F | | 3-F |
| 12 | — | lB | bB | — | 3-E | 1-E | | |
| 13 | — | CrO | bCr | — | 1-E | 3-E | | |
| 14 | — | lB | lB | — | 3-F | 3-F | | |
| 15 | — | GoY | GoY | — | 1-F | 1-E | | |
| 16 | — | l | l | — | 3-F | 1-F | | |
| 17 | — | — | — | L | 3-B | 3-B | 3-B | 3-B |
| 18 | Gy | pBrGr | pBrGr | P | 3-A | 3-A | | 3-A |
| 19 | — | pBrGr | pBrGr | P | 3-D | 3-D | | 3-D |
| 20 | — | — | — | P | 1-F | | | |
| 21 | — | — | — | P | 3-F | | | 1-F |
| 22 | — | l | l | — | 3-E | 3-E | | 3-E |
| 23 | — | dkB | dkB | — | 1-E | | | |
| 24 | — | dkB | dkB | — | 1-E | | | |
| 25 | — | Y | Y | — | 1-E | | | |
| 26 | — | pB | pB | Br/L[b] | 1-F | 3-F | 1-F | 3-F |
| 27 | — | — | — | pP | 1-F | | | |
| 28 | — | — | — | P/P[b] | 3-D | | | 1-D |
| 29 | — | D | pY | | 3-F | 1-D | | |
| 30 | — | D | pY | — | 2-F | | | |
| 31 | — | — | — | P | 2-F | | | 2-F |
| 32 | — | — | — | P | 2-F | | | |
| 33 | — | — | — | P | 2-F | | | 2-F |
| 34 | — | D | D | Gy → BrO | | 3-A | | |
| 35 | — | D | D | — | | 3-A | | |
| 36 | — | D | D | Pk → PkO | | 3-A | | |
| 37 | — | D | D | R → RB | | 3-A | | |
| 38 | — | D | D | pGyP | | 3-E | | |
| 39 | — | D | pYBr | | | 3-F | | |
| 40 | — | D | Pk | — | | 1-F | | |
| 41 | — | D | D | — | | 1-E | | |
| 42 | — | D | — | — | | 1-F | | |
| 43 | — | — | bBGr | — | | 1-E | | |
| 44 | — | — | BBlGr | — | | 1-E | | |
| 45 | — | B | B | — | | 3-F | | |
| 46 | — | — | pB | — | | 1-E | | |
| 47 | — | bBl | bB | — | | 1-E | | |
| 48 | — | D | D | — | | 2-E | | |
| 49 | — | D | D | Br | | 2-E | 3-E | |
| 50 | — | D | D | Br | | 2-E | | |
| 51 | — | B | B | — | | 2-F | | |
| 52 | — | B | B | — | | 2-F | | |
| 53 | — | — | B | — | | 2-F | | |
| 54 | — | — | bB | PkL | | 2-F | 2-F | |
| 55 | — | D | D | pY | | 2-F | | |
| 56 | — | pD | Y | — | | 2-E | | |
| 57 | — | — | lBr | — | | | 1-F | |
| 58 | slPkO | D | D | RBr → bR | | | 3-A | |

TABLE 3. *Continued*

| Number in Fig. 14 | Appearance in various treatments | | | | Occurrence and characteristics | | | |
|---|---|---|---|---|---|---|---|---|
| | Daylight (NH₃) | U.V. | U.V. (NH₃) | Dinitro-aniline | B. leucophaea | B. nutt. | B. sphaer. | B. leucantha |
| 59 | — | D | D | BrG | | | 1-E | |
| 60 | — | — | — | B | | | 1-E | |
| 61 | — | — | — | B | | | 1-F | |
| 62 | — | D | D | slBr | | | 1-F | |
| 63 | — | D | D | PkO | | | 3-F | |
| 64 | — | D | D | PkO | | | 1-F | |
| 65 | — | D | D | — | | | | 1-F |
| 66ᵇ | Y | D | YGr | R | | | 3-A | |
| 67 | — | l | l | — | | | 2-F | |
| 68 | — | Y | YO | lBr | | | 2-A | |
| 69 | — | Y | YO | lBr | | | 2-A | |
| 70 | — | D | Y | — | | 2-E | 2-E | |
| 71 | — | D | Y | — | | | 2-E | |
| 72 | — | B | B | — | | | 2-F | |
| 73 | — | l | l | — | | | 2-F | |
| 74 | — | — | — | B | | | 2-C | |
| 75 | Y | D | dkYGr | G → lBr | | | | 1-A 2-B |
| 76 | Y | D | Y | brY | | | | 1-A 2-B |
| 77 | dull Y | D | dkYGr | Go → lBr | | | | 1-A 2-B |
| 78 | Y | D | Y | pY → lBr | | | | 1-A 2-B |
| 79 | dull Y | lBr | lBr | Gr → lBr | | | | 1-A |
| 80 | — | D | D | pGr → pGyB | | | | 3-B |
| 81 | Y | D | Y | slYBr → pPkO | | | | 3-E |
| 82 | — | GoY | GoY | lBr | | | | 3-A |
| 83 | — | GoY | GoY | lBr | | | 2-B | 3-E |
| 84 | — | GoY | GoY | — | | | | 3-E |
| 85 | — | l | l | — | | | | 1-F |
| 86 | — | l | lBGr | — | | | | 3-F |
| 87 | — | Cr | Cr | — | | | | 1-F |
| 88 | — | — | — | Bl | | | | 1-F |
| 89 | — | D | OY | — | | | | 1-F |
| 90 | — | D | dkY | — | | | | 1-F |
| 91 | — | GoY | GoY | — | | | | 3-F |
| 92 | — | D | D | — | | | | 1-F |
| 93 | — | D | D | — | | | | 1-F |
| 94 | — | B | B | — | | | | 1-F |
| 95 | — | D | D | GoBr | | | | 1-E |
| 96 | — | l | l | — | | | | 3-F |
| 97 | — | D | D | Br | | | | 2-F |
| 98 | — | lB/Gr | — | — | | | | 2-E |
| 99 | — | — | — | Pk | | | | 2-F |
| 100 | — | — | — | Bl | | | | 3-F |
| 101 | — | — | — | Bl | | | | 3-F |
| 102 | — | — | Cr | sldkBl | 3-F | | | |
| 103 | — | B/Gr | bB/Gr | — | 2-D | | | |
| 104 | — | D | pY | — | | | 2-B | |
| 105 | — | D | D | R → RB | | | | 1-F |
| 106 | — | — | — | L | | | | 1-B |

ᵃ *Colors*: B = blue; Br = brown to tan; Cr = cream; D = dark (absorbing UV); Go = gold; Gr = green; Gy = grey; L = lavender; O = orange; P = purple; Pk = pink; R = red; W = white; Y = yellow; BrGo = rust; PkO = salmon; RBr = burnt red.

*Occurrence*: 1 = leaf; 2 = flower; 3 = leaf and flower.

*Descriptive terms*: b = bright; dk = dark; l = light; p = pale (weak); sl = slow color development.

*Characteristics*: A = major, constant spot; B = medium, constant spot; C = minor, constant spot; D = major, variable spot; E = medium, variable spot; F = minor, variable spot.

ᵇ double spot: right side br.Y in NH₃; left side Y-Gr in NH₃-UV.

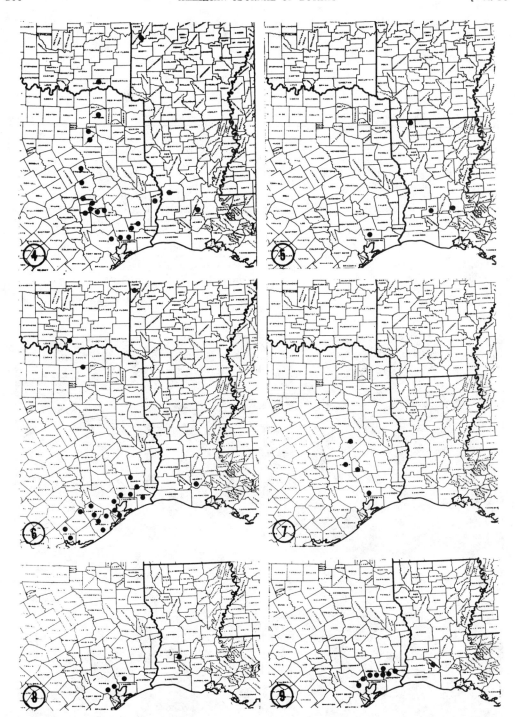

Fig. 4–9. Locations of *Baptisia* hybrids studied. All 6 possible combinations involving the 4 species are shown.—Fig. 4. *B. leucophaea* × *B. nuttalliana.*—Fig. 5. *B. leucantha* × *B. nuttalliana.*—Fig. 6. *B. leucophaea* × *B. sphaerocarpa.*—Fig. 7. *B. nuttalliana* × *B. sphaerocarpa.*—Fig. 8. *B. leucantha* × *B. leucophaea.*—Fig. 9. *B. leucantha* × *B. sphaerocarpa.* (For specific sites see Table 2.)

scattered areas in northeast Texas, Oklahoma, and especially in southeast Texas where large sympatric populations are present. Although no extensive analysis of these populations by chromatography has been attempted, it is evident that backcrossing and more complex genetic exchange occurs in these large hybrid swarms. Since hybridization has apparently been in progress in certain areas for a long period, the periphery of areas of hybridization would be an excellent place to search for biochemical evidence of introgressive hybridization.

Hybrids of *B. leucophaea* × *B. sphaerocarpa* are easily distinguished morphologically, since they are intermediate between the parental types in most characteristics. The reflexional angle of the raceme, bract size, pedicel length, the pattern of pubescence of the ovary, and fruit characters are among the salient features of this hybrid. *Baptisia leucophaea* × *B. nuttalliana* may be distinguished only with difficulty from the hybrid *B. leucophaea* × *B. sphaerocarpa* by one unfamiliar with the various hybrid types.

Chromatographically, the hybrid *B. leucophaea* × *B. sphaerocarpa* is readily identified since a number of major species-specific substances (for this combination) occur together in the hybrid. The hybrid chromatographic pattern is essentially a summation of the parental patterns with some quantitative reduction of individual components. Components especially useful in validating the *B. leucophaea* × *B. sphaerocarpa* hybrids are: 2, 3, 4, 5, 23, 24; 18 and 19 from *B. leucophaea*; and 29, 49, 58, 66, 68, 69, 70, 71 and 83 from *B. sphaerocarpa*.

*Baptisia leucophaea* × *B. nuttalliana*—These species hybridize in scattered locations over a large area of east Texas, western Louisiana, southwest Arkansas and southeast Oklahoma, but the most extensive area of hybridization centers around Huntsville, Texas. Little information is available at this time on the population structure of these hybrid swarms, but it is unlikely that chromatographic techniques alone will be able to provide adequate data for the analysis of backcrossing or related events.

Hybrids of *B. leucophaea* × *B. nuttalliana* may comprise 10% or more of a population in which both species are equally represented. The hybrids are readily distinguished morphologically when

Fig. 10. Photographs of the 4 species of *Baptisia*: (A) *B. leucophaea*; (B) *B. leucantha*; (C) *B. nuttalliana*; and (D) *B. viridis*.

only the 2 species are involved. However, when *B. sphaerocarpa* is present, as noted above, it may be difficult in some instances to distinguish the hybrid *B. leucophaea* × *B. nuttalliana* from the hybrid *B. leucophaea* × *B. sphaerocarpa*. Shorter racemes and a more pubescent foliage are characteristic of the former hybrid.

Validation of a *B. leucophaea* × *B. nuttalliana* hybrid by chromatography is possible, but analysis must be made from stem material. Although a number of leaf components of *B. nuttalliana* are absent from *B. leucophaea*, this latter species has no reliable leaf or flower components absent from *B. nuttalliana*, on the basis of present

Fig. 11. Photographs of the six 2-way hybrid types: (A) *B. leucantha* × *B. leucophaea*; (B) *B. leucantha* × *B. nuttalliana*; (C) *B. leucophaea* × *B. nuttalliana*; (D) *B. nuttalliana* × *B. sphaerocarpa*; (E) *B. leucophaea* × *B. sphaerocarpa*; and (F) *B. leucantha* × *B. sphaerocarpa*.

knowledge. Components 23 and 24 of *B. leucophaea* are potentially useful, but they may be reduced or absent, especially in young plants. However, in the stems of mature hybrid plants, components 23 and 24 of *B. leucophaea* are reliable and distinct. Additionally, components 43 and 44 of *B. nuttalliana* are present in this hybrid. A few other minor distinctions of the hybrid's stem extract added to data from leaves and flowers allow a reasonably certain identification of *B. leucophaea* × *B. nuttalliana* hybrids by chromatography. Components 34, 35, 36, 37, 43 and 44, which are specific for *B. nuttalliana* (in this combination) and which represent major components, often appear reduced and sometimes even absent from hybrids. Since the behavior of these components in the individual hybrids is rather inconsistent, no exact description of these substances in the hybrid is possible at present. There is no difficulty in determining by means of chromatography whether a putative hybrid is derived from *B. leucophaea* × *B. nuttalliana* or *B. leucophaea* × *B. sphaerocarpa*.

*Baptisia leucantha* × *B. leucophaea*—Thus far, only 4 definitive hybrids involving these species have been discovered, each in a separate population in 3 counties in Texas and 1 in Louisiana (Table 1). Since sympatric populations of these 2 species have been examined carefully without yielding any clear evidence of hybridization, it may be assumed that such hybrids are rare. It is significant that of the known hybrids, none was found in large sympatric populations of the 2 parental species; 1 hybrid was found in the complex Dayton site described earlier; 1 was found with a single member of each parental species, 1 was found along a roadside with only a few *B. leucantha* individuals in the adjacent woods and no *B. leucophaea* in the immediate area, and 1 was found in an area with only a few members of each species present.

Hybrids between *B. leucophaea* and *B. leucantha* are morphologically distinctive by virtue of the pale-yellowish flowers, definitely reflexed racemes, and bract and fruit characters. Color of inflorescence, presence of bracts, and fruit form serve to distinguish this hybrid from hybrids of *B. leucantha* × *B. nuttalliana*, but with superficial examination the 2 hybrids may appear similar.

Chromatographically, the hybrid *B. leucantha* × *B. leucophaea* is easily validated since numerous species-specific components are contributed from each parent. Yet, in an area where hybrids of *B. leucantha* × *B. nuttalliana* were also present, as at the Dayton site, chromatographic data alone would not serve to distinguish the 2 types of hybrids. In such combinations, combined morphological and biochemical data rae required. Components useful in validating the hybrid of *B. leucantha* × *B. leucophaea* are: 1, 2, 3, 4, 5, 6, 7, 23, 24 from the former, and 75, 76, 77, 78, 79, 80, 81, 82, 83, and 85 from the latter.

*Baptisia nuttalliana* × *B. sphaerocarpa*—Relatively few hybrids of this origin have been fully validated. One definite hybrid from the Dayton population and 1 from Huntsville, Texas, have been found, and 2 other probable hybrids have been collected. In each instance, except for the plant collected in the complex Dayton population, the hybrid was found together with 1 or a very few individuals of *B. sphaerocarpa*, apparently introduced, among a larger population of *B. nuttalliana*. Large sympatric populations of these 2 species have not yet been found, although they probably exist.

This hybrid is intermediate morphologically between the parents for individual characters but is slightly more *B. sphaerocarpa*-like in gross appearance. The inflorescence is a shortened raceme, and some single axillary flowers occur. The fruit is distinctive in the hybrid, *B. sphaerocarpa*-like in form, but pubescent when young and darkening with age as in *B. nuttalliana*. This hybrid could be confused with backcrosses of *B. leucophaea* × *B. sphaerocarpa* to *B. sphaerocarpa* but is otherwise rather easily recognized.

Chromatographically, this hybrid is easily validated, provided that flowers are available, although there is the usual problem of ascertaining the presence of *B. nuttalliana* rather than *B. leucophaea*. This is best done by means of components 43 and 44 of *B. nuttalliana* and, to some extent, with components 34–37. Flowers provide the *B. sphaerocarpa*-specific substances 70, 71, 69 and 104. It is possible that components 68 (*B. sphaerocarpa*) and 15 (*B. nuttalliana*) are identical. When only *B. nuttalliana* and *B. sphaerocarpa* occur together in an area, hybrids may be identified by components 58, 66, 69, 70, 71 and 104 from *B. sphaerocarpa* and components 2, 3, 4, 43, 44, 18, 19, 34, 35, 36, and 37 (the last 4 with less reliability) from *B. nuttalliana*. Component 29, a major constant spot in *B. sphaerocarpa*, is sometimes present in *B. nuttalliana*.

*Baptisia leucantha* × *B. nuttalliana*—Only a few hybrid plants involving this combination have been identified with certainty. Small sympatric populations of these 2 species occur in several areas of Texas and Louisiana with little evidence of hybridization. These situations, in which relatively little hybridization occurs in sympatric populations, are partly explicable by differences in the times of flowering of the species involved.

This hybrid is distinguished by pale-yellow flowers and much shorter racemes, but it is not greatly different in its general appearance from hybrids of *B. leucantha* × *B. sphaerocarpa*. It is not likely to be easily confused with other hybrid types.

Chromatographically, this hybrid is rather similar to the *B. leucantha* × *B. leucophaea* hybrid, and when all 3 species occur together, a combined morphological and chromatographic analysis is required to prove the identity of a particular plant. Otherwise, the hybrid is quite distinctive and numerous species-specific components are introduced with each parental genome. Components useful in validating this hybrid are 1, 2, 3, 4, 5, 6, 7, 29, 43, and 44 from *B. nuttalliana* and 75, 76, 77, 78, 79, 80, 81, 85, 86, 82, 83 and 84 from *B. leucantha*. Components 34–37, from *B. nuttalliana*, tend to be somewhat obscured in the hybrid and appear to be present in lesser amounts. All hybrids involving *B. leucantha* are potentially useful in the chromatographic analysis of the population structure of hybrid swarms, but only in the instance of *B. leucantha* × *B. sphaerocarpa* hybrids is the incidence of hybrids in the population high enough to warrant an intensive population study.

*Baptisia leucantha* × *B. sphaerocarpa*—Numerous hybrids of this type have been found in southeast Texas and southwestern Louisiana. In sympatric populations, the hybrid may comprise only a small percentage of the total population, but such populations are quite numerous. This hybrid is perhaps the most distinctive of those discussed here. It is a robust plant, more like *B. leucantha* in general habit but having more numerous racemes and intermediate yellow flowers. The fruit is intermediate between the light-brown, small, globose, thick-walled type of *B. sphaerocarpa* and the large, inflated, thin-walled black fruit of *B. leucantha*. It is possible that one unfamiliar with the hybrid types would confuse this hybrid with the *B. leucantha* × *B. nuttalliana* hybrid.

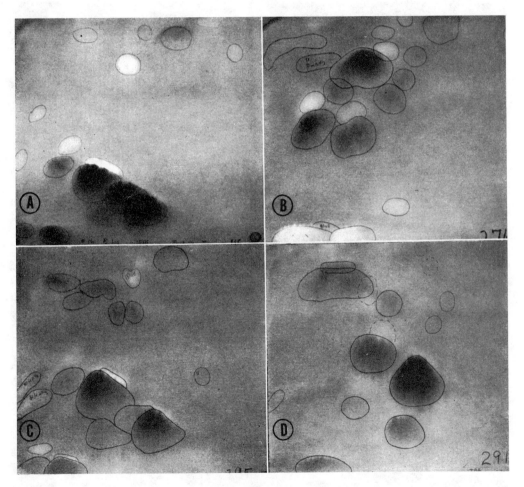

Fig. 12, 13. Chromatograms of leaf extracts photographed in ultraviolet light. Most of the dark spots (absorbing ultraviolet) fluoresce with characteristic color in presence of ammonia vapor. The placement of these chromatograms matches the placement of the corresponding plants of Fig. 10, 11.

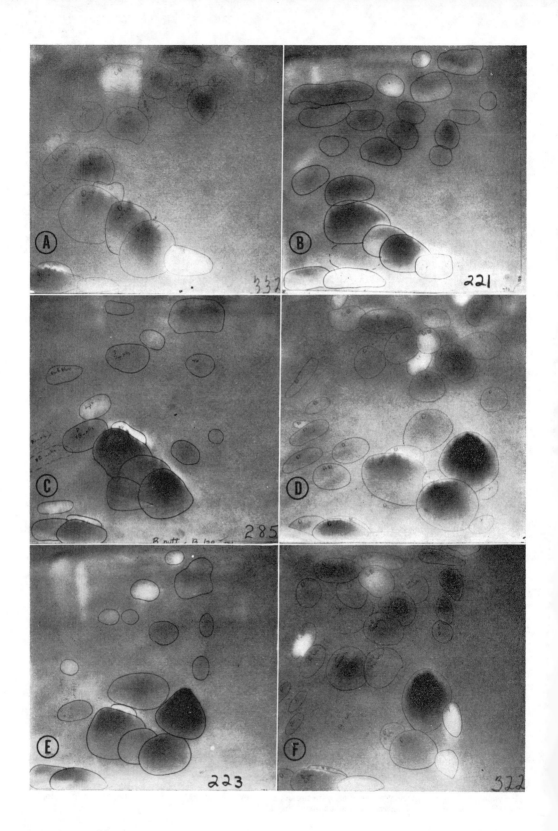

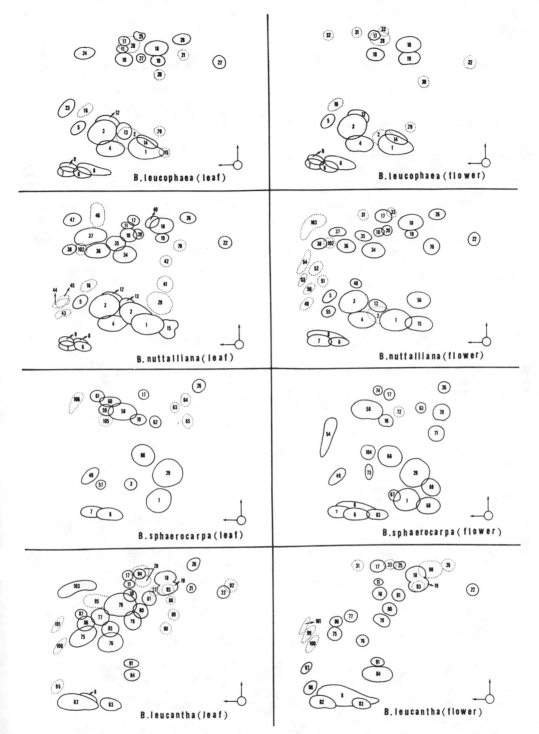

Fig. 14. Composite chromatographic maps of the components of the leaves and the flowers of the 4 species of *Baptisia*. Table 3 contains a description of certain of the properties of the compounds illustrated. Dotted lines refer to compounds which have been selected arbitrarily as inconsistently present following examination of chromatograms of numerous individuals of each species. Absence of one or more of this group of substances may be insignificant.

193

This hybrid is the easiest of all known *Baptisia* hybrids to validate chromatographically, since almost all of the major components are limited to one or the other parental species. Such a large number of species-specific components is available in this hybrid that analysis of backcrossing patterns is quite feasible. At least 1 probable backcross type (to *B. leucantha*) has now been collected. In this hybrid combination the components 68, 69, 70, and 71, which are found in flowers only of *B. sphaerocarpa*, occur in the leaves as well as in the flowers (Alston and Simmons, 1962). These components had previously been considered as "hybrid specific" substances (Alston and Turner, 1962). Components which are useful in the study of this hybrid are 1, 6, 7, 29, 49, 58, 66, 68, 69, 70, 71, and 104, from *B. sphaerocarpa*, and 18, 19, 75, 76, 77, 78, 79, 80, 81, 82, 83, 84, 85 and 86, from *B. leucantha*. This hybrid is so distinctive chromatographically that even a part of 1 leaflet is sufficient to establish its identity with certainty.

CONCLUSIONS—It has been shown that the documentation of natural hybridization by chromatographic techniques is not only a new approach to an old problem but also provides certain types of data not obtainable by morphological criteria alone. The technique is particularly useful in determining the composition of complex hybridizing populations and in the accurate determination of a specific putative hybrid in appropriate circumstances. Or even,

in some instances, as a corollary, there is provided a method for the analysis of introgression in populations peripheral to areas of hybridization. When qualitative intraspecific chemical variation occurs (as in the species *B. nuttalliana* involving particularly component 29, Table 3), this technique allows further population analysis at the species level.

LITERATURE CITED

ALSTON, R. E., AND JANIECE SIMMONS. 1962. A specific and predictable biochemical anomaly in interspecific hybrids of *Baptisia viridis* × *B. leucantha*. Nature 195: 825.
———, AND B. L. TURNER. 1962. New techniques in analysis of complex natural hybridization. Proc. Natl. Acad. Sci. (U.S.) 48: 130–137.
BREHM, B. G. 1962. The distribution of alkaloids, free amino acids, flavonoids and certain other phenolic compounds in *Baptisia leucophaea* Nutt. var. *laevicaulis* Gray and their taxonomic implications. Ph.D. Dissertation. Univ. of Texas, Austin, Texas.
LARISEY, MAXINE M. 1940a. A monograph of the genus *Baptisia*. Ann Missouri Bot. Gard. 27: 119–258.
———. 1940b. Analysis of a hybrid complex between *Baptisia leucantha* and *Baptisia viridis* in Texas. Amer. Jour. Bot. 27: 624–628.
TURNER, B. L. 1959. Legumes of Texas. Univ. of Texas Press, Austin.
———, AND R. E. ALSTON. 1959. Segregation and recombination of chemical constituents in a hybrid swarm of *Baptisia laevicaulis* × *B. viridis* and their taxonomic implications. Amer. Jour. Bot. 46: 678–686.

# 15

## *Drosophila* Hybrids in Nature:
## Proof of Gene Exchange Between Sympatric Species

Abstract. *Genetic studies of two closely related endemic Hawaiian species show that in one area of sympatry about 2 percent of the naturally occurring individuals are hybrids. More than 20 times this many would be expected if the population consisted of a single panmictic unit. Despite hybridization, natural selection appears to maintain the essential integrity of each separate gene pool.*

In the Kahuku area, near the south end of the island of Hawaii, natural interspecific hybridization has been detected between a pair of sympatric endemic species. This circumstance is very unusual for the genus, the other species of which tend to have isolated gene pools which coexist in nature without evidence of gene exchange (*1*). Nevertheless, laboratory crosses between sympatric species are frequently possible and sometimes one or both sexes of F₁ hybrids are fertile (*2*). Accordingly, if reproductive isolation became weak, the potential for gene exchange would exist.

In an extensive literature on *Drosophila* species extending over 50 years, only two cases suggest the possibility of introgression in nature. In both, laboratory crosses give fertile F₁ females and sterile F₁ males. In *D. pseudoobscura* and *D. persimilis*, close genetic scrutiny of the progeny of 27,099 wild-caught flies from sympatric areas yielded four cases of F₁ hybridization (*3*). In each case, a wild female of one species had mated with a male of the other species. Considerable doubt exists, however, as to whether the interspecific matings observed actually took place in nature; since the wild flies were not separated at capture, the interspecific mating could have occurred in the collecting vials on the way to the laboratory. Only a single instance of a backcross hybrid was detected. In *D. metzii* and *D. pellewae*, in-

trogression in nature may be strongly inferred from a face color polymorphism, but critical genetic proof of hybridization in nature is lacking (*4*). We report here a case of hybridization in which one F₁ and three backcross hybrid individuals have been unequivocally recognized in nature.

*Drosophila setosimentum* and *D. ochrobasis* are near-sibling species endemic to the island of Hawaii (*5*). *Drosophila setosimentum* occurs in rain forests between 600 and 1600 m altitude. At five widely separated locations, all above 1000 m, it is accompanied by *D. ochrobasis*. Both species are rare and local, being largely confined to kipukas (islands of vegetation isolated in the midst of newer lava flows) or collapsed lava tubes where their rare host plants grow. Although no reliable means exists for the separation of females on morphological grounds, males of the two species are easily distinguished by a secondary sexual difference in wing pattern.

Wild-caught females were separated from males at capture. In the laboratory, each female was placed in a separate culture vial and allowed to produce offspring. Such a culture (an isofemale line) usually yields F₁ males conforming exclusively to either one species or the other. From each isoline, seven (sometimes fewer) F₁ larvae were taken, and an acetoorcein squash preparation of the salivary gland chromosomes from each was prepared. From each smear the banding order of all five major

chromosome arms (X, 2, 3, 4, and 5) was read directly along the full length of the chromosome, with the use of spaced landmark areas to confirm the sequences. The banding orders of all paired homologs were thereby determined. Inverted sections were read in either homozygous or heterozygous state. The sequence of the hemizygous X in each male larva was also determined.

These cytological data permit inferences about the state of the natural population from which the wild flies were drawn. Except in the Kahuku Ranch population, to be described later, all the F₁ test larvae (195 female and 145 male larvae) examined from 56 wild *D. ochrobasis* females have shown a uniform sequence in chromosome X (*ochrobasis* standard X). In addition, all larvae are homozygous for a chromosome 2 inversion (2k), and all lack an inversion (21) near the opposite end of the same chromosome. This latter inversion is fixed in all but a North Kona, Hawaii, population of *D. setosimentum*. In addition, the commonest gene order of chromosome 4 of *D. ochrobasis* differs from the standard 4th chromosome of *D. setosimentum* by six inversions. These are spread in a roughly tandem manner over the length of the chromosome.

Similarly, 154 female and 164 male test larvae have been recognized from sympatric areas as the progeny of 54 wild *D. setosimentum* flies. In allopatric areas, 221

**195**

*D. setosimentum* were also examined (1062 test larvae). Collections from Kahuku Ranch are excluded from these counts.

Accordingly, from two to nine chromosomal markers serve to distinguish each individual of each species. Their homozygosity and their distribution in the genome is such that three of the five major chromosome pairs are well marked. The lack of heterozygotes in samples of the size reported above means that the probability of incorrect diagnosis of species and of failing to recognize $F_1$ and backcross hybrids between the species is very small. Had we been dealing with one population at Hardy-Weinberg equilibrium, 55 heterozygotes at any one of the autosomal markers would have been expected.

In the laboratory, hybrids between the species are fairly easily made. The resultant chromosomal heterozygotes are those that would be expected from the readings of the two sets of homozygous states. Thus, the *D. ochrobasis* X chromosome in female hybrids is observed in heterozygous state with one of four alternate gene orders known to be polymorphic within *D. setosimentum*. In addition, the expected two-inversion difference (2k 2l$^+$/2k$^+$ 2l) appears in chromosome 2 and the expected six-inversion heterozygote appears in chromosome 4. Further, *D. ochrobasis* adults have a distinctive fixed electrophoretic difference relative to *D. setosimentum*. This may be observed at a β-naphthyl acetate esterase (Est-1) locus (*6*).

Cytological or electrophoretic assays (or both) have been carried out on 180 wild specimens collected from the Kahuku Ranch, Kau District, island of Hawaii. Polytene chromosomes of seven $F_1$ larvae from each isofemale line were scored, and each larval corpse was subjected to electrophoresis. Parallel cytological and electrophoretic data was thus provided for a total of 744 test larvae. Most wild-caught males were subjected to electrophoresis without cytological examination, but some were analyzed by crossing to laboratory virgins from a monomorphic stock and subjected to electrophoresis after $F_1$ larvae had been produced by their mates.

Four exceptional individuals were found. The first was a wild female that transmitted a *D. setosimentum* X chromosome to her sons but showed a *D. ochrobasis–D. setosimentum* heterozygous X in her female progeny. Autosomes 2 and 4 of both sexes showed the hybrid condition in all test larvae. As the female was isolated at capture, this is an unequivocal case (based on heterozygosity for four independent chromosomal markers) of a wild *D. setosimentum* female which had mated with a *D. ochrobasis* male in nature.

The second case was that of a larva collected from a natural oviposition site, a fermenting branch of the plant *Clermontia*. Fourteen third instar larvae from the branch were smeared and the tissues of the same larvae were subjected to electrophoresis. In addition, eight emerging adults from the same branch were subjected to electrophoresis. Of these 22 specimens, 20 were *D. ochrobasis*, one was *D. setosimentum*, and one was a female backcross hybrid. The latter displayed the double polymorphism in chromosome 2 but was homokaryotypic for the *D. ochrobasis* X and 4th chromosomes. Finally, two wild male flies having *D. setosimentum* wing patterns, out of 40 electrophoretically tested but not studied cytologically, proved to be heterozygous for the highly distinctive "null" allele of *D. ochrobasis* at the Est-1 locus.

The foregoing data constitute proof that gene exchange actively occurs between contemporary populations of these species in this locality. Of 180 specimens studied, 4 (2.2 percent) were hybrids. If the population were at Hardy-Weinberg equilibrium, 89 heterozygotes, more than 20 times the observed number, would be expected at the autosomal 2k inversion region alone. The sites at which collections were made are in almost wholly undisturbed rain forest, so that one is not tempted to invoke the idea that man has "hybridized the habitat" (*7*) and thus provided unusual ecological conditions which might have been conducive to the hybridization process or to the biological success of its genetic products.

Discovery of $F_1$ and backcross hybrids is of particular interest in view of Carson and Johnson's suggestion (*5*) that these species have hybridized in the past, as evidenced by the existence of a series of peculiar kinked homologs ("complex chromosomes") which have been found segregating within most sympatric populations of both species. In gene order these kinked chromosomes resemble certain chromosomes of the other species; this appears to have resulted from reciprocal exchange occurring in past generations. Why they have become complexly kinked is not known.

Geographical speciation theory can be invoked to explain the origin of these species in an ecosystem that is continually being fractionated by lava flows (*8*). Nevertheless, the conventionally isolated gene pools of each are not now closed to exchange with the other at places where they come in contact. As in many plants, introgression (*9*) is occurring and apparently has occurred in the past. Despite this, however, the gene pools of the two species give no evidence of flowing together and under-

going free genetic recombination. Accordingly, we hold that the hybridization circumstance is a manifestation of recent (in the geological sense) divergence between these species. Although each species has apparently evolved a recognizable and integrated gene pool of its own, each remains open to the acquisition of small amounts of genetic variability from the other.

This leaky boundary merely represents a stage in what may be a rather common pattern of allopatric speciation in general. Two separated populations may come back into contact after each has virtually completed the integration of its new gene pool. If there is only weak reproductive isolation full recombination might occur. In this, as in many other cases in plants and animals, the reciprocal flow of genes appears to be insignificant probably because of natural selection working against inferior $F_2$ and backcross combinations. The extent of such hybrid zones between closely related species varies greatly; thus the broad zones found in some plants differ greatly from the small zone found in this *Drosophila* case.

A recent theory suggests that a species, as distinguished from an infraspecific population, is characterized by a unique integrated internal system of genetic balance (*10*). So long as closed and balanced genetic systems are not destroyed by recombination, populations can coexist even in the presence of considerable interbreeding. The case we discuss conforms to this view. This is to be expected on a large isolated tropical island which is geologically new and which has rain forests which continue to be dissected by recent lava flows.

H. L. CARSON
*Department of Genetics,*
*University of Hawaii, Honolulu 96822*
P. S. NAIR
*Department of Biology, Southern*
*Illinois University, Edwardsville 62025*
F. M. SENE
*Department of Biology CP11461,*
*University of São Paulo,*
*São Paulo, 05421 Brazil*

**References and Notes**

1. J. T. Patterson and W. S. Stone, *Evolution in the Genus* Drosophila (Macmillan, New York, 1952).
2. E. Craddock, *Evolution* **28**, 593 (1974); H. Carson, *ibid.* **8**, 148 (1954).
3. Th. Dobzhansky, *Am. Nat.* **107**, 312 (1973).
4. S. Pipkin, *Evolution* **22**, 140 (1968).
5. H. Carson and W. Johnson, *ibid.* **29**, 11 (1975).
6. P. Nair, F. Sene, H. Carson, *Genetics* **80**, s60 (1975).
7. E. Anderson, *Evolution* **2**, 1 (1948).
8. H. Carson, *Stadler Symposia* **3**, 51 (1971).
9. V. Grant, *Plant Speciation* (Columbia Univ. Press, New York, 1971).
10. H. Carson, *Am. Nat.* **109**, 83 (1975).
11. Supported by NSF grants GB27586 and GB29288 to the University of Hawaii and by Office of Research and Projects, Southern Illinois University. F.M.S. is postdoctoral fellow of Fund. Amparo a Pesquisa do Est, S. Paulo, Brasil. We thank K. Y. Kaneshiro and L. T. Teramoto for help.

1 May 1975; revised 3 June 1975

Part V

# HYBRID ZONES

# Editor's Comments
# on Papers 16, 17, and 18

**16   MOORE**
*An Evaluation of Narrow Hybrid Zones in Vertebrates*

**17   HUNT and SELANDER**
*Biochemical Genetics of Hybridisation in European House Mice*

**18   BLOOM**
*Multivariate Analysis of the Introgressive Replacement of* Clarkia Nitens *by* Clarkia Speciosa Polylantha *(Onagraceae)*

In many animal and plant genera, population systems of common ancestry diverge following the emergence of geographical barriers, and then come into contact along broad fronts after the barrier recedes. Remington (1968) describes "suture zones" where major biotic assemblages have established contact in the past few thousand years. The zones contain several plant and animal taxa and belts of hybridization between congeners. In some instances, the populations that hybridize have moderate reproductive barriers and have been accorded species status. In most instances, hybridization involves population systems designated as subspecies. There are contact zones between taxa that are not located in Remington's suture zones.

Mayr (1942) refers to zones of extensive hybridization and transition from one taxon to another as secondary intergradation: "Secondary intergradation refers to cases in which the two population systems now connected by a steeply sloping character gradient were isolated at one time and now have come into contact again, after a number of differences have evolved." The zone of intergradation (or hybridization) may be a few to over a hundered miles wide, vary in width from one part of the zone to another, may expand or contract in time, or may shift positions over time (Mayr, 1963). The presence of zones of intergradation are the result of gene flow restriction. The greater the slope, irregularity, and discordance of the single character gradients, the stronger are genetic barriers or the greater is isolation by distance. The width of a zone of secondary intergradation depends on the period the taxa have been in contact, their crossing and fertility relationships, the selective status of immigrants in alien habitats, dispersal parameters, and the reproductive capacity of zone populations relative to "pure" populations. The reproductive capacity may be defined as the number of offspring left by a given individual. The lower the reproductive capacity of organisms within the zone, the narrower will be the zone, because reproductive deficits therein will be counterbalanced by input from the parental taxa (Crosby, 1966).

Moore (Paper 16) reviews the major contributions on secondary intergradation in vertebrates. Within the context of hybrid zones, he challenges the notion that hybridization disrupts coadapted gene complexes and that the unity of a species is maintained by the bonds of mating. He postulates that hybrid zones coincide with ecotones, and that hybrids are not inferior to the parental taxa therein. Here is the crux of the issue: Under what circumstances are hybrids superior to their parents? Although there are ideas on the subject (e.g., Stebbins, 1959), data are scant. A comprehensive genetic analysis of a hybrid zone was conducted by Hunt and Selander (Paper 17) involving *Mus musculus musculus* and *M. m. domesticus* in Denmark. The subspecies are quite distinct; this is confirmed by allozyme data. Of particular interest is the fact that six of seven polymorphic enzymatic loci form step clines centered along or near the same line, but the geographical pattern of gene flow is distinctive for each locus. Also, introgression at most loci is more extensive into ssp. *musculus* than into ssp. *domesticus*. This type of asymmetry also has been described in hybrid zones of birds (Short, 1965; Yang and Selander, 1968). In contrast to the *Mus* situation, allozymic variation at nineteen to twenty loci showed that hybridization along a zone between population systems in *Scelophorus grammicus* yielded only $F_1$ and $BC_1$ hybrids, and that there is little introgression at best (Hall and Selander, 1973).

Description of zones of secondary intergradation in plants is much less common than in animals. In view of the distribution patterns and corresponding patterns of intraspecific differentiation (Ehrendorfer, 1968; J. Heslop-Harrison, 1964), it seems that the topic simply has not attracted much interest from botanists. Subspecific intergradation has taken place near the Illinois–Indiana border in woodland *Phlox divaricata* (Levin, 1967) and prairie *P. pilosa* (Levin and Levy, 1971). Both species have an Ozarkian and Appalachian element that spread into glaciated regions after the hypsithermal maximum. Both species have the same pollination systems and seed-dispersal capabilities. These similarities notwithstanding, the outcome of contact varies between the species. In *P. divaricata*, the zone is over 200 miles wide, whereas in *P. pilosa* it is only 40 miles wide. The character gradients in *P. divaricata* are much less steep and irregular than those in *P. pilosa*. The differences apparently are due to greater genome incompatibility in *P. pilosa* than in *P. divaricata*. In *Asclepias incarinata* and *A. tuberosa*, one finds Ozarkian and Appalachian subspecies with zones of intergradation hundreds of miles wide (Woodson, 1947a). Woodson (1947b, 1962) demonstrated a shift in the genocline between the subspecies of *A. tuberosa* over fourteen years, and suggested that there has been a strong and consistent selection of the western genotypes over the eastern. In view of a generation time of at least several years, stability in population size, and restricted pollen and seed movement, the shift in leaf character expression more likely reflects a plastic rather than genetic response to different environmental conditions.

In contrast to the compatibility of the genomes in the aforementioned subspecies pairs, intergradation may occur between taxa with strong reproductive barriers. An excellent example of this phenomenon has been described in *Clarkia* by Bloom (Paper 18). Not only does gene flow occur across a complex chromosomal barrier, but the zone of phenotypic change does not coincide with the zone of chromosomal changes.

An excellent example of a widespread interspecific hybrid zone is afforded by *Juniperus virginiana* and *J. scopulorum* in the Missouri River Basin of the north central United States. Van Haverbeke (1968) described morphological evidence for allopatric hybridization, but with indecision. Recently, Flake, Urbatsch, and Turner (1978) studied forty volatile chemical characters from populations in the basin area and found overwhelming evidence that the enigmatically variable populations along the bispecific contact zone was the result of introgression. Studies on the volatile constituents of terpenes also have been used to document what was considered the classical case of interspecific introgression or secondary intergradation between *Juniperus virginiana* and *J. ashei* (Anderson, 1953). In this instance, the use of over sixty chemical characters failed to reveal any evidence of gene exchange or even $F_1$ hybrids even where both species grew intermixed (Flake et al., 1969; Adams and Turner, 1970; Flake et al., 1973).

# 16

Copyright © 1977 by the Quarterly Review of Biology
Reprinted from *Quart. Rev. Biol.* **52**:263–277 (1977)

# AN EVALUATION OF NARROW HYBRID ZONES IN VERTEBRATES

WILLIAM S. MOORE

*Department of Biology, Wayne State University, Detroit, Michigan 48202*

ABSTRACT

*A review of the literature on vertebrate hybridization reveals the existence of a number of narrow hybrid zones. Three hypotheses have been suggested to explain the occurrence of these zones. The ephemeral-zone hypothesis states that hybridization will end either in speciation or fusion of the hybridizing taxa by means of introgression. The dynamic-equilibrium hypothesis allows the possibility that narrow hybrid zones might be stable: where hybrids are confined to a small area by steep selection gradients, "crystallization" of an antihybridization mechanism might be prevented by naive immigrants from the parental populations even though hybrids are selected against. The hybrid-superiority hypothesis states that hybrids are more fit than parental phenotypes in some environments.*

*The ephemeral-zone hypothesis fails to explain the antiquity and apparent stability of several hybrid zones. The dynamic-equilibrium hypothesis does not adequately explain the persistence of hybrid populations that do not receive a substantial influx of genes from both parental populations. The hybrid-superiority hypothesis is consistent with the various sizes, shapes, and positions reported for stable hybrid zones because, under this hypothesis, the range of a hybrid population is determined by the range of environmental conditions within which the hybrids are superior.*

*Although there are exceptions, most vertebrate hybrid zones are, in fact, narrow. The hybrid-superiority hypothesis must accommodate this fact. The additional hypothesis is offered that hybrids, in some cases, can succeed in environments where competition from parental phenotypes is weak. Thus, hybrid populations are often found in areas devoid of stable ecological communities. Ecotones are one such area, and I suggest that stable hybrid zones are often narrow because they tend to occur in ecotones which are themselves narrow.*

## INTRODUCTION

T HE EVENTS which ensue secondary contact between two morphologically and presumably genetically distinct vertebrate populations are obscure. Certainly, hybridization is to be expected if the taxa in question have not diverged too far, but the final outcome is unpredictable. It is widely believed that natural hybridization is ephemeral, leading ultimately to either speciation — i.e., the perfection of reproductive isolation — or to fusion of the two races through introgressive hybridization (e.g., Dobzhansky, 1940; Sibley, 1957; Wilson, 1965; Remington, 1968). This hypothesis no longer seems tenable, however, because there are several examples of hybrid zones which appear to be stable (see Short, 1970, and below — p. 266 ff.).

Short (1969) made a distinction between a hybrid zone and a zone of overlap and hybridization; both are subcategories of zones of secondary intergradation. Although this could be an important distinction, it can rarely be made with regard to published reports of secondary intergradation zones. In this article I have used the term hybrid zone interchangeably with zone of secondary intergradation in order to reduce verbiage.

The hypothesis that hybrid zones are always ephemeral is based on two concepts: the concept of coadapted gene complexes; and the concept that gene flow is a strong cohesive force in maintaining the unity of a species. It can be reasonably argued that an organism's phenotype is a singular manifestation of the synergistic actions of numerous gene loci. A successful organism must, therefore, have an integrated genetic system whose component genes work harmoniously to produce a physiologically homeostatic organism capable of reproduction. When a species population is divided, differential mutation and coadaptation would be expected to cause divergence between the isolates. If divergence progressed to the extent that hybridization would disrupt the harmony of the distinctly coadapted gene complexes, secondary contact would lead to speciation because natural selection would eliminate those individuals prone to hybridization. On the other hand, if divergence and differential coadaptation had not progressed to the extent that hybrids were less fit, fusion of the races would be expected, because the viable and fertile hybrids would serve as a bridge for introgressive hybridization. In fact, Wilson (1965) said that when two Mendelian populations with imperfect intrinsic isolating mechanisms come into contact, one of these two outcomes is inevitable and no other equilibrium state seems possible.

Another hypothesis, which I will refer to as the dynamic-equilibrium hypothesis (or model), reconciles the existence of narrow stable hybrid zones with the concept of coadapted gene complexes. Given that two populations had diverged to the point that hybrids suffer depressed fitness, Bigelow (1965) postulated that gene flow through the hybrid zone into the parental populations would be inhibited by selection. Where selection gradients were steep, intergradation would be restricted to a narrow zone between the parental populations. Although hybrids might be inherently less fit than parental phenotypes, only a few individuals in or near the zone of secondary contact would be exposed to selection disfavoring hybridization, while a much larger proportion of the parental phenotypes would never experience this selection pressure. Gene flow from the parental populations into the zone of secondary intergradation could "swamp" alleles which cause individuals to avoid hybridizing and, thus, hinder the evolution of an antihybridization mechanism.

A third hypothesis which, like the dynamic-equilibrium hypothesis, could account for stable hybrid zones is that the hybrids are actually more fit than the parental phenotypes in the restricted regions where they occur. This hypothesis has been espoused by botanists for years (e.g., Anderson, 1949; Muller, 1952; Grant, 1971), but it is not often given serious consideration as an alternative to the ephemeral hybrid zone and dynamic-equilibrium hypotheses by students of vertebrate hybridization. Hagen (1967), however, argued in favor of hybrid superiority as an explanation of the narrow hybrid zones between the anadromous form "trachurus" and the freshwater form "leiurus" of three-spine sticklebacks (Gasterosteus aculeatus). More recently Short (1970) pointed out that ephemeral hybrid zones are the exception rather than the rule in avian hybrids and concluded (Short, 1972) that these hybrids are more fit than parental phenotypes in zones of antiquity, although strong selection against alien genes is occurring in the parental populations. And, finally, Littlejohn and Watson (1973) concluded that the most likely explanation of a stable narrow hybrid zone between two closely related anurans Geocrinia laevis and G. victoriana in Victoria was that the hybrids were more fit than either parental species within a restricted region at the interface of the ranges of the two species populations.

The hybrid-superiority hypothesis has largely been ignored because the occurrence of adapted hybrids would be inconsistent with the central dogma of animal speciation theory, namely, that the integrity of a species, at least in its infancy, is maintained by coadaptation of the species gene pool. That coadaptation exists is unequivocal (e.g., Wallace and Vetukhiv, 1955), but its universal importance as a cause of speciation is a presumption and not a demonstrated fact.

My present purpose is, first, to advocate the hypothesis that hybrids between morphologically, genetically, and ecologically distinct phenotypes are in many cases more fit than either of the parental phenotypes; and, second, to develop a hypothesis which explains the occurrence of stable hybrid zones. My argument is premised primarily on the common properties of several such zones which have been described in the literature, but my initiative to explore this literature was prompted by an investigation of fitness components in unisexual fishes (Moore, 1976). The basis of this initiative will be briefly presented also.

### THE CASE OF *Poeciliopsis monacha-occidentalis*

*P. monacha-occidentalis* is an all-female fish species. As is characteristic of unisexual vertebrates, it is also a hybrid; that is, a population of *P. monacha-occidentalis* is genetically equivalent to a population of female $F_1$ hybrids between *P. monacha* and *P. occidentalis* (see Schultz, 1969, 1971, 1973 for reviews of unisexuality in *Poeciliopsis*). Moore (1976) analyzed the fitness of this hybrid in the context of a mathematical model. The results indicated that the broad distributional success of *P. monacha-occidentalis* is primarily attributable to the simple fact that it produces approximately twice as many female offspring as does *P. occidentalis*, with which it is sympatric (Fig. 1A). When all-femaleness is discounted, however, *P. occidentalis* is better adapted than the hybrid over the entire range of sympatry with the exception of a narrow zone (the Rio Mayo), where the ranges of the parental species overlap slightly, and of a single locale in the next northerly drainage.

The model can also be used to deduce the distribution of *P. monacha-occidentalis* under the hypothetical condition that assumes it to be not a unisexual species but, rather, a *P. monacha* × *P. occidentalis* bisexual hybrid. This distribution is illustrated in Figure 1B.

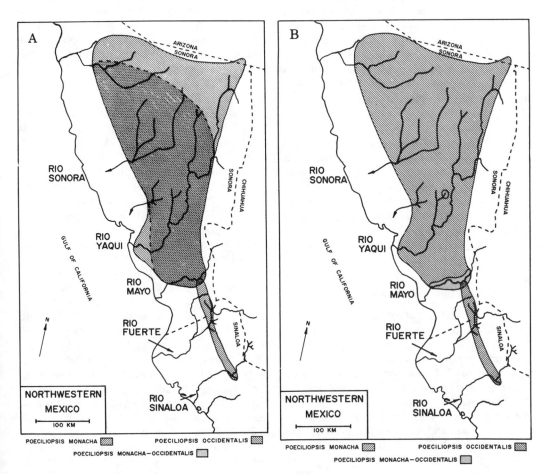

FIG. 1. DISTRIBUTION OF THE COMPLEX *Poeciliopsis occidentalis* — *P. monacha-occidentalis*
A, the actual distribution of the component species; B, the predicted ranges of the component species if *P. monacha-occidentalis* were an independent, sexually reproducing hybrid.

Although some reservations need to be registered regarding the applicability of this result to the problem of narrow hybrid zones, two features of the hypothetical distribution are remarkable — viz., that the hybrids are more fit than one of the parental species anywhere, and that the region of hybrid superiority is, in the main, restricted to a narrow zone at the interface of the ranges of the parental species.

The major reservations are twofold. First, the result was obtained from a mathematical model, and therefore, the inference can be no more valid than the set of simplifying assumptions made when the biological problem was translated into a set of mathematical relationships. Second, all *P. monacha-occidentalis* individuals are F$_1$ hybrids, whereas a population of bisexual *P. monacha* × *P. occidentalis* hybrids would consist of numerous recombined phenotypes. Although heterosis does not appear to be an important component of fitness in *Poeciliopsis* (Thibault, 1974; Moore, 1976), it is possible that the adaptive superiority of *P. monacha-occidentalis* in the Rio Mayo would be lost if the F$_1$ hybrid condition of this species were eroded by recombination.

These reservations notwithstanding, the situation in *P. monacha-occidentalis* provides the insight that hybrids between some vertebrate species may be more fit than the parental species, under some circumstances. In any case, discarding the presumption that hybrids are invariably inferior allows one to explain readily much of the otherwise enigmatic and inexplicable data on hybrid zones.

### HYBRID ZONES IN OTHER VERTEBRATES

The hypothesis that all hybrid zones are ephemeral is untenable because several ancient zones do exist. In addition to the often cited zone of hybridization between the Hooded and Carrion Crows, *Corvus corone* and *C. cornix*, which traverses central Europe (Mayr, 1963), it has been reasonably argued that the following hybrid zones are stable and of remote origin: domestic mice, *Mus musculus musculus* × *M. m. domesticus*, on Jutland (Hunt and Selander, 1973); iguanid lizards, *Sceloporus woodi* × *S. undulatus undulatus*, in Florida (Jackson, 1973); and warblers, *Dendroica coronata* × *D. auduboni* in Alberta and British Columbia (Hubbard, 1969). The domestic mice were thought to have

been hybridizing for 5000 years, *Sceloporus* for up to 100,000 years, and *Dendroica* for "several thousand years" (5,500-6,500 — Hubbard, 1969). There is no evidence of assortative mating in any of these hybrid zones nor evidence of introgression beyond the narrow confines of the zones.

In addition, there are numerous reports of hybridizing specific or subspecific pairs where stable hybrid zones are at least indicated. These reports variously include evidence of hybrid fertility, backcrossing, morphological intermediacy, hybrid viability, constancy in historical times, and random mating in the secondary contact zone. In some cases, inferences based on geological data or events recorded in historical times, were made as to the time when secondary contact was established. These complexes include two ecologically distinct morphs of sticklebacks, *Gasterosteus aculeatus*, in British Columbia (Hagen, 1967); toads, *Bufo woodhousei* × *B. fowleri*, in Texas (Meacham, 1962); anurans, *Geocrinia laevis* × *G. victoriana*, in Victoria (Littlejohn, Watson, and Loftus-Hills, 1971; Littlejohn and Watson, 1973, 1974, 1976); anurans, *Litoria ewingi* and *L. paraewingi*, in Victoria (Watson, Loftus-Hills, and Littlejohn, 1971; Watson, 1972; Gartside, 1972; Littlejohn, 1976; whiptail lizards, *Cnemidophorus tigris gracilis* × *C. t. marmoratus*, in Arizona and New Mexico (Zweifel, 1962; Dessauer, Fox, and Pough, 1962); leopard lizards, *Crotaphytus wislizenii* × *C. silus*, in California (Montanucci, 1970); chromosomal races of the *Sceloporus grammicus* complex of iguanid lizards in Mexico (Hall and Selander, 1973); orioles, *Icterus galbula galbula* × *I. g. bullockii* (Sibley and Short, 1964; Rising, 1970, 1973), flickers, *Colaptes auratus auratus* × *C. a. cafer* (Short, 1965), and towhees, *Pipilo erythrophthalmus erythrophthalmus* × *P. e. arcticus* (Sibley and West, 1959), on the Great Plains; grackles, *Quiscalus quiscula quiscula* × *Q. q. versicolor*, in eastern and southeastern United States (Huntington, 1952; Yang and Selander, 1968); titmice, *Parus bicolor* × *P. atricristatus*, in Texas (Dixon, 1955); towhees, *Pipilo erythrophthalmus* × *P. ocai*, in Mexico (Sibley, 1954); cactus wrens, *Campylorynchus rufinucha humilis* × *C. r. nigricaudatus*, in Mexico (Selander, 1965); toucans, *Pteroglossus torquatus torquatus* × *P. (t.) sanguineus*, and jacamars, *Galbula ruficanda ruficanda* × *G. (r.) melanogenia*, in Colombia (Haffer 1967); and pocket gophers,

*Thomomys bottae* and *T. townsendii*, in California (Thaeler, 1968).

Whereas the hybrid-superiority hypothesis accommodates all of these stable hybrid zones, circumstances in some of the zones appear to be inconsistent with predictions made by the dynamic-equilibrium hypothesis. To see the discrepancies, it is necessary to explore in greater depth the latter hypothesis and its implications.

An implicit requirement for stability in the dynamic-equilibrium hypothesis is that the zone be sandwiched between the parental species populations, or at least so situated that the hybrid population receives a substantial influx of genes from both parental species. Imagine that species A and B come into secondary contact and begin to hybridize. Although the hybrids are less fit than either A or B, a hybrid zone is established and maintained according to the dynamic-equilibrium hypothesis. Now, suppose that the range of species B contracts, thus breaking contact between the hybrid population and species B. As the remnant genes of species B combine with species A genes they are eliminated by selection from the hybrid population. There is no longer a source to replace dwindling species B genes, and the hybrid population would be swamped by species A.

Narrowness also has been implied to be a necessary condition for stability in the dynamic-equilibrium model. Bigelow (1965) posed the question: "If selection can maintain effective reproductive isolation for several thousand generations despite interbreeding in a zone of contact, one might ask how selection could fail to perfect ethological or (other) mechanisms to prevent interbreeding in that zone of contact" (p. 454). Later he answers this question as follows: "Actually, the evolution of mechanisms that prevent interbreeding is not likely to occur in a narrow hybrid zone. The 'zones of contact' cited by Mayr and others usually involve but a small fraction of the total range of the component populations. In every case the vast bulk of each population lives *outside* [Bigelow's italics] the zone of contact and is derived from a long line of ancestors that have seen neither the zone nor an individual from the population beyond the zone. . . . The evolution of mechanisms to inhibit interbreeding appears to have taken place as a direct result of selection *in a narrow hybrid zone of contact*

[Bigelow's italics] rather infrequently" (p. 454). Jackson (1973) echoed this view when he interpreted the hybrid zone in iguanid lizards: "Continuance of hybridization between *undulatus* and *woodi* probably results from the very sharp boundary between the two plant associations. This sharpness allows such a small part of each species population to be sympatric with the other that gene flow to the ecotone swamps any tendency toward development of ethological reproductive isolating mechanisms" (p. 67). This also seems to be the implication of the following statement from Watson's (1972) abstract: "The presumed antiquity of the contact, and the narrowness of the zone suggest a stable situation" (referring to the *Litoria* hybrid zone).

The postulate that stable hybrid zones must be narrow actually supports the position I advocate, namely, that the dynamic-equilibrium model does not explain the observed variation in breadth of some of the zones cited above. Nevertheless, I am uncertain that narrowness is, in fact, a necessary condition for stability in this model. Suppose that a hybrid zone has gained considerable breadth relative to the dispersal capabilities of the species involved. "Naive" migrants from allopatric portions of the parental populations would rarely reach the center of the hybrid zone. Presumably in each generation of random mating, however, a few nearly pure parental genotypes would segregate in this central hybrid swarm. These would be strongly favored by selection, particularly if they were also fortuitously endowed with a complement of genes that caused them to mate with similar genotypes, thus reducing the number of maladapted hybrid offspring produced. Since the central portion of the zone would be effectively isolated from "naive" migrants, an antihybridization mechanism might "crystallize" there. Crosby (1970) actually simulated a comparable situation. Crosby's plant "subspecies" were defined by eight independent loci; in addition, the two "subspecies" were identically polymorphic at three loci that determined flowering time and two loci that determined duration of flowering. Hybrid genotypes were selected against in proportion to their admixture of genes. When the hybrid swarm was initiated, the identity of the parental phenotypes was rapidly lost. By generation 44, however, bimodality was again apparent, and by generation 80 "speciation" had occurred. A marked divergence in flowering time between "species" concomitant with reductions in flowering duration in both species effectively prevented miscegenation by generation 80. In theory, then, resegregation of parental genotypes should occur in a hybrid swarm or, equivalently, near the center of a broad hybrid zone, and it follows that narrowness is, indeed, a necessary condition to prevent the evolution of reproductive isolation in a zone where hybrids are less fit than parental phenotypes. Whether these events would occur in reality is uncertain. It would depend upon the probability of parental, or near parental, genotypes resegregating. This, in turn, would depend on population size, the number of loci that distinguish the parental genotypes, and the number of loci that determine an antihybridization mechanism. Data of these sorts are yet to be collected.

To summarize these predictions of the dynamic-equilibrium hypothesis: it is clear that a hybrid zone in contact with one parental species would be unstable, and it seems likely that a broad hybrid zone would be unstable also. Furthermore, if Crosby's (1970) result is generally true, speciation would be expected in a hybrid swarm if the hybrids were maladapted.

A comparison of the predictions of the dynamic-equilibrium hypothesis with the list of apparently stable hybrid zones cited above reveals several discrepancies. The hybrid populations of leopard lizards, *Crotaphytus wislizenii* × *C. silus*, apparently are in contact only with *C. silus* at some locales, and with neither parental species at others. Thus, the persistence of these hybrids does not depend upon an influx of genes from both parental populations. Furthermore, not all of these hybrid zones are uniformly narrow. For example, the zone of hybrid mice in Denmark flares in width along the west coast of Jutland; the zone of grackles is as much as 320 km wide in the southern Appalachians (Huntington, 1952) although it is much narrower in Louisiana (Yang and Selander, 1968). The narrow hybrid zone between the anurans *Geocrinia laevis* and *G. victoriana* also has a feature that is not adequately explained by the dynamic-equilibrium hypothesis. Littlejohn and Watson (1973) reported a salient protrusion of the range of *G. laevis* into that of *G. victoriana* near the northeastern end of this hybrid zone. The protrusion is sufficiently narrow that gene flow from the clearly allopatric portion of the *G. laevis* population to the tip of the protrusion ought to be very limited. Since the flow of genes to the distal region of the protrusion would not be expected to be of sufficient magnitude to disrupt the evolution of an antihybridization mechanism, the dynamic-equilibrium model would predict that reproductive isolation would be more highly developed at the tip of the protrusion than in other parts of the hybrid zone where the parental populations interface along a broad front. Littlejohn and Watson (1973), however, reported that hybridization at the tip of the protrusion was not discernibly different from that occurring elsewhere in the zone.

A final piece of evidence that casts doubt on the validity of the dynamic-equilibrium hypothesis also comes from Crosby's (1970) simulation studies. Crosby simulated the essential circumstances of the dynamic-equilibrium hypothesis (see Crosby, 1970, pp. 288-289, *Centrifugal* gene flow), and the outcome was "speciation" rather than stabilization of the hybrid zone. Again, there is some doubt as to whether Crosby's result can be generalized, but it does raise the possibility that the postulates of the dynamic-equilibrium hypothesis do not lead to an equilibrium at all, or if they do, the equilibrium is an unstable one.

Implicit in the hypothesis that hybrids can be more fit than parental phenotypes is the postulate that ecological factors are primary in determining the fitness of these hybrids. Accordingly, the breadth of a zone is determined by the geographical range of ecological conditions to which the hybrid is adapted — or more plausibly, to which the parental phenotypes are less adapted. The complex array of hybrid populations of towhees in Mexico is readily accommodated by the hypothesis that ecological factors determine the relative fitness of hybrid phenotypes. At some locales, *Pipilo erythrophthalmus* and *P. ocai* are sympatric and do not hybridize. Elsewhere, however, partially isolated populations comprised entirely of intermediates are known. At still other locales step clines in hybrid index occur (Sibley, 1954; Sibley and West, 1958; Sibley and Sibley, 1964).

Although the hybrid-superiority model can account for irregularities in the structure of secondary intergradation zones, the fact remains that most of these zones are narrow and occur at the interface of the ranges of parental populations. To survive, the hybrid-superiority hypothesis must account for this fact. A simple explanation is suggested by the zoogeography of primary fitness values in *P. monacha-occidentalis*, viz., hybrids are better adapted than parental species in narrow geographical regions. This explanation, however, only raises the question: why would hybrids be more fit in such restricted areas?

A plausible answer is suggested by the botanical literature on hybridization and, again, by the unisexual vertebrates. Systematic botanists have long recognized that natural plant hybrids are often restricted to man-disturbed environments; i.e., they are weeds in an ecological sense (e.g., see Grant, 1971, Chap. 11). Wright and Lowe (1968) borrowed this concept and the term "weed" habitat to describe collectively the habitats that support parthenogenetic species of *Cnemidophorus*. Under this rubric they included ecotones, disclimax communities, marginal habitats, and perpetually disturbed habitats. The common property of these habitats is that there is no bisexual analogue with which the parthenospecies must compete. Now, there is substantial evidence that the parthenogenetic species of *Cnemidophorus* have hybrid phenotypes (Lowe and Wright, 1966; Neaves and Gerald, 1968; Fritts, 1969; Neaves, 1969; McKinney, Kay, and Anderson, 1973; Parker and Selander, 1976). So the question is: are parthenogenetic whiptail lizards relatively well adapted to "weed" habitats because they are parthenogenetic, or because they are hybrids?

Most of the stable hybrid zones appear to occur in ecological conditions which conform to Wright and Lowe's (1968) definition of a "weed" habitat (see below). Since both sexual and asexual hybrids share a common type of habitat, it is probable that hybridity and not asexuality is the adaptation to these habitats. Particularly relevant is the frequent ecotonal occurrence of unisexual *Cnemidophorus* as opposed to bisexual species which occupy more stable widespread habitats. For example, the bisexual species *C. inornatus* occurs in grassland communities whereas the unisexual *C. uniparens* occurs in desert-grassland ecotones.

Most of the better documented narrow secondary intergradation zones occur in ecotones also. The two subspecies of grackles, *Quiscalus quiscula quiscula* and *Q. q. versicolor*, have diverged ecologically. In Louisiana, the more northerly *Q. q. versicolor* inhabits pine forest and mixed pine-hardwood forest where *Q. q. quiscula* inhabits cypress-tupelogum swamp and coastal marshes (Yang and Selander, 1968). The narrow zone of hybridization between these subspecies occurs along the interface of the two vegetational zones. The zone is 24 km wide where the two associations are in contact, but it is 64 km wide where they are separated by disturbed bottomland forest.

The hybrids of Audubon's Warbler and the Myrtle Warbler occur in ". . . the meeting and mingling place of the northern boreal forests, and the western, or cordilleran forest" in Alberta and British Columbia (Hubbard, 1969). The antiquity and narrowness of this zone imply that selection gradients inhibit gene flow through the zone. Although the precise nature of these gradients is obscure, the Myrtle Warbler is favored in boreal forest while Audubon's Warbler is favored in cordilleran forest (Hubbard, 1969).

The Eastern Tufted and Western Black-crested Titmouse hybrid zone occurs in a narrow ecotone between an eastern deciduous forest assemblage and a more xeric woody assemblage including live oak, juniper, and mesquite. The abrupt floristic and faunistic transition is correlated with a humidity gradient and with changes in edaphic conditions (Dixon, 1955).

*Bufo fowleri* and *B. woodhousei* are parapatric and hybridize freely in a narrow south-north zone slightly to the east of that involving titmice in Texas. *B. fowleri* and *B. woodhousei* are ecologically quite distinct. *B. fowleri* has a proclivity for forested regions and sandy soil; *B. woodhousei* occurs in a wider range of habitats including blackland prairie and oak-hickory forest, but not pine-oak forest which is inhabited by *B. fowleri*. The hybrid populations occur primarily in the narrow belt of transitional oak-hickory forest sandwiched between pine-oak forest to the east and blackland prairie to the west (Meacham, 1962).

Zweifel (1962) noted that the narrow secondary intergradation zone between *Cnemidophorus tigris gracilis* and *C. t. marmoratus* is correlated with the specific and subspecific range limits of several forms with which these two lizards are geographically and ecologically associated. Lowe (1955) described the general vicinity of these hybrid populations as a complex ecotone between the Sonoran Desert and more eastern biotas.

Montanucci (1970) described the occurrence of natural hybrids between *Crotaphytus wislizenii* and *C. silus* in a narrow ecotone between the California Steppe and Great Basin pinyon-juniper woodland and noted that the distribution of hybrids closely coincides with the limits of the ecotone.

Several subspecific pairs of woodland birds hybridize on the Great Plains. Three pairs which are of particular interest here are the orioles, *Icterus galbula galbula* and *I. g. bullockii* (Sibley and Short, 1964; Rising, 1970, 1973); the flickers, *Colaptes auratus auratus* and *C. a. cafer* (Short, 1965); and the towhees, *Pipilo erythrophthalmus erythrophthalmus* and *P. e. arcticus* (Sibley and West, 1959). The congruence in position and width of the secondary intergradation zones in orioles and flickers is remarkable (see maps in Sibley and Short, 1964, p. 131; Short, 1965, p. 325). Particularly striking is the similarity of plots of the respective hybrid indices along the Platte River transect (Sibley and Short, 1964, p. 138; Short, 1965, p. 327). Rising (1970) noted that hybrid index in the orioles is correlated with precipitation and suggested that the western *I. g. bullockii* is adapted to a warmer, drier environment than is the eastern *I. g. galbula*. The potential vegetational map of Kansas (Kuechler, 1974) shows a transition from bluestem-grama prairie to northern grama-buffalograss prairie where the oriole and flicker hybrid zones cross Kansas, but a correlation between the center trace of these zones and a vegetational transition is not apparent in northern Nebraska and South Dakota (Kuechler, 1964). Whether a grassland ecotone could be a causative factor in the occurrence of hybridization in woodland birds is doubtful. Sibley and West (1959), however, noted that the riparian woodland becomes narrower and drier, cottonwoods more scattered, and the understory thinner, west of the junction of the North and South Platte rivers. It may be significant that this is where the sharp transition in hybrid indices begins for both orioles and flickers.

The center trace of the secondary intergradation zone in towhees is somewhat more to the east than those of the orioles and flickers, particularly along the Niobrara River where the western flora extends further to the east. The distributions of the Spotted (*Pipilo eryth-rophthalmus arcticus*) and the Unspotted (*P. e. erythrophthalmus*) Towhees appear to be determined by vegetational patterns. Both races require a dense woody undergrowth, but the Spotted Towhee usually occupies a chaparral formation of woody shrubs without an arboreal cover as is more typical of the western plains, whereas the Unspotted Towhee occupies the understory shrubbery of the eastern deciduous forest. Sibley and West (1959) suggested that the spotting pattern affords camouflage in chaparral habitats where the absence of an arboreal canopy allows sunlight to dapple in low dense shrubbery.

It is likely that all three zones on the Great Plains resulted from secondary contacts (see Short, 1965, pp. 407-411, for a discussion of Pleistocene climatic conditions on the Plains) and that these zones originated with the recession of the Wisconsin Glaciers. Although European man has altered the availability of habitat for orioles and flickers (Sibley and Short, 1964; Short, 1965; Anderson, 1971), Rising (1970) has argued that the oriole zone is stable. Although introgression may be more extensive in the flickers, the zone within which birds are undoubtedly hybrids is little, if any, wider than the oriole zone. In view of the antiquity of the flicker zone, it is likely that it is stable also. The shrubby undergrowth inhabited by towhees has not been increased significantly by man (Sibley and West, 1959). If man's activities have not increased the opportunity for hybridization in towhees, this zone is also probably stable.

To summarize the situation on the Great Plains: It is likely that at least three subspecific pairs of birds hybridize in stable secondary intergradation zones. The occurrence of the zones appears to correlate with a change in climatic conditions, viz., precipitation. It is also possible that this region represents an interface between distinct floristic assemblages, i.e., an ecotone. In any case, it seems possible that hybrids are more fit than parental phenotypes in these zones.

Perhaps the sharpest ecotone supporting a stable hybrid zone is that between sand-pine and sandhill plant associations where the iguanid lizard hybrid, *Sceloporus woodi* × *S. undulatus undulatus*, occurs in Florida (Jackson, 1973). The antiquity of the *Sceloporus* hybrid zone is intriguing. It is likely that these species have been hybridizing for 100,000 years, and yet there is no evidence of reproductive isolation nor extensive introgression into the parental populations. Obviously selection gradients must inhibit introgression, but an explanation for the failure of a premating reproductive isolating mechanism to evolve is not obvious. It is possible that the two species lack the genetic basis for the evolution of a premating reproductive isolating mechanism, but this seems very unlikely since such mechanisms are known in congeneric species (Jackson, 1973). The dynamic-equilibrium hypothesis could explain the situation in *Sceloporus* and cannot be discounted. The equally viable and more parsimonious hypothesis that I advocate is that the hybrids of *Sceloporus* are not inferior in the ecotonal areas, as was presumed by Jackson (1973).

Littlejohn, Watson, and Loftus-Hills (1971) and Littlejohn and Watson (1973) described the region in Victoria where *Geocrinia laevis* and *G. victoriana* hybridize as a subtle ecotone. The hybrid zone is not correlated with any "man-made ecotone," but it does seem to be correlated with a steep precipitation gradient and with the periphery of the pristine range of a forest since obliterated by the activities of European man. Littlejohn and Watson (1976) subsequently described this hybrid zone as occurring near the ecotone between dry sclerophyll eucalypt forest and open grassland of the basaltic plains of southwestern Victoria.

Another anuran hybrid zone, this one involving *Litoria ewingi* and *L. paraewingi*, occurs approximately 300 km northeast of the *Geocrinia* zone (Watson, Loftus-Hills, and Littlejohn, 1971; Watson, 1972; Gartside, 1972; Littlejohn, 1976). The *Litoria* zone is clearly associated with a forest-grassland ecotone. When the considerable data on this zone are synthesized, however, the picture that emerges does not clearly favor one hybrid-zone hypothesis over the other two. Watson (1972) thought it probable that the *Litoria* zone was established at the end of the last period of Pleistocene glaciation (about 12,000 years B. P.). Genetic incompatibilities between the parental species are manifest in developmental aberrations when *L. ewingi* females are crossed to *L. paraewingi* males, but offspring from the reciprocal cross develop normally (Watson, Loftus-Hills, and Littlejohn, 1971). Surprisingly, there is no evidence that mating call has diverged between these species in terms of characteristics which are known to effect

mate discrimination within the *Litoria ewingi* complex (Littlejohn, 1976), nor do females of either species demonstrate any ability to discriminate between the mating calls of allospecific and conspecific males (Watson, Loftus-Hills, and Littlejohn, 1971). Littlejohn (1976) did discern a slight displacement in scatter plots of two call parameters when near-allopatric and distant-allopatric males were compared. The displacement was slight, however, and did not involve the critical parameter of pulse repetition rate which effects mate discrimination in at least some congeneric species.

The *Litoria* hybrid zone, more than any other, does satisfy the dynamic-equilibrium model: it is narrow, it is probably stable, and there is clear evidence of fitness loss in hybrids. Furthermore, both developmental (Watson, 1972) and mating-call (Littlejohn, 1976) studies indicate the presence of significant numbers of the parental species in the hybrid zone. The latter fact is important because a postulate of the dynamic-equilibrium model is that there are enough "naive" individuals moving into the hybrid zone to prevent crystallization of an antihybridization mechanism. The dynamic-equilibrium model is unclear with regard to the association of narrow hybrid zones with ecotones. Bigelow (1965, p. 452) recognized that coadaptation of a species gene pool is, at least to a slight extent, forged by an intricate pattern of ecological factors. It is possible, then, that distinct ecological communities on either side of a hybrid zone act as the selection gradients which are postulated to prevent fusion in the dynamic-equilibrium model. In this model, a disturbed habitat is perhaps a sufficient condition for the establishment of a narrow hybrid zone; but it is not a necessary condition, since a disruption of physiological homeostasis could also serve as the selection gradient against gene flow.

In the case of the *Litoria* hybrid zone, the observed fitness loss is physiological in nature, and therefore it is unlikely that this has caused the zone to occur in an ecotone. It is possible that the association of this zone with an ecotone is coincidental, but it is also plausible that the hybrids are more fit than the parental phenotypes in the ecotone despite some developmental aberrations (Littlejohn, 1976). For example, a very large portion of *Poeciliopsis monacha-occidentalis* eggs go unfertilized in Rio

Mayo populations, and yet this hybrid is immensely successful there (Moore, 1976). High survivorship of *P. monacha-occidentalis* in the Rio Mayo is apparently more than enough to compensate for the large number of unfertilized eggs. By analogy, it is plausible that high survivorship beyond metamorphosis in *Litoria* hybrids compensates for a high level of larval mortality. Finally, since the observed developmental abnormalities manifest themselves only in crosses between *L. ewingi* females and *L. paraewingi* males, it is possible that they are caused by a cytoplasmic reaction and that hybrids suffer no fitness loss once this initial hindrance is cleared.

The most thorough genetical analysis of a hybrid zone is that of *Mus musculus musculus* × *M. m. domesticus* in Denmark (Hunt and Selander, 1973). This zone traverses central Europe from south to north and extends across Jutland. *M. m. musculus* and *M. m. domesticus* have diverged morphologically to the extent that they probably would be classified as distinct species by typological criteria. Allozyme analysis shows a genetic similarity index comparable to sibling species of other rodents. Allele frequencies at six of seven enzymatic loci studied formed step clines centered along or near the same line. Hunt and Selander (1973) concluded, as had Ursin (1952), that the zone lies in a region where climatic gradients, particularly in precipitation, create environments to which *M. m. domesticus* and *M. m. musculus* are equally well adapted. They also concluded that selection pressures resulting from disruption of the internal genetic environment must be a factor because the major shift in allele frequencies for all loci studied occurs along the same line.

The latter conclusion is perplexing because there is no evidence of reproductive isolation or assortative mating between the two subspecies. Coadaptation of gene complexes means that alleles which substitute at a given locus interact with alleles which substitute at other loci in such a way as to produce fit phenotypes. It is impossible to have coadaptation in gene pools of parental forms without the production of less fit phenotypes by hybridization. It is difficult, therefore, to reconcile the postulate that *M. m. domesticus* and *M. m. musculus* have gene complexes coadapted to distinct internal genetic environments with the facts that the hybrid zone is old and a premating isolating mechanism has not evolved.

Alternatively, the congruent step clines could be caused by a change in a common environmental variable. To test this hypothesis, Hunt and Selander (1973) did a step-wise multiple linear regression of hybrid index for a limited sample taken from the center of the hybrid zone on several climatic and geographical variables. The only good predictor of hybrid index was latitude, although July precipitation seemed weakly associated with hybrid index also. In spite of no clear correlation between hybrid index and environmental variables in the limited sampling region, there is evidence that the positions and slopes of these step clines are environmentally determined. First, the clines are invariably more gradual and have a more northerly center along the western edge of Jutland. Second, the allele frequencies on the islands of Fyn and Als are invariably more similar to the corresponding frequencies on northeastern Jutland than on more proximal southern Jutland. Precipitation levels on these islands are comparable to that of northeastern Jutland but less than those on nearby southern Jutland. Soil type on Als and Fyn also resembles that of northeastern Jutland. When all of Jutland and the adjacent islands where data were collected are considered, hybrid index does appear to be related to precipitation, although the relationship is not linear.

To summarize the situation in *Mus: M. musculus musculus* and *M. m. domesticus* have diverged to the level of distinct species by morphological and enzymatic criteria. Natural hybrids occur in a narrow stable hybrid zone in Jutland. Overall, hybrid index appears to correlate with precipitation and soil type, although a linear correlation analysis on a limited sample from the area where allele frequency clines were steepest did not detect a good environmental predictor of hybrid index. No correlation analysis was made between hybrid index and biotic variables; so, it is unknown whether this region is an ecotone.

### DISCUSSION

Exploring the ramifications of coadaptation in a gene pool requires that we distinguish coadaptation to an endogenous regime of selection pressures from coadaptation to an exogenous regime of selection pressures. Minimally, the alleles at all loci in a fit organism must interact in such a way as to produce a physiologically homeostatic individual capable of reproduction. In nature, a homeostatic phenotype must also be successful in the ranges of physical parameters it encounters (e.g., temperature, humidity, soil type) as well as in its biotic interactions in an ecological community if it is to survive to reproduce. The latter set of demands constitutes an exogenous regime of selection pressures, whereas the demands for physiological homeostasis are endogenous. When an allele enters a natural population it is not only selected against if it perturbs physiological homeostasis but also if it renders its carriers less successful in terms of intraspecific and interspecific competition, predator avoidance, pathogen resistance, and the like.

When hybridization ensues secondary contact, a large infusion of new alleles into the parental populations is initiated. If the gene complexes of the parental populations were distinctly coadapted to endogenous regimes of selection pressures, hybird offspring would be selected against and the evolution of an antihybridization mechanism would be expected regardless of ecological conditions. On the other hand, if the gene complexes of the parental phenotypes were coadapted to distinct exogenous regimes of selection pressures, the outcome of hybridization would depend upon ecological conditions. If hybridization occurred in a zone of marginal habitat for both parental phenotypes, the hybrids would not have to overcome rigorous competition from parental phenotypes and they could, therefore, persist if physiologically homeostatic.

I suggest that the several evidently stable hybrid zones cited and discussed above are the result of coadaptation of parental gene pools to distinct exogenous selection regimes. Introgression is limited and fusion prevented by ecological selection gradients. Antihybridization mechanisms do not evolve because the hybrids are physiologically homeostatic and because they occur in regions where the parental phenotypes are no better adapted than the hybrid phenotypes. The narrowness common in stable hybrid zones is attributable to the fact that the hybrids often occur in ecotones which are themselves narrow. Hybrids are probably not preadapted to these zones but, rather, their success depends on the parental forms being less adapted. When geographical barriers divide the gene pool of a species it is obvious that it remains but one species in a whole ecological

community divided into isolates. It is thought that during prolonged isolation coadaptation of specific gene pools occurs. It is logical to recognize that whole communities must undergo divergence and coadaptation also. Because all species interact in a community through food webs, competition, patterns of protective coloration, commensalism, mutualism, and the like, a simple change in species structure should alter the regimes of selection pressures operating on many species in the community. Consequently, allele frequencies which are sensitive to biotic selection pressures would adjust in many species in the community. As an interpretation of vertebrate hybrid zones this is a novel idea; however, it is far from original. Grant (1971, p. 158) summarized the belief held by many botanists regarding the occurrence of natural plant hybrids as follows:

The explanation of the correlation between hybridization and habitat disturbance which is favored by most students is an extension of one proposed by Kerner in 1891. In a closed stable community no habitat is available for such hybrid zygotes as are formed from time to time. Stabilizing selection eliminates them from the scene almost as soon as they arise. But, where the natural community has been broken into by road building, overgrazing, or the like, so that new open habitats are created, the hybrids can and do become established. In other words, environmental isolation operates to suppress hybridization between intercompatible species in a stable, closed community, but ceases to be fully effective in an open habitat.

The subspecific occurrence of spotting patterns in towhees is a simple example of adaptation to distinct communities. If the spotting pattern is camouflaging, as Sibley and West (1959) suggested, then hybrid phenotypes would be more susceptible to predation in either deciduous forest or chaparral than the parental phenotypes, regardless of the vigor or fertility of the hybrid. In an ecotone, however, the intermediate phenotype may be no less cryptic than the parental phenotypes and would not be selected against.

*Sceloporus woodi* and *S. undulatus undulatus* occupy similar niches but in distinct ecological communities. The two have diverged in morphological characteristics that are apparent adaptations to their respective communities. Although speculative, alternatives to the hypothesis that hybrids of *Sceloporus* are inferior should be considered. For example, it is possible that hybrids, because of their intermediacy, are able to compete moderately well in either community whereas the more specialized parental phenotypes compete well in their own community but poorly in the other. The success of a parental phenotype at the ecotone depends upon finding habitat available in its specific community. The hybrid, in contrast, might be an opportunist utilizing whatever habitat is available at least moderately well. Presumably, a large number of genes contribute to the ecologically adaptive morphology of both the towhees and *Sceloporus*. The important point is that a regime of ecological and not physiological selection pressures enforces the integrity of this kind of adaptive gene complex.

To cast this analysis in a final perspective, it is useful to recall the method of deductive testing described by Popper (1959). If a theory withstands a scrutiny " . . . then the theory has, for the time being passed its test: we have found no reason to discard it. But if the decision is negative or in other words, if the conclusions have been falsified, then their falsification also falsifies the theory from which they were logically deduced" (p. 33). The hybrid-superiority hypothesis has been subjected to less scrutiny than the other hypotheses, and this may well account for its vigor at this juncture. An a priori assumption of most hybrid-zone studies is that the hybrids are inferior to parental phenotypes; as a consequence, the kinds of data that could falsify the hybrid-superiority hypothesis have usually not been sought. The hybrid-superiority hypothesis does make predictions that are testable. The prediction which, if invalidated, would directly falsify this hypothesis is that hybrids should be more fit than parental phenotypes in stable hybrid zones. Predictions that would less directly test the hypothesis but are more testable from a practical standpoint include the following: (1) all stable hybrid zones should be restricted to regions devoid of, or at least at the periphery of, stable ecological communities; (2) mating should be random in stable hybrid zones; (3) there should be no character displacement associated with a stable hybrid zone, particularly in features that could effect reproductive isolation (such as mating call). In addition, the genetic structure of the population through the hybrid zone would probably

manifest some distinctive features; however, these are difficult to deduce without the aid of a mathematical model. Mathematical modelling would be very useful at this juncture to generate predictions about genetic structure, and, also, to test the internal consistency of the hypothesis. That is, would a geographical region of hybrid superiority suffice to cause the stabilization of a hybrid population between the ranges of the parental populations, given some reasonable assumptions about the genetics of the parental differences and of a potential isolating mechanism?

Comprehensive studies of individual hybrid zones thought to be stable could test the "mettle" of the hybrid-superiority hypothesis and further test the other two. These studies should include comparative data on survivorship, fecundity, fertility, mating behavior, gene frequencies, ecology, and zoogeography. Several recent studies are commendable in this respect. But, these data would be of little consequence if they were not used to choose between plausible hypotheses. Popper (1959) expressed part of his method in a metaphor borrowed from the late 18th and early 19th century philosopher Novalis. Popper: "Theories are nets cast to catch what we call the 'world': to rationalize, to explain and to master it." Novalis' germinal thought seems an apropos terminus to this discussion: "Hypotheses are nets: only he who casts will catch."

## SUMMARY

Three alternative hypotheses have been proposed to explain the occurrence of vertebrate hybrid zones: (1) the ephemeral-zone hypothesis states that hybridization following secondary contact will lead to either speciation or fusion of the populations. Speciation is to be expected if the populations have diverged to the extent that the hybrids are less fit than the parental phenotypes; otherwise, the hybrids would serve as a bridge for introgressive hybridization. (2) The second hypothesis accounts for the evident stability of some hybrid zones despite selection against hybrids. Bigelow (1965) proposed that stable narrow hybrid zones might result from a dynamic balance between gene flow into the hybrid zone and selection against hybrids. He suggested that steep selection gradients might inhibit introgression and that the evolution of an antihybridization mechanism in the restricted zone of contact might be disrupted by migrants moving into the contact zone from the more extensive areas of allopatry. (3) The third hypothesis states that hybrids are actually more fit than the parental phenotypes in the restricted regions where hybrid zones occur. The latter hypothesis has not gained strong advocacy because it is inconsistent with the concepts that hybridization disrupts coadapted gene complexes and that the unity of a species is maintained by gene flow.

Neither of the first two hypotheses explains the full range of observed hybrid-zone phenomena. The failure of the ephemeral zone model is that there are several zones which are old and evidently stable. The dynamic-equilibrium model requires that the hybrid zone be sandwiched between the parental populations, or so situated that genes flow from both parental populations into the hybrid zone. Since some apparently stable hybrid zones do not satisfy these conditions, the dynamic-equilibrium model does not adequately explain all stable hybrid zones.

The geography of fitness in the unisexual hybrid fish *Poeciliopsis monacha-occidentalis* supports the hypothesis that hybrids are in some circumstances more fit than parental phenotypes. Furthermore, to discard the presumption that hybrids are invariably less fit allows one to explain readily much of the otherwise inexplicable data on hybrid zones.

Although the hybrid-superiority model can account for a stable hybrid zone of almost any shape or size, the fact remains that most stable hybrid zones are narrow. The nearly universal association of stable hybrid zones with ecotones suggests that ecological factors are often important in determining the success of hybrids. It is suggested here that coadaptation within a species gene pool can be a response to two regimes of selection pressures, one endogenous, the other exogenous. Hybrids between races that have gene complexes distinctly coadapted as responses to endogenous selection regimes are likely to be selected against in any environment because disruption of the coadapted gene complexes would perturb physiological homeostasis. In contrast, hybrids between races that have gene complexes distinctly coadapted as responses to exogenous but not endogenous selection regimes might survive in nature provided they did not have to compete in a stable community with species that were well adapted to that community.

Ecotones probably provide marginal habitats for many of the species that comprise the ecological communities on either side of the ecotone. Homologous species or races which have diverged in response to the peculiar exogenous selection regimes exerted by the respective communities may produce hybrids which are physiologically homeostatic and are no less adapted to the transitional habitat than are the parental phenotypes. In this circumstance, a stable hybrid population would be established in the ecotone. Selection gradients exerted by the distinctly integrated ecological communities on either side of the ecotone would prevent expansion of the hybrid zone; and reproductive isolation would not evolve because, where the opportunity to hybridize occurs, there is no selection against hybrids.

ACKNOWLEDGMENTS

I would like to thank the following individuals for critically reading one or more drafts of this paper: A. Bradley Eisenbrey; Gerard R. Joswiak, Leo S. Luckinbill, William L. Thompson, and Thomas Uzzell. I would also like to acknowledge two anonymous reviewers whose criticisms guided a revision of the manuscript. I recognize that the people who have reviewed this paper disagree to varying extents with its arguments and conclusions; therefore, I assume sole responsibility for its content. This research was supported in part by NSF Grant 74-19894.

## LIST OF LITERATURE

ANDERSON, B. W. 1971. Man's influence on hybridization in two avian species in South Dakota. *Condor,* 73: 342-347.

ANDERSON, E. 1949. *Introgressive Hybridization.* John Wiley, New York.

BIGELOW, R. S. 1965. Hybrid zones and reproductive isolation. *Evolution,* 19: 449-458.

CROSBY, J. L. 1970. The evolution of genetic discontinuity: Computer models of the selection of barriers to interbreeding between subspecies. *Heredity,* 25: 253-297.

DESSAUER, H. C., W. FOX, and F. H. POUGH. 1962. Starch-gel electrophoresis of transferrins, esterases and other plasma proteins of hybrids between two subspecies of whiptail lizard (*Cnemidophorus*). *Copeia*, 1962: 767-774.

DIXON, K. L. 1955. An ecological analysis of the interbreeding of crested titmice in Texas. *Univ. Calif. Publ. Zool.*, 54: 125-206.

DOBZHANSKY, T. 1940. Speciation as a stage in evolutionary divergence. *Am. Nat.* 74: 312-321.

FRITTS, T. H. 1969. The systematics of the parthenogenetic lizards of the *Cnemidophorus cozumela* complex. *Copeia*, 1969: 519-535.

GARTSIDE, D. F. 1972. The *Litoria ewingi* complex (Anura: Hylidae) in south-eastern Australia. III. Blood protein variation across a narrow hybrid zone between *L. ewingi* and *L. paraewingi. Aust. J. Zool.*, 20: 435-443.

GRANT, V. 1971. *Plant Speciation.* Columbia Univ. Press, New York.

HAFFER, J. 1967. Speciation in Colombian forest birds west of the Andes. *Am. Mus. Novit.*, No. 2294: 1-57.

HAGEN, D. W. 1967. Isolating mechanisms in threespine sticklebacks (*Gasterosteus*). *J. Fish. Res. Bd. Canada*, 24: 1637-1692.

HALL, W. P., and R. K. SELANDER. 1973. Hybridization of karyotypically differentiated populations in the *Sceloporus grammicus* complex (Iguanidae). *Evolution*, 27: 226-242.

HUBBARD, J. P. 1969. The relationships and evolution of the *Dendroica coronata* complex. *Auk*, 86: 393-432.

HUNT, W. G., and R. K. SELANDER. 1973. Biochemical genetics of hybridisation in European house mice. *Heredity*, 31: 11-33.

HUNTINGTON, C. E. 1952. Hybridization in the Purple Grackle, *Quiscalus quiscula. Syst. Zool.*, 1: 149-170.

JACKSON, J. F. 1973. The phenetics and ecology of a narrow hybrid zone. *Evolution*, 27: 58-68.

KUECHLER, A. W. 1964. Potential natural vegetation of the conterminous United States. Am. Geogr. Soc. Spec. Publ. no. 36; New York.

———. 1974. A new vegetation map of Kansas. *Ecology*, 55: 586-604.

LITTLEJOHN, M. J. 1976. The *Litoria ewingi* complex (Anura: Hylidae) in south-eastern Australia IV. Variation in mating-call structure across a narrow hybrid zone between *L. ewingi* and *L. paraewingi. Aust. J. Zool.*, 24: 283-293.

LITTLEJOHN, M. J., and G. F. WATSON. 1973. Mating-call variation across a narrow hybrid zone between *Crinia laevis* and *C. victoriana* (Anura: Leptodactylidae). *Aust. J. Zool.*, 21:277-284.

———, and ———. 1974. Mating call discrimination and phonotaxis by females of the *Crinia laevis* complex (Anura: Leptodactylidae). *Copeia*, 1974: 171-175.

———, and ———. 1976. Effectiveness of a hybrid mating call in eliciting Phonotaxis by females of the *Geocrinia laevis* complex (Anura: Leptodactylidae). *Copeia*, 1976: 76-79.

LITTLEJOHN, M. J., G. F. WATSON, and J. J. LOFTUS-HILLS. 1971. Contact hybridization in the *Crinia laevis* complex (Anura: Leptodactylidae). *Aust. J. Zool.*, 19: 85-100.

LOWE, C. H. 1955. The eastern limit of the Sonoran Desert in the United States with additions to the known herptofauna of New Mexico. *Ecology*, 36: 343-345.

LOWE, C. H., and J. W. WRIGHT. 1966. Evolution of parthenogenetic species of *Cnemidophorus* (whiptail lizards) in western North America. *J. Ariz. Acad. Science*, 4: 81-87.

MAYR, E. 1963. *Animal Species and Evolution.* Belknap Press, Cambridge.

McKINNEY, C. O., F. R. KAY, and R. A. ANDERSON. 1973. A new all-female species of the genus *Cnemidophorus. Herpetologica*, 29: 361-366.

MEACHAM, W. R. 1962. Factors affecting secondary intergradation between two allopatric populations in the *Bufo woodhousei* complex. *Am. Midl. Nat.*, 67: 282-304.

MONTANUCCI, R. R. 1970. Analysis of hybridization between *Crotaphytus wislizenii* and *Crotaphytus silus* (Saura: Iguanidae) in California. *Copeia*, 1970: 104-123.

MOORE, W. S. 1976. Components of fitness in the unisexual fish *Poeciliopsis monacha-occidentalis. Evolution*, 30: 564-578.

MULLER, C. H. 1952. Ecological control of hybridization in *Quercus:* A factor in the mechanism of evolution. *Evolution*, 6: 147-161.

NEAVES, W. B. 1969. Gene dosage at the lactate dehydrogenase b locus in triploid and diploid lizards (*Cnemidophorus*). *Science*, 164: 557-559.

NEAVES, W. B. and P. S. GERALD. 1968. Lactate dehydrogenase isozymes in parthenogenetic teiid lizards (*Cnemidophorus*). *Science*, 160: 1004-1005.

PARKER, E. D., and R. K. SELANDER. 1976. The organization of diversity in the parthenogenetic lizard *Cnemidophorus tesselatus. Genetics*, 84: 791-805.

POPPER, K. R. 1959. *The Logic of Scientific Discovery.* Harper and Row, New York.

REMINGTON, C. L. 1968. Suture-zones of hybrid interaction between recently joined biotas. In T. Dobzhansky, M. K. Hecht, and W. C. Steere (eds.), *Evolutionary Biology*, Vol. 2, p. 321-428. Appleton-Century-Crofts, New York.

RISING, J. D. 1970. Morphological variation and evolution in some North American Orioles. *Syst. Zool.*, 19: 315-351.

———. 1973. Morphological variation and status of the orioles, *Icterus galbula, I. bullockii*, and *I. abeillei*, in the northern Great Plains and in Durango, Mexico. *Can. J. Zool.*, 51: 1267-1273.

SCHULTZ, R. J. 1969. Hybridization, unisexuality, and polyploidy in the teleost *Poeciliopsis* (Poeciliidae) and other vertebrates. *Am. Nat.*, 103: 605-619.

———. 1971. Special adaptive problems associated with unisexual fishes. *Am. Zool.*, 11: 351-360.

———. 1973. Origin and synthesis of a unisexual fish. In J. H. Schroeder (ed.), *Genetics and Mutagenesis of Fish*, p. 207-211. Springer-Verlag, Berlin.

SELANDER, R. K. 1965. Hybridization of Rufous-naped Wrens in Chiapas, Mexico. *Auk*, 82: 206-214.

SHORT, L. L. 1965. Hybridization in the flickers (*Colaptes*) of North America. *Bull. Am. Mus. Nat. Hist.*, 129: 307-428.

———. 1969. Taxonomic aspects of avian hybridization. *Auk*, 86: 84-105.

———. 1970. A reply to Uzzell and Ashmole. *Syst. Zool.*, 19: 199-202.

———. 1972. Hybridization, taxonomy and avian evolution. *Ann. Missouri Bot. Gard.*, 59: 447-453.

SIBLEY, C. G. 1954. Hybridization in the red-eyed towhees of Mexico. *Evolution*, 8: 252-290.

———. 1957. The evolutionary and taxonomic significance of sexual dimorphism and hybridization in birds. *Condor*, 59: 166-191.

SIBLEY, C. G., and L. L. SHORT. 1964. Hybridization in the orioles of the Great Plains. *Condor*, 66: 130-150.

SIBLEY, C. G., and F. C. SIBLEY. 1964. Hybridization in the red-eyed towhees of Mexico: the populations of the southeastern plateau region. *Auk*, 81: 479-504.

SIBLEY, C. G., and D. A. WEST. 1958. Hybridization in the red-eyed towhees of Mexico: the eastern plateau populations. *Condor*, 60: 85-104.

———, and ———. 1959. Hybridization in the rufous-sided towhees of the Great Plains. *Auk*, 76: 326-338.

THAELER, C. S. 1968. An analysis of three hybrid populations of pocket gophers (genus *Thomomys*). *Evolution*, 22: 543-555.

THIBAULT, R. E. 1974. The ecology of unisexual and bisexual fishes of the genus *Poeciliopsis:* A study in niche relationships. Ph.D. Dissertation, Univ. of Conn.

URSIN, E. 1952. Occurrence of voles, mice, and rats (Muridae) in Denmark, with a special note on a zone of intergradation between two subspecies of the House Mouse (*Mus musculus* L.). *Vid. Medd. Dansk Naturhist. Foren.*, 114: 217-244.

WALLACE, B., and M. VETUKHIV. 1955. Adaptive organization of the gene pools of Drosophila populations. *Cold Spring Harbor Symp. Quant. Biol.*, 20: 303-309.

WATSON, G. F. 1972. The *Litoria ewingi* complex (Anura: Hylidae) in south-eastern Australia. II. Genetic incompatibility and delimitation of a narrow hybrid zone between *L. ewingi* and *L. paraewingi*. *Aust. J. Zool.*, 20: 423-433.

WATSON, G. F., J. J. LOFTUS-HILLS, and M. J. LITTLEJOHN. 1971. The *Litoria ewingi* complex in south-eastern Australia. I. A new species from Victoria. *Aust. J. Zool.*, 19: 401-416.

WILSON, E. O. 1965. The challenge from related species. In H. G. Baker and G. L. Stebbins (eds.), *The Genetics of Colonizing Species*, p. 7-25. Academic Press, New York.

WRIGHT, J. W., and C. H. LOWE. 1968. Weeds, polyploids, and the geographical and ecological distribution of all-female species of *Cnemidophorus*. *Copeia*, 1968: 128-138.

YANG, S. Y., and R. K. SELANDER. 1968. Hybridization in the Grackle *Quiscalus quiscula* in Louisiana. *Syst. Zool.*, 17: 107-143.

ZWEIFEL, R. G. 1962. Analysis of hybridization between two subspecies of the desert Whiptail Lizard, *Cnemidophorus tigris*. *Copeia*, 1962: 749-766.

## 17

Copyright © 1973 by the Genetical Society of Great Britain

Reprinted from *Heredity* 31:11–33 (1973)

# BIOCHEMICAL GENETICS OF HYBRIDISATION IN EUROPEAN HOUSE MICE

W. GRAINGER HUNT and ROBERT K. SELANDER

*Department of Zoology, University of Texas, Austin, Texas 78712, U.S.A.*

Received 27.vii.72

## SUMMARY

Techniques for demonstrating allozymic variation in seven enzymes (four esterases, isocitrate dehydrogenase, malic enzyme, and malate dehydrogenase) were employed to study genic variation in a narrow zone of hybridisation between allopatric semispecies of the house mouse (*Mus musculus musculus* and *M. m. domesticus*) on the Jutland Peninsula of Denmark. Material consisted of 2696 mice collected at 152 farms representing 44 sample areas on the peninsula and adjacent islands. The history of movements of early farming cultures with which mice were associated as commensals suggests that *musculus* and *domesticus* have been in contact and hybridising in northern Europe since 3000 B.C. The zone in Jutland and another in Germany lie in regions transitional between Atlantic and continental climates, the two parental forms meeting where they are equally well adapted to ecological conditions.

The zone in Jutland has not shifted since Ursin defined its position in 1952 on the basis of morphological characters. An analysis of genotypic proportions in populations in the zone of hybridisation failed to demonstrate assortative mating, thus supporting laboratory evidence of free interbreeding between the semispecies. Genic heterozygosity levels are " normal " on the large islands of Falster, Fyn, and Als but reduced on the small islands of Alrø and Hjaelm, presumably through the founder effect or genetic drift. The zone is strongly asymmetrical north to south, with extensive introgression of *domesticus* alleles into *musculus*, but little introgression in the other direction. A marked increase in width of the zone in western Jutland is associated with a more extensive gradient of environmental factors, particularly precipitation. In the narrow eastern part of the zone, 90 per cent. of the transition in genetic character (as measured by a hybrid index) occurs over a distance of 20 km. The extent of introgression varies markedly among loci. Linkage between the loci studied is not a major factor affecting patterns of introgression. The extreme steepness of the gradient of transition in genetic character and the occurrence of major changes in frequencies at all loci along the same line are cited as evidence that the selective values of alleles are determined in part by the internal genetic environment. The "new" genetic environment created by introgression of *domesticus* alleles into *musculus* populations apparently favours the occurrence of minor alleles at the *Es-2* and *Es-3* loci. The failure of the *musculus* and *domesticus* gene pools to fuse despite long-standing hybridisation argues that genetic isolation cannot be equated with reproductive isolation. Selection against introgression of the genes studied (or the chromosomal segments that they mark) is presumed to involve reduced fitness in backcross generations caused by disruption of co-adapted parental gene complexes.

## 1. INTRODUCTION

THE remarkable narrowness and apparent temporal stability of many zones of allopatric hybridisation present an intriguing problem in evolutionary genetics. Mayr (1963, p. 378) has suggested that there is in such zones " a vigorous selection against the infiltration of genes from one balanced gene complex into the other, but without the development of any isolating mechanisms as a by-product of this selection ". Restriction of gene flow is thus attributed to an interdependence of genes and a consequent genetic imbalance between differentiated gene pools. While Huxley (1943), Mayr (1963), Bigelow (1965) and Ford (1971) maintain that zones of allopatric hybridisation often reach stable equilibrium, with introgression balanced by selection, Wilson (1965) and Remington (1968) believe that they are generally transitory, being eliminated with the perfection of reproductive isolating mechanisms or fusion of the parental gene pools. All hypotheses invoking hybrid inferiority predict that introgression involves something more than a simple diffusion of alleles (Mayr, 1963).

The present study of variation in allele frequencies at seven gene loci through a narrow zone of hybridisation between two allopatric forms of the house mouse (*Mus musculus musculus* and *M. m. domesticus*) on the Jutland Peninsula of Denmark was motivated by the following questions: (1) Is introgression occurring beyond the geographic limits detected in an earlier analysis of morphology (Ursin, 1952)? (2) What environmental, historical, or other factors are important in determining the position and width of the zone? (3) Do patterns of introgression vary among loci, and is introgression symmetrical? (4) Do patterns of introgression provide evidence relevant to the current controversy regarding the selective neutrality or non-neutrality of allozymic variants?

## 2. Systematic and ecological background

### (i) *Differences between the parental forms*

European house mice were assigned to two subspecies groups by Schwarz and Schwarz (1943), the *spicilegus* group represented by *M. m. musculus* and *M. m. spicilegus*, and the *wagneri* group including *M. m. domesticus* and *M. m. brevirostris* (fig. 1). *Domesticus* and *musculus* hybridise on the Jutland Peninsula and on a front extending from Kiel Bay on the Baltic Sea south through Germany (Zimmermann, 1949). The positions of these disjunct zones roughly correspond to those of similar zones of hybridisation involving the carrion and hooded crows (*Corvus c. corone* and *C. c. corax*), as described by Meise (1928).

Morphological and ecological differences between *musculus* and *domesticus* were discussed by Zimmermann (1949) and Ursin (1952). *Musculus* is a short-tailed, light-bellied mouse with a brownish yellow lateral line, whereas *domesticus* is long-tailed and dark-bellied, with no obvious discontinuity of colour on the sides. *Domesticus* is slightly larger than *musculus* and has more tail rings.

Schwarz and Schwarz (1943, pp. 70-72) maintained that *domesticus* is a specialised commensal, while *musculus* has not progressed beyond " a primitive stage of commensalism ". But Ursin's (1952) and our field work in Denmark indicates that *musculus* is fully as successful as *domesticus* in exploiting farm buildings and other man-made " habitats ". Zimmermann (1949) reported that *musculus* is more resistant to cold and has a greater tendency to winter outdoors, but populations of either form rarely overwinter in fields in Denmark (Ursin, 1952). The long tail of *domesticus* was regarded by Schwarz and Schwarz as an adaptation for commensalism, but Zimmermann questioned this interpretation by demonstrating that variation in tail length in Europe follows Allen's rule.

In a study of the behaviour of laboratory-raised individuals of both forms and their $F_1$ hybrids, Brubaker (1970) found that *domesticus* showed greater wheel-running, swimming and " open field " activity, while *musculus* had a greater tendency to hoard. We have noted that wild-caught *domesticus* struggle vigorously and bite when handled, whereas *musculus* are relatively docile and infrequently bite. *Musculus* is also more active than *domesticus* in the daytime.

*Musculus* and *domesticus* differ more in appearance and behaviour than the average run of subspecies designated by taxonomists, and they are quite distinct genetically. From an analysis of allozymic variation in Jutland

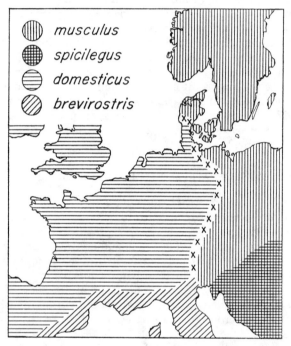

*musculus*
*spicilegus*
*domesticus*
*brevirostris*

Fig. 1.—European distribution of four subspecies of *Mus musculus*. Zones of allopatric hybridisation between *musculus* and *domesticus* are indicated by crosses.

populations of *musculus* and *domesticus*, Selander, Hunt and Yang (1969) derived a coefficient of genetic similarty of 0·79, a value close to those obtained for sibling species of the rodents *Sigmodon* and *Peromyscus* (Selander and Johnson, 1973). Considering the relatively strong differentiation of *musculus* and *domesticus* and, more importantly, the apparent isolation of their gene pools despite hybridisation (see Discussion), it is realistic to consider them semispecies rather than subspecies.

### (ii) *Establishment of the zone of hybridisation*

Zimmermann (1949) suggested that *musculus* was derived from *spicilegus* and reached the Baltic Sea area during the post-glacial optimum, when several other dry-land vertebrates now represented in northern Europe by relic populations ranged north to the Baltic Sea. (House mice have been reported from an early post-glacial forest fauna in central Germany by

Brunner, 1941.) Following its establishment in northern Europe, *musculus* is presumed to have changed from an entirely aboriginal to a partially commensal form and subsequently to have accompanied the spread of grain-farming into north-central France, establishing itself over much of northern Europe by 4200 B.C.

*Domesticus* is believed to have reached Europe later than *musculus*, arriving as a commensal associated with the neolithic grain culture that spread from North Africa into Spain and southern France (Zimmermann, 1949). By 3000 B.C. the long-isolated farming traditions of the western Mediterranean and northern and eastern Europe made contact, and at this time it is likely that *domesticus* came to occupy Britain, northern France, and western Germany. The probable position of the zone of contact of these cultures, as described by Waterbolk (1968), lies several hundred miles west of the present boundary between *domesticus* and *musculus*. Because the European distribution of *musculus* corresponds to the drier " continental " climate, and that of *domesticus* to the wetter " Atlantic " climate, the shift in the zone of contact may have been caused by an eastward movement of the climatic boundary since 3000 B.C. Alternatively, *domesticus* simply may have replaced populations of *musculus* that earlier had occupied the eastern margin of the Atlantic climatic region. In any event, it is likely that interbreeding occurred when the two forms first met and that a zone of hybridisation has existed in some form in central-northern Europe for several thousand years.

The present disjunction of the northern Jutland and German populations of *musculus* presumably was established as the zone of contact moved eastward, thus permitting *domesticus* to occupy the southern Jutland Peninsula. This invasion did not displace *musculus* from the larger islands in the western Baltic Sea, including Als and Fyn.

### (iii) *The environment of Jutland*

The Jutland Peninsula is 435 km. long and varies in width from 50 km. at the German border to 178 km. in the central region. The topography is principally the result of Quaternary glaciation, and particularly the last glacial period, when ice sheets covered the northern and eastern parts (Davies, 1944). Ice emanating from Scandinavia retreated earlier and more rapidly than that which had spread westward from the Baltic. Consequently, eastern Jutland has a greater and more varied relief than northern Jutland because of the longer period available for the accumulation of morainic deposits. In northern Jutland the soil is mainly sandy loam of relatively low fertility, and the terrain is bare and windswept. The landscape of eastern Jutland is characterised by numerous valleys and rounded hills (up to 172 m. in elevation), and soils are mainly clay loam and sandy loam. Because glacial drainage flowed westward, western Jutland is a flat, sandy outwash plain with scattered low sandhills formed in the second glacial period. In geology and landscape, the Danish islands are similar to eastern Jutland.

Because of the small size and gentle relief of Jutland, there are no large rivers. Many small lakes occur on the peninsula, and there is a large estuary, the Ringkøbing Fjord, in central-eastern Jutland. The northern and central regions are separated by the Limfjord, a waterway extending from the North Sea to the Baltic Sea.

The climate of Jutland and adjacent islands is fairly uniform. However, mean yearly precipitation decreases from south to north, and the transition occurs much more gradually in western than in eastern Jutland (fig. 2). The lowest temperatures occur inland in the central and northern half of the peninsula, but the actual range of geographic variation in temperature is very small.

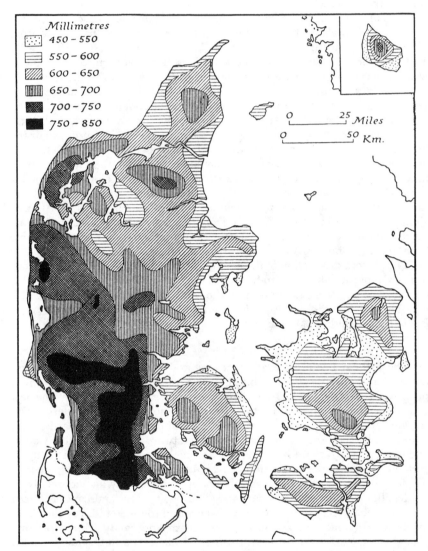

*Millimetres*

450 – 550

550 – 600

600 – 650

650 – 700

700 – 750

750 – 850

FIG. 2.—Total yearly precipitation in Denmark. From Davies (1944) after Danske Meteorol. Inst. (1933, plate 40).

The distribution of native vegetation and agricultural crops (predominantly cereals) corresponds fairly well to that of soil types. Eastern Jutland and the islands have been under cultivation for centuries, with forest largely confined to the higher land and steeper slopes. Before 1800 large regions of sandy soil in western and central Jutland were covered by a native heath. Although farming has increased in the west and north in the last 100 years, the density of towns and farms has remained much lower in these regions than in the east. Agriculture is now very uniform throughout Denmark; approximately four-fifths of the land area is agricultural, and of this more than 80 per cent. is arable (Davies, 1944).

## 3. MATERIALS AND METHODS

### (i) *Samples*

A total of 2696 house mice was collected at 152 farms on the Jutland Peninsula and on five islands east of the peninsula from August 1968 to February 1969 (fig. 3). Because attempts to collect mice in grainfields were generally unsuccessful, almost all mice were live-trapped in farm buildings or captured when grain ricks were threshed. All were adults or independent subadults.

Studies of allozymic variation in house mouse populations in North America have demonstrated marked heterogeneity in allele frequencies among barns on the same farm and among farms in local regions (Selander and Yang, 1969; Selander, 1970), resulting from the polygynous, tribal and territorial social system of the species (Anderson and Hill, 1965; Reimer and Petras, 1967; DeFries and McClearn, 1972). To reduce the confusing effect of local differentiation in our analysis of geographic variation, the farms were grouped into 44 sample areas (fig. 3), and estimates of allele frequencies for sample areas were based on pooled samples from all farms. With three exceptions (insular sample areas 42, 43 and 44), each sample area includes two or more farms. (A complete list of farms and numbers of mice collected is provided by Hunt [1970].) Farms were coded alphabetically within sample areas. Sample areas on the mainland were numbered along three south-north transects as follows: eastern, 1-16; central, 17-30; and western, 31-39 (fig. 3).

### (ii) *Allozymic analysis*

In an earlier study (Selander, Hunt and Yang, 1969), allozymic variation in 36 proteins controlled by 41 loci was assessed in two samples of *domesticus* from southern Jutland and four samples of *musculus* from northern Jutland and islands east of the peninsula. At 13 of the 17 variable loci there were marked differences in allele frequencies between the semispecies; and at several loci alternate alleles were fixed or nearly so in northern and southern populations. On the basis of this survey seven polymorphic enzymes (each encoded by a separate locus) were selected for detailed analysis: four esterases (*Es-1*, *Es-2*, *Es-3* and *Es-5*), supernatant isocitrate dehydrogenase (*Idh-1*), supernatant malic enzyme (*Me-1*) (*Mdh-1* of Henderson [1966], Selander, Hunt and Yang [1969], and others; and *Mod-1* of Shows, Chapman and Ruddle [1970]), and mitochondrial malate dehydrogenase (*Mdh-2*) (*Mor-1* of Shows, Chapman and Ruddle [1970]).

Techniques of tissue preparation, electrophoresis and protein staining are described by Selander and Yang (1969). Because of the expense of staining for IDH, ME and MDH, the numbers of mice typed for these enzymes were reduced for most areas represented by samples exceeding 30. Mean sample sizes per area were 60 for esterases and 28 for other enzymes.

The distribution of allozymic variants in the house mouse has been studied in natural populations in North America (Petras *et al.*, 1969; Ruddle *et al.*,

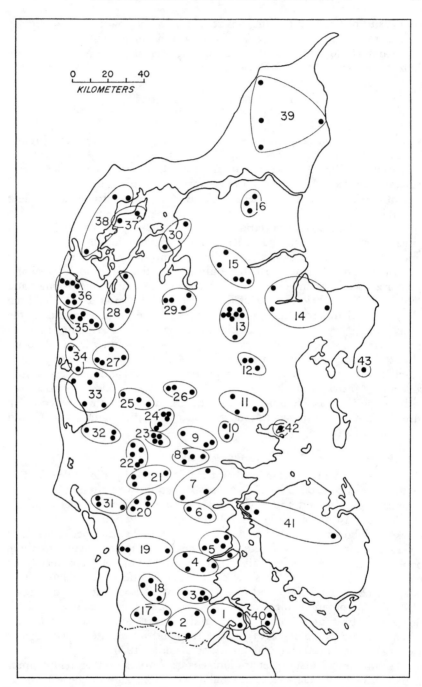

FIG. 3.—Sample localities for specimens of *Mus musculus*. Samples from individual farms (dots) are grouped into numbered sample areas, as indicated. The insular populations are as follows: 40, Als; 41, Fyn; 42, Alrø; and 43, Hjaelm. Falster Island (44) lies east of the area shown.

1969; Selander and Yang, 1969; Selander, Yang and Hunt, 1969), the Hawaiian Islands (Wheeler and Selander, 1972), parts of Britain (Berry and Murphy, 1970), and Jutland (Selander, Hunt and Yang, 1969). The distribution of alleles in inbred strains was summarised by Roderick *et al.* (1971) and Taylor (1972). The literature on breeding experiments demonstrating the generally codominant inheritance of allozymic variants is cited in these papers and in those mentioned in the following paragraph. A genetic analysis of variants in Jutland populations was presented by Wheeler (1972).

The linkage relationships of the loci considered are as follows: *Es-1*, *Es-2* and *Es-5* are in linkage group XVIII (Petras and Biddle, 1967; Popp, 1967; Ruddle, Shows and Roderick, 1969; Wheeler and Selander, 1972). Average recombination frequencies from several studies are 0·100 for *Es-1* and *Es-2*, 0·080 for *Es-1* and *Es-5*, and 0·009 for *Es-2* and *Es-5*, suggesting the linear arrangement *Es-1*, *Es-5*, *Es-2*, with only *Es-5* and *Es-2* tightly linked. *Es-3* is in group VII (Roderick, Hutton and Ruddle, 1970); *Idh-1* is in group XIII (Hutton and Roderick, 1970; Chapman, Ruddle and Roderick, 1971); and *Me-1* is in group II (Shows, Chapman and Ruddle, 1970). The linkage group for *Mdh-2* has not been determined.

## 4. RESULTS

### (i) *Polymorphic variation*

Alleles are of two types, " major " alleles in moderate or high frequency in one or both semispecies and " minor " alleles in low frequencies with generally localised distribution. Geographic variation in frequencies of major alleles is shown in figs. 4 to 7. (Genotypic and allelic frequencies for all samples are presented by Hunt [1970].)

### (a) *Esterase-1*

Populations of *musculus* in northern Jutland are monomorphic for the *Es-1^a* allele, and those of *domesticus* near the German border are monomorphic for *Es-1^b* (fig. 4). Introgression is more or less symmetrical north to south but is more extensive in western than in eastern Jutland. The distance between the 0·10 and 0·90 isofrequency lines for the *Es-1^a* allele is 100 km. in the west but only 30 km. in the east. On Als Island (sample area 40), *Es-1^a* has a frequency of 0·04, presumably reflecting introgression from the adjacent mainland, but other insular populations (Fyn, 41; Alrø, 42; Hjaelm, 43) are monomorphic for *Es-1^b*.

A minor null allele, *Es-1^c*, occurs at a frequency of 0·18 in sample area 39 in northern Jutland.

### (b) *Esterase-2*

Six alleles are represented, two of which, *Es-2^b* and *Es-2^c*, are major. Shown in fig. 4 are the adjusted (relative) frequencies of the major alleles, calculated by eliminating minor alleles. *Es-2^b* and *Es-2^c* are essentially fixed in *domesticus* and *musculus*, respectively. Introgression is asymmetrical north to south: *Es-2^b* penetrates to the Limfjord, but there is little introgression of *Es-2^c* into southern populations. Again, the zone of transition in allele frequencies is broader in the west than in the east; distances between the 0·10 and 0·90 isofrequency lines for *Es-2^c* are 120 and 45 km. respectively.

Populations on Als and Fyn islands are slightly introgressed from the mainland.

*Es-2$^a$*, a null allele, was detected in homozygous state in five sample areas in north-central Jutland. This allele may also occur in low frequency elsewhere on the peninsula, but, because it is undetectable in heterozygous condition, we could not determine its frequency directly. Maximum likelihood methods of estimating frequencies of null alleles from heterozygote

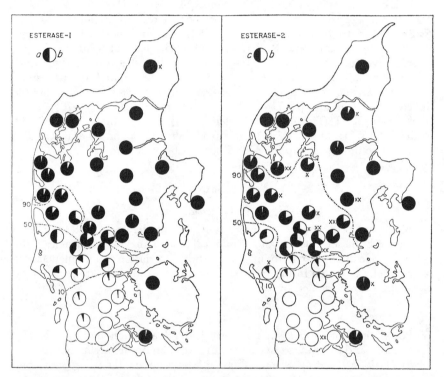

FIG. 4.—Geographic variation in frequency of major alleles at the *Es-1* and *Es-2* loci. Isofrequency lines for the *Es-1$^a$* and *Es-2$^c$* alleles are indicated. The occurrence of minor alleles is indicated by an X.

deficiencies in samples were considered inappropriate because of complications of the Wahlund effect (Selander, Yang and Hunt, 1969). *Es-2$^e$* occurs in low frequency in eight sample areas, *Es-2$^f$* in five, and *Es-2$^g$* in two.

(c) *Esterase-3*

Populations in 43 of the 44 areas sampled are polymorphic for *Es-3$^b$* and *Es-3$^c$* (fig. 5). There is no conspicuous geographic variation in allele frequencies, but *Es-3$^b$* occurs in slightly higher frequency, on the average, in the southern and western parts of the peninsula than in the north-eastern region occupied by relatively " pure " populations of *musculus*. This pattern is consistent with evidence from other loci of a more extensive introgression in the west than in the east. The frequency of *Es-3$^b$* is surprisingly low (0·09) in sample area 1, being only slightly higher than that (0·06) on the adjacent Als Island.

A null allele, *Es-3^e*, was recorded in three sample areas, and *Es-3^f* was found in four areas.

### (d) *Esterase-5*

Phenotypic variation in esterase-5 involves the presence or absence of a band of enzymatic activity. Presence results from a dominant allele, *Es-5^b*, while absence reflects homozygosity of the null *Es-5^a* allele (Petras and Biddle, 1967). *Es-5^a* is fixed or nearly so in *musculus*, but *domesticus* is polymorphic, with *Es-5^b* in frequencies of approximately 0·20 to 0·30. In fig. 5, variation at the *Es-5* locus is expressed in terms of the frequency of the homozygous

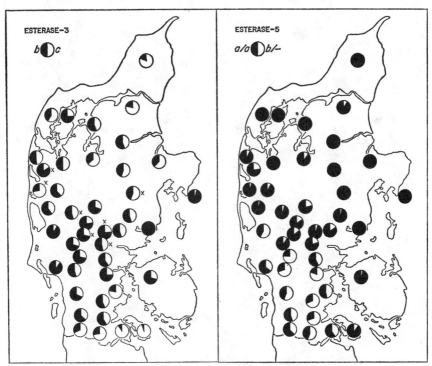

FIG. 5.—Geographic variation in frequency of major alleles at the *Es-3* locus and of genotypes at the *Es-5* locus. The occurrence of minor *Es-3* alleles is indicated by an X.

*Es-5^a* phenotype. The introgressive pattern at the *Es-5* locus is similar to that at *Es-2*. *Es-5^b* penetrates far into *musculus*, particularly in the west, but the relative uniformity of frequencies in southern populations indicates that little introgression is occurring from *musculus* to *domesticus*. Populations on Als and Fyn islands apparently are slightly introgressed from the mainland.

### (e) *Isocitrate dehydrogenase-1*

*Idh-1^a* and *Idh-1^b* are, with few exceptions, alternately fixed in populations of *domesticus* and *musculus*, respectively (fig. 6). Introgression is symmetrical north to south, as in the case of *Es-1*, but asymmetrical east to west. If the presence of *Idh-1^b* in southern Jutland is in fact the result of introgression, this is the only locus at which a *musculus* allele has penetrated all the way to the German border. Introgression of *Idh-1^a* from *domesticus* also appears unusually extensive, with this allele being represented in areas beyond the Limfjord (38 and 39). The presence of *Idh-1ᵇ* at a frequency of 0·12 in sample area 1 suggests gene flow from Als Island, where populations appear also to be introgressed from the mainland. *Idh-1ᵇ* is fixed in other insular populations.

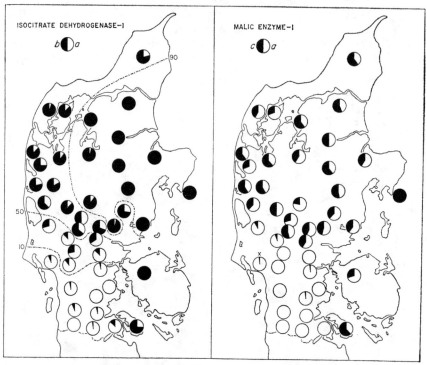

Fig. 6.—Geographic variation in frequency of major alleles at the *Idh-1* and *Me-1* loci. The occurrence of the minor allele *Me-1ᵈ* is indicated by an X.

### (f) *Malic enzyme-1*

*Domesticus* is monomorphic for *Me-1ᵃ*, while *musculus* is polymorphic for *Me-1ᵃ* and *Me-1ᶜ* (fig. 6). The transition in central Jutland is very sharp, and there is little evidence of introgression of *Me-1ᵃ* into northern populations, even on the western side of the peninsula. Some penetration of *Me-1ᶜ* into southern populations is apparent, however.

A third allele, *Me-1ᵈ*, was found in sample area 31 in south-western Jutland.

### (g) *Malate dehydrogenase-2*

The pattern of variation at *Mdh-2* is similar to that at *Me-1* in that *domesticus* is monomorphic (*Mdh-2ᵇ*) and *musculus* is polymorphic for two alleles (*Mdh-2ᵇ* and *Mdh-2ᶜ*). Introgression from *musculus* into *domesticus* is not apparent, but the relatively high frequency of *Mdh-2ᵇ* in north-western Jutland presumably reflects extensive introgression from the south (fig. 7). A third allele, *Mdh-2ᵃ*, was found in two sample areas in northern Jutland.

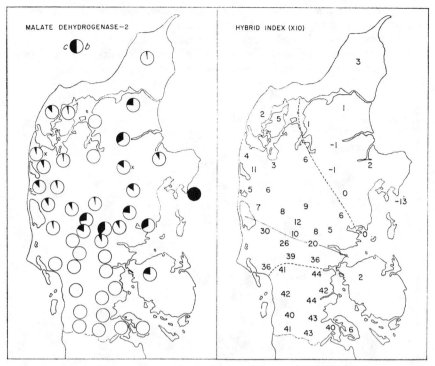

FIG. 7.—Left: Geographic variation in frequency of major alleles at the *Mdh-2* locus. The occurrence of the minor allele *Mdh-2ª* is indicated by an X. Right: Geographic variation in hybrid index score ($I \times 10$) based on six polymorphic loci. Dashed lines indicate limits of introgression, as reflected by $I$, and the continuous line indicates the centre of the zone of hybridisation.

## (ii) *Hybrid index*

To measure overall genetic character, a hybrid index score ($I$) for each sample area was computed as follows:

$$I = f_{Es-1b} + f_{Es-2b} + f_{Idh-1a} + (1 - f_{Es-5a/Es-5a}) + (f_{Me-1a} - 0.487)$$
$$+ (f_{Mdh-2b} - 0.818).$$

Frequencies of alleles are adjusted ones, and $f_{Es-5a/Es-5a}$ is the frequency of the null phenotype. The value 0·487 is the mean frequency of *Me-1ª* in sample areas 12 through 15, representing minimally introgressed *musculus*, and 0·818 is the mean frequency of the *Mdh-2b* allele in these areas. The index is designed to yield highest scores ($\sim 4.2$) for " pure " populations of *domesticus* (2, 3, 17 and 18) and lowest scores ($\sim 0.0$) for those of *musculus* (12 to 15). *Es-3* was excluded because it provides little information on introgression.

The centre of the zone of hybridisation can be represented as the 2·1 isofrequency line, which lies between sample areas 7 and 8 in eastern Jutland, between 22 and 23 in the centre of the peninsula, and between 32 and 33 in the west (fig. 7). North-south and east-west introgressive asymmetries are readily apparent.

To illustrate the transition from *musculus* to *domesticus* in greater resolution, we have in fig. 8 shown hybrid index values for individual farms in central Jutland for which relatively large samples are available. Most of the transition in the eastern region occurs over a distance of less than 20 km. Interfarm heterogeneity within sample areas also is demonstrated in fig. 8.

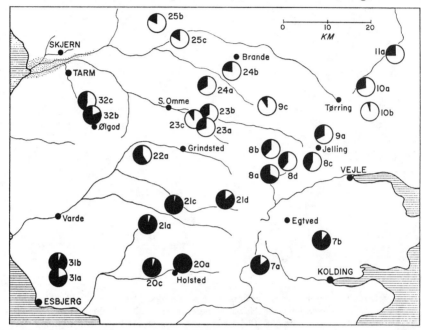

Fig. 8.—Interfarm variation in overall genic character at six polymorphic loci in central Jutland. Each circle represents a farm. Black areas of circles are proportional to hybrid index values, expressed as percentages of 4·2, the average value for samples of "pure" *domesticus*. Marshland between Skjern and Tarm is indicated by stippling.

### (iii) *Degree of interbreeding*

Evidence from the field and laboratory indicates that, if interbreeding between *musculus* and *domesticus* is not entirely free, at least there is no strong tendency for assortative mating. In samples from individual farms within the zone we have not detected evidence of selective mating, the proportions of genotypes in samples from individual farms being similar to those expected in panmictic populations (data in Hunt, 1970; see also table 2).

### 5. Discussion

#### (i) *Comparison with Ursin's earlier analysis*

The centre of the zone of hybridisation defined by Ursin's (1952) study of pelage colour and skin and skeletal dimensions extended from Fredericia (near Vejle) west through Grindsted to a point between Tarm and Skjern (see localities in fig. 8). That this is also the line defined by our analysis indicates that the centre has not shifted in the 20 years since Ursin's study.

Because the tail and hindfoot were slightly shorter in *musculus* populations north of the Limfjord than in those south of this channel, Ursin concluded that " weak gene-flow " from *domesticus* could be traced north to the Limfjord, an inference confirmed by our demonstration of introgression in the western part of the peninsula at least as far north as Mors Island (37) in the Limfjord (fig. 7). (In the case of *Idh-1*, introgression may extend to the northern tip of the peninsula.) Ursin's data suggested a north-south asymmetry, for he could not detect with certainty introgressive effects south of the Vejle-Grindsted-Skjern line. We have shown that introgression is relatively limited south of the centre of the zone (fig. 7), especially in the east, although at *Es-1* and *Idh-1* it can be traced almost to the German border. The east-west asymmetry apparent from our study was also suggested by Ursin's data.

Populations on islands east of the Jutland Peninsula were assigned to *musculus* by Ursin, with the comment that they are similar to those north of the Limfjord. Our sample from Falster, a large island 120 km. east of the peninsula, is " pure " *musculus* ($I = -0.53$). Similarly, the Fyn population is not appreciably introgressed ($I = 0.22$), but the Als population ($I = 0.55$) shows slight introgression at most loci. The clearest indication of gene flow from Als to the adjacent mainland is provided by the relatively high frequency of *Idh-1^b* in sample area 1 (fig. 6).

Because the narrow channels separating Fyn and Als from the mainland are bridged by highways along which there is a heavy traffic of agricultural products in which mice could be carried, the severe restriction of gene flow is surprising. It may be significant that in climate and soil type Fyn and Als resemble the north-eastern area of Jutland occupied by *musculus* rather than the adjacent mainland inhabited by *domesticus* (see fig. 2).

Genic heterozygosity levels are " normal " on the large islands of Falster, Fyn, and Als. On the small island of Alrø, *Es-3*, a locus polymorphic in all mainland populations, is monomorphic (fig. 5). Hjaelm Island is occupied by a small population presumably founded by a few individuals and dependent for its existence on the lighthouse station located there. *Me-1* and *Mdh-2*, which are polymorphic in *musculus* on the adjacent mainland, have been fixed on Hjaelm (figs. 6 and 7), presumably through the founder effect or genetic drift, but *Es-3* remains weakly polymorphic (fig. 5).

### (ii) *Position and width of the zone*

#### (a) *Position*

The question of whether hybridisation between *musculus* and *domesticus* is stable or transient pertains to both the location and width of the zone. As noted earlier, circumstantial evidence of an eastward shift in the zone from its original position is provided by its present disjunction in Jutland and Germany. The fact that these zones lie in regions of transition between Atlantic and continental climates suggests that climatic factors influence their positions (see also beyond). Since *musculus* and *domesticus* are almost entirely commensal in habit, their distributions and densities also undoubtedly have been influenced by historical changes in agricultural practices and the distribution of human populations. If, as suggested by Ursin (1952), the two forms meet where they are equally well adapted to ecological conditions, temporal variation in these conditions will preclude long-term stability of position of the zones.

How long *musculus* and *domesticus* have been in contact along their present front in central Jutland is unknown, but the western part of the zone cannot have existed in its present form prior to the mid-nineteenth century. Through prehistoric and early Christian times, much of Jutland was covered with oak and beech forest (Davies, 1944). This habitat and the scattered farming communities would have supported populations of *Mus*, but the non-forested sandy outwash plains and marshy areas in the west probably would have been only sparsely and locally inhabited. In eastern Jutland the zone could have existed in much its present form and position for centuries, following the extensive destruction of forests in medieval times, but only tenuous avenues of contact between *musculus* and *domesticus* could have existed in the west prior to the 1850s, when a major programme of reclamation of heathlands and improvement of soil conditions was initiated, permitting an extensive development of agriculture on the sandhill islands and plains. Similarly, prior to the programme of reclamation of marshlands initiated in south-western Jutland in the mid-nineteenth century, human settlement was concentrated on scattered islands of drier land rising from meadows and bogs. The extensive peat bogs of northern Jutland and marshlands adjacent to the Limfjord, now partly reclaimed, also may have affected the distribution of *musculus* in earlier times.

Ursin emphasised the abruptness of the transition between *musculus* and *domesticus*, noting (1952, p. 235) that " populations with conspicuous hybrid characters apparently are met with only within an area the breadth of which is about 50 km. . . . Only within a zone of at most a few kilometres are the hybrid characters so pronounced that it is doubtful to what subspecies the population is most nearly related." The transition in overall genetic character in central Jutland is shown in fig. 9, in which hybrid index scores for sample areas along three transects are plotted against distance from south to north on the peninsula. Along the Vejle transect in the east (sample areas 1 to 16), " pure " *musculus* and " pure " *domesticus* populations are found within 80 km. of one another, and 90 per cent. of the transition in hybrid index occurs over a distance of 20 km. But along the Skjern transect in the west introgression is evident for a distance of at least 50 km. south of the centre of the zone and may extend northward up the coast beyond the Limfjord, a distance of more than 200 km.

Minor variation among loci in position of the mid-line of the zone probably is unimportant, considering the large part played by stochastic processes in determining allele frequencies at individual farms. Recall that we are dealing with disjunct, semi-isolated populations on farms, and that marked genetic heterogeneity exists among farms within small regions, largely as a result of founder effects (Selander, 1970). Founder effects and sampling error are probably also responsible for the few obvious " reversals " in the gradient of character transition detected in our study. As shown in fig. 7, sample area 23 has a lower hybrid index than area 24, owing to an unexpectedly high frequency of the *musculus* alleles *Idh-1$^b$* and *Me-1$^c$* (fig. 6). (Allele frequencies at other loci in area 23 are " properly " intermediate between those of areas 22 and 24.) Another reversal occurs in the west, where the hybrid index for sample area 35 is unexpectedly high (1·1) as a consequence of variation at *Es-2, Es-5, Idh-1* and *Me-1*.

The centre of the zone does not correspond to any sharp natural or man-made ecological discontinuity or to any obvious barrier to dispersal; and at least in the central and western parts of the peninsula it does not correspond to a steepening of gradients of environmental factors. The centre in the east lies near Vejle, where total yearly precipitation decreases northward from 800 to 625 mm. over a distance of 30 km. (figs. 2 and 9). Elevation is also

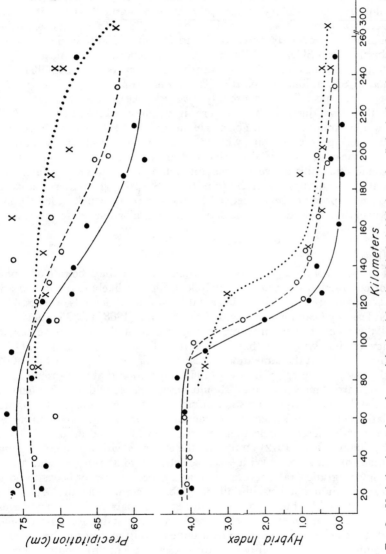

Fig. 9.—Variation in total yearly precipitation (above) and in hybrid index value (below) along three south-to-north transects through the Jutland Peninsula. Dots and solid line, Vejle transect (sample areas 1-16); circles and dashed line, Grindsted transect (17-30); Xs and dotted line, Skjern transect (31-39). Distance is measured in kilometres from the German border.

increasing in this region, so that areas above 100 m. occur just north of the mid-line. In the west, however, environmental correlates with the centre of the zone are less apparent. If precipitation alone were determining the position of the zone, we might expect the mid-line to lie near the south-western border of the Limfjord. Rather, it falls between Skjern and Tarm, where there is no abrupt change in precipitation, elevation, or soil type.

As shown in fig. 8, it is not apparent that the streams in Jutland affect gene flow in the central and eastern parts of the peninsula. For example, the characters of populations north (23b) and south (23a) of the Grindsted River are similar, as previously noted by Ursin (1952). However, the extensive marshes between Skjern and Tarm, where the Skjern and Grindsted rivers flow into the Ringkøbing Fjord, may impede dispersal. Ursin identified *musculus* at Skjern and *domesticus* at Tarm, only about 5 km. distant. Our material from near Tarm is strongly *domesticus*-like in character ($I = 3 \cdot 0$), while that from just north of Skjern is strongly *musculus*-like ($I = 0 \cdot 7$). However, a similar change in $I$ occurs in an equally short distance ($\sim 20$ km.) in the eastern part of the zone, where there are no obvious barriers to migration. In the east, the partial barrier to dispersal formed by the Vejle Fjord and the forested hills between Vejle and Jelling does not seem sufficient to account for the abruptness of the transition. Surprisingly, patterns of variation do not appear to be influenced by the Limfjord, which at least under more primitive conditions would have been a major barrier to dispersal of mice.

(b) *Width*

If the width of the hybrid zone in Jutland is not stable, the rate of expansion or contraction must be extremely low, for it is likely that *musculus* and *domesticus* have been hybridising in cental Europe for thousands of years. For *Mus* at least, we cannot agree with Remington's (1968, p. 375) suggestion that " the amount of natural hybridizing is commonly more a measure of recency of first sympatry than a measure of the amount of genetic dissimilarity . . .". At the same time, we cannot judge whether the width of the zone is slowly changing. Bigelow's (1965) suggestion that in transient zones trends toward " incorporation " should be detectable within a few generations may be an overly optimistic assessment of such situations. Computer simulations of zones of secondary contact or of clines of variation maintained by the opposing forces of selection and gene flow (Jain and Bradshaw, 1966; Crosby, 1969; Cook, 1972) emphasise the slowness with which equilibrium conditions are attained. With weak selection against introgressant alleles and a low migration rate, several thousand generations may be required to reach equilibrium at a locus initially fixed for different alleles in the parental forms. As noted by Fincham (1972) in regard to balanced polymorphisms, " most land environments may not be sufficiently stable for long enough periods for true equilibria ever to be attained ". In any event, the question of absolute or relative stability is in the present case of little consequence in attempting to understand factors determining the extent of introgression at different loci.

A notable feature of the zone is that introgression at most loci is more extensive into *musculus* than into *domesticus*. This type of asymmetry is a common feature of allopatric hybrid zones (Short, 1965; Yang and Selander, 1968). In the case of *Mus* it could reflect one or a combination of conditions, including (1) greater interfarm migration in *musculus* than in *domesticus*, or a net northward dispersal; and (2) lesser intensity of selection against intro-gressant alleles in *musculus* populations than in those of *domesticus*.

Variation in width is also a common feature of allopatric hybrid zones, as early demonstrated by Meise (1928) for *Corvus* in Europe, and in some cases such variation is clearly associated with habitat features (Yang and Selander, 1968). The east-west asymmetry in width of the *Mus* zone corresponds to variation in steepness of environmental gradients, particularly precipitation (fig. 9), but also to elevation and, to some degree, soil type. Introgression of *domesticus* into *musculus* is relatively extensive in the west, where climatic conditions similar to those of southern Jutland extend up the coast even to the north-western part of the peninsula beyond the Limfjord. This feature of variation argues that selection by environmental factors has some part in limiting introgression.

Theoretically, the observed east-west asymmetry in width could be caused by regional differences in rate of migration. For example, a greater density of farms supporting mouse populations might facilitate interfarm migration, thus producing a broader transition in genetic character. But we have no reason to suspect that migration rate is in fact higher in the west than in the east.

The lesser extent of introgression where extensive contact and hybridisa-tion presumably have existed for the longest period is not the result of develop-ment of incipient reproductive isolating " mechanisms " (see theory of Dobzhansky, 1940, 1970), since, as noted earlier, we have no evidence of non-random mating or of sterility barriers in any part of the zone. It is also unlikely that earlier chance establishment of parental populations north and south of the present centre of the zone is responsible for the asymmetry, since the zone is narrowest where there has been the longest period of time avail-able for colonisation.

In an attempt to identify particular environmental variables regulating introgression, we performed a step-wise multiple linear regression of hybrid index values for 17 sample areas near the mid-line of the zone against the following independent variables: altitude, total yearly precipitation, July precipitation, February precipitation, mean yearly temperature, mean July temperature, and mean February temperature (data from Danske Meteorol. Inst., 1933, representing averages for the period 1886-1925). The fraction ($R^2$) of the sum of squares of $I$ attributable to the regression is only 0·50, of which almost half (0·23) is associated with July precipitation. Repeating the analysis with latitude as one of the independent variables, we obtained an $R^2$ of 0·88, of which 0·77 results from the effect of latitude. Hence, con-sidered apart from latitude, with which they are strongly correlated, precipi-tation, temperature, and altitude are not very useful " predictors " of the overall genetic character of populations in the hybrid zone. Either we have not identified all the important environmental variables or there are other factors influencing patterns of introgression.

### (iii) *Differential introgression and selection*

If selection is a factor controlling the extent of introgression, it is, as Mayr (1963) has suggested, unlikely that it will affect all loci with the same inten-sity. Rather, it is to be expected that the progress of some alien alleles will be checked near the centre of the hybrid zone, while the flow of others will proceed varying distances into regions occupied by the parental populations.

Our analysis suggests that the geographic pattern of gene flow is distinctive for each locus. Variation is especially apparent among the three loci (*Es-1*, *Es-2* and *Idh-1*) at which alternate alleles are fixed or nearly so in the two semispecies. *Idh-1*$^a$ has penetrated into *musculus* populations beyond the Limfjord, whereas *Es-1*$^b$ can be detected only as far north as the south-western edge of the Limfjord. Introgression of *Mdh-2*$^b$ to regions north of the Limfjord is also suggested by the pattern of allele frequencies shown in fig. 7.

Introgression of *musculus* alleles into *domesticus* is relatively limited but similarly variable. *Idh-1*$^b$ reaches the German border, and *Es-1*$^a$ goes further south than *Es-2*$^c$. There is a slight south-western introgression of *Es-5*$^a$, but no apparent introgression of *Mdh-2*$^c$.

Although linkage disequilibrium is difficult to assess in *Mus* populations because of their subdivided structure and the occurrence of null alleles at several of the loci, a comparison of allele frequencies at the linked loci *Es-1*,

TABLE 1

*Allele frequency correlations in central Jutland**

|  | Es-1$^b$ | Es-2$^b$ | Es-3$^b$ | Es-5$^a$/$^a$ | Idh-1$^a$ | Me-1$^a$ | Mdh-2$^b$ |
|---|---|---|---|---|---|---|---|
| Es-1$^b$ | 1·00 | 0·86 | 0·39 | −0·70 | 0·88 | 0·73 | 0·36 |
| Es-2$^b$ |  | 1·00 | 0·51 | −0·83 | 0·84 | 0·84 | 0·25 |
| Es-3$^b$ |  |  | 1·00 | −0·51 | 0·52 | 0·60 | −0·08 |
| Es-5$^a$/$^a$ |  |  |  | 1·00 | −0·61 | −0·76 | −0·39 |
| Idh-1$^a$ |  |  |  |  | 1·00 | 0·81 | 0·21 |
| Me-1$^a$ |  |  |  |  |  | 1·00 | 0·32 |
| Mdh-2$^b$ |  |  |  |  |  |  | 1·00 |

* 20 sample areas (7-11, 21-29, and 31-36).

*Es-2* and *Es-5* suggests that linkage between the loci studied is not an important factor affecting patterns of introgression. Shown in table 1 are correlations of allele frequencies between pairs of loci in the primary area in which introgression is occurring (bounded by dashed lines in fig. 7). There are strong correlations for combinations of *Es-1*, *Es-2* and *Es-5*, but correlation is equally strong for the unlinked pairs *Es-1*$^b$—*Idh-1*$^a$, *Es-2*$^b$—*Idh-1*$^a$ and *Es-2*$^b$—*Me-1*$^a$. Near the centre of the hybrid zone individual farm populations are in equilibrium with regard to *Es-1* and *Es-5*, and this is generally so also for *Es-1* and *Es-2* (table 2). But mild disequilibrium between the closely linked *Es-2* and *Es-5* loci is seen at about half the farms.

In the absence of relevant experimental data, our discussion of how selection operates in the hybrid zone is necessarily speculative. Following Ursin (1952), we conclude that the zone lies in a region where gradients of climatic factors, particularly precipitation, create environments to which the semispecies are equally well adapted. The east-west asymmetry in width of the zone reflects variation in extent of the area in which the critical set of intermediate environmental conditions is distributed. Direct selective pressures of the external environment may influence some aspects of allele frequency variation, but it is likely that selection resulting from the disruption of the respective co-adapted gene complexes of the parental forms is also involved. Particularly pertinent to the argument that the selective value of alleles at individual loci is determined in part by the internal genetic environment are the extreme steepness of the gradient of transition in genetic character and the fact that frequencies at all loci have a major shift along the same mid-point line. If the loci were simply responding to external environmental gradients, the mid-points of frequency would vary in position, although sigmoidal curves would result from the interaction of migration and selective gradients (Haldane, 1948; Fisher, 1950). Apparently selective coefficients for all alleles change sign along the Vejle-Grindsted-Skjern line. Our failure to find close relationships between geographic patterns of external environmental variables and introgressive patterns at individual loci also points to

TABLE 2

*Genotypic combinations at paired loci in linkage group XVIII at farms in central Jutland*

| Genotypic combination | Farm: Observed (expected) | | | | | |
|---|---|---|---|---|---|---|
| | 22a | 23a | 25c | 27b | 32b | 32c |
| *Es-1/Es-2* | | | | | | |
| aa/bb | 1 (2) | 1 (1) | 0 (0) | 0 (0) | 8 (7) | 2 (2) |
| aa/cc | 3 (3) | 11 (11) | 4 (7) | 14 (14) | 0 (0) | 20 (16) |
| aa/bc | 3 (2) | 17 (18) | 11 (8) | 3 (3) | 0 (1) | 0 (4) |
| bb/bb | 8 (5) | 0 (0) | 0 (0) | 0 (0) | 26 (25) | 0 (0) |
| bb/cc | 7 (8) | 0 (0) | 3 (2) | 0 (0) | 0 (0) | 0 (0) |
| bb/bc | 2 (5) | 1 (1) | 2 (3) | 0 (0) | 2 (3) | 0 (0) |
| ab/bb | 1 (3) | 0 (0) | 0 (0) | 0 (0) | 25 (27) | 2 (2) |
| ab/cc | 6 (5) | 3 (3) | 11 (9) | 6 (6) | 0 (0) | 6 (10) |
| ab/bc | 5 (3) | 5 (5) | 8 (10) | 2 (2) | 5 (3) | 6 (2) |
| $\chi^2_{(4)}$ | 7·54 | 0·94 | 3·67 | 0·19 | 2·41 | 12·54* |
| *Es-1/Es-5* | | | | | | |
| aa/aa | 6 (6) | 21 (23) | 14 (14) | 16 (15) | 4 (2) | 17 (15) |
| aa/b- | 1 (1) | 8 (6) | 1 (1) | 1 (2) | 4 (6) | 5 (7) |
| bb/aa | 16 (16) | 1 (1) | 5 (5) | 0 (0) | 5 (8) | 0 (0) |
| bb/b- | 1 (1) | 0 (0) | 0 (0) | 0 (0) | 23 (20) | 0 (0) |
| ab/aa | 11 (11) | 8 (6) | 18 (18) | 6 (7) | 9 (8) | 7 (9) |
| ab/b- | 1 (1) | 0 (2) | 1 (1) | 2 (1) | 21 (22) | 7 (5) |
| $\chi^2_{(2)}$ | 0·42 | 3·16 | 0·37 | 1·71 | 3·39 | 2·78 |
| *Es-2/Es-5* | | | | | | |
| bb/aa | 9 (9) | 1 (1) | 0 (0) | 0 (0) | 12 (16) | 1 (3) |
| bb/b- | 1 (1) | 0 (0) | 0 (0) | 0 (0) | 47 (43) | 3 (1) |
| cc/aa | 14 (15) | 14 (11) | 18 (17) | 20 (18) | 0 (0) | 20 (17) |
| cc/b- | 2 (1) | 0 (3) | 0 (1) | 0 (2) | 0 (0) | 6 (9) |
| bc/aa | 10 (9) | 15 (18) | 19 (20) | 2 (4) | 6 (2) | 3 (4) |
| bc/b- | 0 (1) | 8 (5) | 2 (1) | 3 (1) | 1 (5) | 3 (2) |
| $\chi^2_{(2)}$ | 1·33 | 6·73* | 1·73 | 13·64** | 13·58** | 5·30 |

the genetic environment as a factor determining the distribution of alleles. We are thus suggesting that recombination products of the parental *musculus* and *domesticus* genotypes are selected against by a combination of external and internal environmental factors. A similar argument was advanced by Clarke (1968) and Clarke and Murray (1971) to explain certain aspects of the distribution of colour-morph frequencies in *Partula* and *Cepaea* snails. And Carson (1972) has emphasised the fact that one-to-one correlations between genes and external environmental parameters are unlikely because " the object of selection is an interacting system rather than a single gene ".

Disruption of the coadapted structure of the *musculus* gene pool may be causally related to the relatively widespread occurrence of minor *Es-2* and

*Es-3* alleles in the northern part of the hybrid zone (figs. 4 and 5). Introgression from *domesticus* could modify the *musculus* pool in such a way that selective barriers to incorporation of new alleles are relaxed, or the minor alleles could even be favoured in low frequency in the " new " genetic environment created by introgression (see Stebbins, 1971).

Despite a long period of interbreeding not only in Jutland but also in Germany, the *musculus* and *domesticus* gene pools have not fused. Hence, our studies support the view that genetic isolation cannot be directly equated with reproductive isolation, since strongly integrated and co-adapted gene pools may be effectively protected against introgression even in the absence of reproductive isolation (Stebbins, 1950; Bigelow, 1965; Hagen and McPhail, 1970). Comparable cases have been analysed by Hagen (1967) in fishes and by Hall and Selander (1973) in lizards.

The behaviour of alleles at the structural loci examined in the present study does not seem readily compatible with the hypothesis that allozymic variants in populations and in closely allied species are selectively neutral (King and Jukes, 1969; Kimura and Ohta, 1971). However, the confirmed " neutralist " can point to the possibility that selection is affecting the chromosomal segments that the loci mark rather than the loci themselves. At present we see no way of resolving the problem of whether selection acts on the loci studied or on others with which they are very tightly linked. In any event, our study suggests that selection against potentially introgressive genes, presumably involving reduced fitness in backcross generations, is very strong, and that the degree of disruption of the coadapted parental gene complexes caused by introgression is not equal over all loci.

*Acknowledgments.*—H. M. Thamdrup and F. W. Braestrup provided laboratory facilities and assistance at the Zoological Institute of the University of Aarhus. Valuable assistance in the field or laboratory was also provided by C. Schrøder, A. Munk and A. M. Sørensen, and problems arising in connection with the work in Denmark were capably handled by H. Walhovd, P. Valentin-Jensen, P. Bang and M. Lund. W. E. Johnson and S. Y. Yang provided helpful suggestions relating to electrophoresis, and C. S. Pankratz assisted in computer data processing. The contribution of R. S. Ralin to preparation of the manuscript and figures is gratefully acknowledged.

This research was supported by NSF Grants GB-6662 and GB-15664 and NIH Grant GM-15769. Support from NIH Training Grants GM-00337 and MH-12476 is acknowledged by the senior author.

## 6. References

ANDERSON, P. K., AND HILL, J. L. 1965. *Mus musculus*: experimental induction of territory formation. *Science, 148*, 1753-1755.

BERRY, R. J., AND MURPHY, H. M. 1970. The biochemical genetics of an island population of the house mouse. *Proc. roy. Soc. B, 176*, 87-103.

BIGELOW, R. S. 1965. Hybrid zones and reproductive isolation. *Evolution, 19*, 449-458.

BRUBAKER, L. L. 1970. A behavior-genetic study of race and deme differences in *Mus musculus*. Ph.D. thesis, University of Texas at Austin.

BRUNNER, G. 1941. Die Kreuzgrotte bei Pottenstein (Ofr.) und das Peterloch bei Woppental (Opf.). *Abh. Nat. Hist. Ges. Nürnberg, 27* (*Fide* Zimmermann, 1949).

CARSON, H. L. 1972. Reorganization of the gene pool during speciation. Paper presented at *Workshop on Population Structure*, University of Hawaii, July, 1972.

CHAPMAN, V. M., RUDDLE, F. H., AND RODERICK, T. H. 1971. Linkage of isozyme loci in the mouse: phosphoglucomutase-2 (*Pgm-2*), mitochondrial NADP malate dehydrogenase (*Mod-2*), and dipeptidase-1 (*Dip-1*). *Biochem. Genet., 5*, 101-110.

CLARKE, B. 1968. Balanced polymorphism and regional differentiation in land snails. In *Evolution and Environment*, Ed. E. T. Drake, 351-368. Yale Univ. Press, New Haven.

CLARKE, B., AND MURRAY, J. 1971. Polymorphism in a Polynesian land snail *Partula suturalis vexillum*. In *Ecological Genetics and Evolution*, Ed. R. Creed, 51-64. Blackwell Scientific Publ., Oxford.

COOK, L. M. 1972. *Coefficients of Natural Selection*. Hutchinson, London.

CROSBY, J. L. 1969. The evolution of genetic discontinuity: computer models of the selection of barriers to interbreeding between subspecies. *Heredity*, 25, 253-297.

DANSKE METEOROLOGISKE INSTITUT. 1933. *Danmarks Klima*. Copenhagen.

DAVIES, E. (Ed.) 1944. *Denmark: Geographic Handbook Series*. [British] Naval Intelligence Division.

DEFRIES, J. C., AND MCCLEARN, G. E. 1972. Behavioral genetics and the fine structure of mouse populations: a study in microevolution. In *Evolutionary Biology*, Vol. 5, Ed. T. Dobzhansky, M. K. Hecht and W. C. Steere, 279-291. Appleton-Century-Crofts, New York.

DOBZHANSKY, T. 1940. Speciation as a stage in evolutionary divergence. *Amer. Nat.*, 74, 312-321.

DOBZHANSKY, T. 1970. *Genetics of the Evolutionary Process*. Columbia Univ. Press, New York.

FINCHAM, J. R. S. 1972. Heterozygous advantage as a likely general basis for enzyme polymorphisms. *Heredity*, 28, 387-391.

FISHER, R. A. 1950. Gene frequencies in a cline determined by selection and diffusion. *Biometrics*, 6, 353-361.

FORD, E. B. 1971. *Ecological Genetics*, Third Ed. Chapman and Hall, London.

HAGEN, D. W. 1967. Isolating mechanisms in threespine sticklebacks (*Gasterosteus*). *J. Fish. Res. Bd. Canada*, 24, 1637-1692.

HAGEN, D. W., AND MCPHAIL, J. D. 1970. The species problem within *Gasterosteus aculeatus* on the Pacific Coast of North America. *J. Fish. Res. Bd. Canada*, 27, 147-155.

HALDANE, J. B. S. 1948. The theory of a cline. *J. Genet.*, 48, 277-284.

HALL, W. P., AND SELANDER, R. K. 1973. Hybridization in karyotypically differentiated populations of the *Sceloporus grammicus* complex (Iguanidae). *Evolution*, 27, in press.

HENDERSON, N. S. 1966. Isozymes and genetic control of NADP-malate dehydrogenase in mice. *Arch. Biochem. Biophys.*, 117, 28-33.

HUNT, W. G. 1970. Biochemical variation in hybridizing subspecies of the house mouse (*Mus musculus*) in Europe. Ph.D. thesis, University of Texas at Austin.

HUTTON, J. J., AND RODERICK, T. H. 1970. Linkage analyses using biochemical variants in mice. III. Linkage relationships of eleven biochemical markers. *Biochem. Genet.*, 4, 339-350.

HUXLEY, J. 1943. *Evolution: The Modern Synthesis*. Harper, New York.

JAIN, S. K., AND BRADSHAW, A. D. 1966. Evolutionary divergence among adjacent plant populations. I. The evidence and its theoretical analysis. *Heredity*, 21, 407-441.

KIMURA, M., AND OHTA, T. 1971. *Theoretical Aspects of Population Genetics*. Princeton Univ. Press, Princeton.

KING, J. L., AND JUKES, T. H. 1969. Non-Darwinian evolution. *Science*, 164, 788-798.

MAYR, E. 1963. *Animal Species and Evolution*. Harvard Univ. Press, Cambridge.

MEISE, W. 1928. Die Verbreitung der Aaskrähe (Formenkreis *Co.vus corone* L.). *J.f. Ornithol.*, 76, 1-203.

PETRAS, M. L., AND BIDDLE, F. G. 1967. Serum esterases in the house mouse, *Mus musculus*. *Canad. J. Genet. Cytol.*, 9, 704-710.

PETRAS, M. L., REIMER, J. D., BIDDLE, F. G., MARTIN, J. E., AND LINTON, R. S. 1969. Studies of natural populations of *Mus*. V. A survey of nine loci for polymorphisms. *Canad. J. Genet. Cytol.*, 11, 497-513.

POPP, R. A. 1967. Linkage of *Es-1* and *Es-2* in the mouse. *J. Hered.*, 58, 186-188.

REIMER, J. D., AND PETRAS, M. L. 1967. Breeding structure of the house mouse, *Mus musculus*, in a population cage. *J. Mammal.*, 48, 88-99.

REMINGTON, C. L. 1968. Suture-zones of hybrid interaction between recently joined biotas. In *Evolutionary Biology*, Vol. 2, Ed. T. Dobzhansky, M. K. Hecht and W. C. Steere, 321-428. Appleton-Century-Crofts, New York.

RODERICK, T. H., HUTTON, J. J., AND RUDDLE, F. H. 1970. Linkage of esterase-3 (*Es-3*) and *Rex* (*Re*) on linkage group VII of the mouse. *J. Hered.*, 61, 278-279.

RODERICK, T. H., RUDDLE, F. H., CHAPMAN, V. M., AND SHOWS, T. B. 1971. Biochemical polymorphisms in feral and inbred mice (*Mus musculus*). *Biochem. Genet.*, 5, 457-466.

RUDDLE, F. H., RODERICK, T. H., SHOWS, T. B., WEIGL, P. G., CHIPMAN, R. K., AND ANDERSON, P. K. 1969. Measurements of genetic heterozygosity by means of enzyme polymorphisms in wild populations of the mouse. *J. Hered.*, 60, 321-322.

RUDDLE, F. H., SHOWS, T. B., AND RODERICK, T. H. 1969. Esterase genetics in *Mus musculus*: expression, linkage and polymorphism of locus *Es-2. Genetics, 62*, 393-399.

SCHWARZ, E., AND SCHWARZ, H. K. 1943. The wild and commensal stocks of the house mouse, *Mus musculus* Linnaeus. *J. Mammal., 24*, 59-72.

SELANDER, R. K. 1970. Behavior and genetic variation in natural populations. *Amer. Zool., 10*, 53-66.

SELANDER, R. K., AND JOHNSON, W. E. 1973. Genetic variation among vertebrate species. *Ann. Rev. Ecol. Syst., 4*, in press.

SELANDER, R. K., AND YANG, S. Y. 1969. Protein polymorphism and genic heterozygosity in a wild population of the house mouse (*Mus musculus*). *Genetics, 63*, 653-667.

SELANDER, R. K., HUNT, W. G., AND YANG, S. Y. 1969. Protein polymorphism and genic heterozygosity in two European subspecies of the house mouse. *Evolution, 23*, 379-390.

SELANDER, R. K., YANG, S. Y., AND HUNT, W. G. 1969. Polymorphism in esterases and hemoglobin in wild populations of the house mouse (*Mus musculus*). *Studies in Genetics V*, Univ. Texas Publ. 6918, 271-338.

SHORT, L. L., JR. 1965. Hybridization in the flickers (*Colaptes*) of North America. *Bull. Amer. Mus. Nat. Hist., 129*, 307-428.

SHOWS, T. B., CHAPMAN, V. M., AND RUDDLE, F. H. 1970. Mitochondrial malate dehydrogenase and malic enzyme: Mendelian inherited electrophoretic variants in the mouse. *Biochem. Genet., 4*, 707-718.

STEBBINS, G. L. 1950. *Variation and Evolution in Plants.* Columbia Univ. Press, New York.

STEBBINS, G. L. 1971. *Processes of Organic Evolution*, Second Ed. Prentice-Hall, Englewood Cliffs, N.J.

TAYLOR, B. A. 1972. Genetic relationships between inbred strains of mice. *J. Hered., 63*, 83-86.

URSIN, E. 1952. Occurrence of voles, mice, and rats (*Muridae*) in Denmark, with a special note on a zone of intergradation between two subspecies of the house mouse (*Mus musculus* L.). *Vid. Medd. Dansk Naturhist. Foren., 114*, 217-244.

WATERBOLK, H. T. 1968. Food production in prehistoric Europe. *Science, 162*, 1093-1102.

WHEELER, L. L. 1972. Inheritance of allozymes in subspecific $F_1$ hybrids of *Mus musculus* from Denmark. *Studies in Genetics VII*, Univ. Texas Publ. 7213, 319-326.

WHEELER, L. L., AND SELANDER, R. K. 1972. Genetic variation in populations of the house mouse, *Mus musculus*, in the Hawaiian Islands. *Studies in Genetics VII*, Univ. Texas Publ. 7213, 269-296.

WILSON, E. O. 1965. The challenge from related species. In *The Genetics of Colonizing Species*, Ed. H. G. Baker and G. L. Stebbins, 7-27. Academic Press, New York.

YANG, S. Y., AND SELANDER, R. K. 1968. Hybridization in the grackle *Quiscalus quiscalus* in Louisiana. *Syst. Zool., 17*, 107-143.

ZIMMERMANN, K. 1949. Zur Kenntnis der mitteleuropäischen Hausmäuse. *Zool. Jahrb. (Syst.), 78*, 301-322.

# 18

Reprinted from *Evolution* **30**:412–424 (1976)

## MULTIVARIATE ANALYSIS OF THE INTROGRESSIVE REPLACEMENT OF *CLARKIA NITENS* BY *CLARKIA SPECIOSA POLYANTHA* (ONAGRACEAE)

William L. Bloom

*Botany Department, University of Kansas, Lawrence, Kansas 66045*

Received August 1, 1975

Revised January 16, 1976

Introgressive hybridization has received considerable attention following the initial delineation of the concept by Anderson and Hubricht (1938). Introgression was viewed as ". . . the introduction of a comparatively few genes from one species into the germplasm of another . . ." (Anderson, 1949, p. 18). As such, the process offers the recipient species the unique opportunity of testing the adaptive value of genes which have already survived the gauntlet of selection in a closely related donor species. However, no consensus has been reached as to the evolutionary importance of introgression. At one extreme the hybridization required for introgression is viewed as merely "evolutionary noise" (Wagner, 1970) with no biological significance while at the other extreme it is considered of importance in producing totally new genetic combinations, even capable of yielding new genes (cf. Watt, 1972). Further, there is no agreement on the frequency of introgression in natural populations. Certainly the number of strongly documented cases for introgression is very low (Heiser, 1973). It is not clear whether this deficiency is due to a rarity of introgression or rather due to the difficulty of recognizing this process in natural populations. The study reported here indicates the latter may be the case.

*Clarkia speciosa* and *C. nitens* are outcrossing annuals of central California. *C. speciosa* consists of several ecogeographic races in the Coast Ranges and one in the foothills of the southern Sierra Nevada, *C. s. polyantha*. *C. s. polyantha* is replaced northward in the sierra foothills by *C. ni-*

*tens*, which was originally recognized as distinct from *C. speciosa* because of numerous floral and vegetative differences and the very low fertility of artificial hybrids (Lewis and Lewis, 1955). The low fertility was due to the fact that *C. nitens* and *C. speciosa* differ in chromosome arrangement. The former species has the "North" arrangement and the latter the "South" arrangement, and these differ by seven reciprocal translocations involving eight of the nine haploid chromosomes. However, additional work revealed one major exception to this simple correspondence between the morphological and chromosomal differentiation in the two species. The populations immediately south of *C. nitens* for a distance of about 100 miles appear to resemble *C. s. polyantha* most closely but have the North chromosome arrangement characteristic of *C. nitens* (Mosquin and Lewis, 1959). The polyantha phenotype, therefore, in the southern half of its range is associated with the South chromosome arrangement (typical of *C. speciosa*) but the northern populations of *C. s. polyantha* have the North chromosome arrangement which is otherwise restricted to *C. nitens*.

The North and South chromosome arrangements have been found to be separated by a 20 mile wide boundary (within the *C. s. polyantha* phenotype) in which populations have a minimum of ten different chromosome arrangements (Bloom and Lewis, 1972). Each of these boundary arrangements appears to be distributed quite locally; they differ from the North and South arrangements by as many as six translocations. Chromosomal studies of

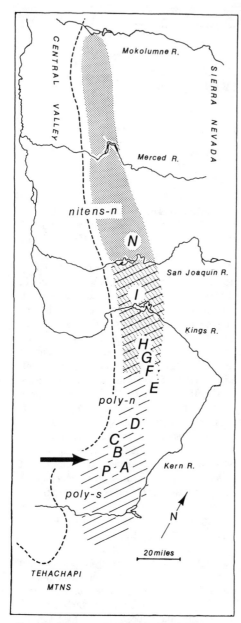

FIG. 1. Major patterns of chromosomal and external morphological variation in polyantha and nitens in the Sierra Nevada foothills of central California. Hatching indicates the polyantha phenotype, shading indicates nitens, and the combination pattern indicates morphological hybridity as shown in Fig. 2. Boundary region between the North (n) and South (s) chromosomal races is indicated by the arrow. Note the polyantha phenotype extending northward across and beyond the chromosomal boundary.

this boundary region suggest the following: (1) the boundary chromosome arrangements arose within a zone of hybridization between the South and North chromosome races; (2) the new arrangements and their spatial distribution through the boundary area eliminated the hybrid infertility barrier between the North and South chromosome races; and (3) the sequence of boundary arrangements is geographically stable, thus marking the area of initial contact between South and North races.

The apparent discrepancy between chromosomal and morphological boundaries in the *C. s. polyantha-C. nitens* group, when considered together with chromosomal evidence from the boundary region, suggests that introgression has occurred from polyantha into nitens. The present study is a statistical evaluation of the morphological relationships between *C. polyantha, C. nitens* and the putative introgressant populations located north of the chromosomal boundary (Fig. 1).

## METHODS

Seeds were collected from natural populations, germinated on wet filter paper and viable seedlings were transplanted to a common garden where the plantings were randomized. Each seed parent collected from the natural population was represented by one plant to maximize the number of individuals sampled from each population. An attempt was made to collect seeds from a minimum of 20 randomly spaced seed parents from each population; if fewer than 20, then seeds were collected from every individual found in the population. Eleven natural populations were sampled (Table 1), one from each of the two parental taxa, *C. nitens* (N) and *C. s. polyantha* (P), eight (B to I) from between the area of initial hybridization and *C. nitens,* and one (A) population located at the northern edge of the South chromosome race. The two parental populations sampled were located at least 15 air miles from populations exhibiting any chromosomal

TABLE 1. *Location of populations of* Clarkia s. polyantha *and* C. nitens *from which seeds were collected. Populations are arranged from top to bottom in a South to North geographic order, elevation is expressed in feet.*

| Population | Location |
|---|---|
| P | Tulare Co., 5.7 miles east of bridge at White River on M-109 to Posey, 3.95 miles west of junction with Posey–Glennville road, elevation 2800′ (Bloom 429). |
| A | Tulare Co., .1 mile east of California Hot Springs on road (J22) to Pine Flat, elevation 3390′ (Bloom 530). |
| B | Tulare Co., Deer Creek road, 10.0 miles south from junction with Tule River Indian Reservation Road, elevation 1100′ (Bloom 410). |
| C | Tulare Co., Deer Creek Road, 2.1 miles south from junction with Tule River Indian Reservation Road, elevation 500′ (Bloom 520). |
| D | Tulare Co., Springville to Exeter road, 2.1 miles south of junction with Balch Park Road, elevation 1200′ (Bloom 426). |
| E | Tulare Co., Hwy. 198 at junction with road to Mineral King, 1 mile north of Three Rivers, elevation 1100′ (Bloom 521). |
| F | Tulare Co., Lemon Cove to Badger road (J21) along Dry Creek, 2.5 miles north of Homer Ranch, elevation 950′ (Bloom 522). |
| G | Tulare Co., Lemon Cove to Badger road (J21) along Dry Creek, 6.3 miles south of Badger, elevation 1200′ (Bloom 523). |
| H | Tulare Co., 2 miles south of Badger at junction of J21 and Hartland Road, elevation 2655′ (Bloom 524). |
| I | Fresno Co., Watts Valley Road, 6.6 miles west of Pittman Hill Road, elevation 920′ (Bloom 528). |
| N | Madera Co., San Joaquin Experimental Range south of Coarse Gold, elevation 1000′ (Bloom 71-115). |

or morphological variation attributable to hybridization.

Individuals in full flower were scored for color determined visually against a neutral gray background of 18% reflec-tance with illumination from two "high in-tensity" lamps having approximately day-light color temperature. A reference set of individuals selected to represent the full range of color variation present in the en-tire sample was used as the standard and each individual was scored by comparison with this set. Three major colors (hues) could be discerned—purple, yellow and green. Purple was separated into six classes of intensity (maximum intensity = 3.0) yellow into three (maximum intensity = 1.5) and green into two classes. Frequent repeat scoring of individuals showed high reliability of color classification with this system. The purpleness of a particular flower region was considered one character, yellowness another and greenness another character. The color "white" (no discern-able hue) was given a value of 0 for all three color characters. Hence, the color variation of eight morphological regions was coded as 15 characters for statistical analysis (Table 2).

Metric variation of nine leaf and flower characters (Table 3) was determined from plants in full flower. Three stem and fruit characters were measured from plants with all seed fully mature. Internode length was obtained by averaging the three adja-cent internodes above the oldest (lowest) fruit on the major flowering axis. Flower measurements were taken from the second and third oldest flowers on the major flow-ering axis and the values averaged. The character "hypanthium width" is actually half the circumference of the top of the hypanthium obtained by flattening the hy-panthium laterally and measuring the width. Leaf character scores are averages of measurements of the two leaves subtend-ing the second and third oldest flower on the major flowering axis. Raceme length was obtained after flowering from the long-est inflorescence of each plant and the num-ber of capsules on this axis included every capsule with at least one mature seed.

Canonical variate analysis (program BMDO7M) was used to evaluate the

TABLE 2. *Flower color scores of eleven* Clarkia *populations—one* Clarkia nitens *(N), one* C. speciosa polyantha *(P) and nine geographically intermediate populations. Mean and standard deviation in arbitrary units.*

| Character | Population | | | | | |
|---|---|---|---|---|---|---|
| | P (N = 22) | A (N = 21) | B (N = 17) | C (N = 16) | D (N = 21) | E (N = 20) |
| Stigma purple | 1.91 ± 0.4 | 2.07 ± 0.3 | 2.00 ± 0.2 | 2.06 ± 0.5 | 2.29 ± 0.5 | 1.88 ± 0.8 |
| Stigma yellow | .00 | .00 | .00 | .00 | .00 | .00 |
| Style top purple | .02 ± 0.1 | .02 ± 0.1 | .15 ± .02 | .22 ± 0.5 | .36 ± 0.4 | .50 ± 0.5 |
| Style top green | .07 ± 0.2 | .05 ± 0.2 | .00 | .00 | .00 | .00 |
| Style top yellow | .00 | .00 | .00 | .00 | .00 | .00 |
| Style bottom purple | 1.18 ± 0.6 | 1.21 ± 0.6 | 1.71 ± 0.6 | 1.44 ± 0.6 | 1.81 ± 0.5 | 2.20 ± 0.5 |
| Style bottom yellow | .00 | .00 | .00 | .00 | .00 | .00 |
| Inner hypanthium purple | .14 ± 0.4 | .00 | .50 ± 0.6 | 1.84 ± 0.8 | .67 ± 0.3 | 1.23 ± 0.8 |
| Inner hypanthium green | .77 ± 0.4 | .67 ± 0.2 | .24 ± 0.3 | .00 | .10 ± 0.3 | .00 |
| Filament purple | .36 ± 0.4 | .12 ± 0.3 | .79 ± 0.3 | .75 ± 0.4 | 1.02 ± 0.2 | .88 ± .04 |
| Filament yellow | .00 | .00 | .00 | .00 | .00 | .00 |
| Petal base purple | .73 ± 0.3 | .62 ± 0.2 | .94 ± 0.2 | .84 ± 0.2 | 1.00 ± 0.0 | 1.08 ± 0.2 |
| Petal base yellow | .00 | .00 | .00 | .00 | .00 | .00 |
| Petal spot purple | 1.98 ± 0.1 | 2.00 ± 0.0 | 2.00 ± 0.0 | 2.06 ± 0.3 | 2.00 ± 0.3 | 2.20 ± 0.3 |
| Petal top purple | 1.00 ± 0.0 | 1.00 ± 0.0 | 1.00 ± 0.0 | 1.00 ± 0.0 | 1.00 ± 0.0 | 1.08 ± 0.2 |

| Character | Population | | | | |
|---|---|---|---|---|---|
| | F (N = 13) | G (N = 19) | H (N = 18) | I (N = 30) | N (N = 24) | Grand mean |
| Stigma purple | .27 ± 0.7 | .34 ± 0.7 | .03 ± 0.1 | .00 | .00 | 1.12 |
| Stigma yellow | .73 ± 0.4 | .58 ± 0.4 | .86 ± 0.3 | .97 ± 0.1 | 1.02 ± 0.1 | .40 |
| Style top purple | .23 ± 0.3 | .21 ± 0.3 | .06 ± 0.2 | .10 ± 0.2 | .00 | .16 |
| Style top green | .00 | .00 | .00 | .00 | .00 | .01 |
| Style top yellow | .00 | .00 | .42 ± 0.2 | .42 ± 0.2 | .50 ± 0.0 | .14 |
| Style bottom purple | 1.85 ± 0.4 | 1.89 ± 0.4 | 1.81 ± 0.5 | 2.48 ± 0.5 | .63 ± 0.4 | 1.66 |
| Style bottom yellow | .00 | .00 | .00 | .00 | .06 ± 0.2 | .01 |
| Inner hypanthium purple | .15 ± 0.2 | .32 ± 0.2 | .17 ± 0.2 | .72 ± 0.4 | .04 ± 0.2 | .51 |
| Inner hypanthium green | .58 ± 0.3 | .58 ± 0.2 | .75 ± 0.2 | .22 ± 0.3 | .54 ± 0.2 | .40 |
| Filament purple | .81 ± 0.3 | .89 ± 0.4 | .83 ± 0.3 | .70 ± 0.4 | .00 | .63 |
| Filament yellow | .00 | .00 | .00 | .00 | .06 ± 0.2 | .01 |
| Petal base purple | .96 ± 0.1 | .95 ± 0.2 | 1.00 ± 0.0 | 1.14 ± 0.3 | .00 | .83 |
| Petal base yellow | .00 | .00 | .00 | .02 ± 0.1 | .50 ± 0.0 | .06 |
| Petal spot purple | 2.12 ± 0.2 | 2.08 ± 0.2 | 2.00 ± 0.0 | 2.12 ± 0.3 | 2.04 ± 0.1 | 2.05 |
| Petal top purple | 1.00 ± 0.0 | 1.00 ± 0.0 | 1.05 ± 0.2 | 1.02 ± 0.1 | .75 ± 0.3 | .99 |

phenotypic relationships of the eleven populations studied (see Blackith and Reyment, 1971, for theory and applications of this method). The radius of the 95% confidence circles for the canonical centroids was calculated using the formula $1.96/\sqrt{N}$, where $N$ is the sample size of the population (Seal, 1964).

Hereafter for brevity *C. nitens* and *C. s. polyantha* will be referred to as nitens and polyantha, respectively.

## RESULTS

*Flower color variation.*—Nitens flowers are invariably characterized by a distinctly yellow stigma, style top and petal base in

TABLE 3. *Metric variation of leaf, stem and flower parts of eleven* Clarkia *populations—one* C. nitens *(N), one* C. s. polyantha *(P) and nine geographically intermediate populations (A through I). Mean and standard deviation shown in mm.*

| Character | Population | | | | | |
|---|---|---|---|---|---|---|
| | P (N = 19) | A (N = 19) | B (N = 17) | C (N = 14) | D (N = 21) | E (N = 20) |
| Leaf length | 50.21 ± 10.8 | 50.05 ± 9.5 | 55.82 ± 12.2 | 54.43 ± 11.8 | 43.71 ± 13.3 | 56.45 ± 10.4 |
| Leaf width | 6.47 ± 1.5 | 7.53 ± 1.5 | 6.94 ± 1.7 | 7.07 ± 2.2 | 6.24 ± 1.7 | 5.65 ± 1.3 |
| Length to max. leaf width | 23.68 ± 5.9 | 22.11 ± 4.7 | 22.94 ± 5.6 | 25.00 ± 6.7 | 20.10 ± 8.4 | 26.60 ± 6.6 |
| Raceme internode length | 21.84 ± 6.8 | 21.16 ± 4.6 | 25.76 ± 5.4 | 24.79 ± 5.7 | 21.14 ± 7.7 | 27.30 ± 6.7 |
| Petal length | 20.63 ± 1.7 | 24.53 ± 2.3 | 20.47 ± 1.9 | 19.50 ± 2.5 | 19.33 ± 3.2 | 22.40 ± 3.2 |
| Petal width | 24.00 ± 3.6 | 28.05 ± 3.8 | 23.66 ± 2.9 | 21.93 ± 3.3 | 21.05 ± 3.6 | 25.50 ± 3.3 |
| Hypanthium length | 8.26 ± 1.0 | 8.95 ± 1.8 | 8.41 ± 1.5 | 8.86 ± 1.2 | 8.33 ± 1.4 | 8.95 ± 1.1 |
| Hypanthium width | 6.53 ± 0.8 | 7.58 ± 1.2 | 6.29 ± 1.0 | 6.07 ± 0.7 | 6.05 ± 0.7 | 6.35 ± 0.8 |
| Style length | 18.47 ± 1.3 | 20.26 ± 2.3 | 17.53 ± 2.2 | 19.64 ± 2.3 | 18.38 ± 2.6 | 18.65 ± 2.0 |
| Inflorescence length | 29.77 ± 11.2 | 25.68 ± 13.0 | 36.22 ± 17.0 | 34.34 ± 11.1 | 25.91 ± 11.5 | 33.28 ± 13.2 |
| Bud number | 20.32 ± 6.9 | 18.58 ± 6.6 | 20.65 ± 7.6 | 21.64 ± 7.8 | 20.05 ± 5.5 | 19.10 ± 7.3 |
| Capsule number | 19.05 ± 7.0 | 16.68 ± 7.0 | 17.59 ± 7.0 | 20.29 ± 7.3 | 18.52 ± 5.4 | 18.20 ± 6.6 |

| Character | Population | | | | | |
|---|---|---|---|---|---|---|
| | F (N = 15) | G (N = 13) | H (N = 14) | I (N = 29) | N (N = 40) | Grand mean |
| Leaf length | 47.40 ± 6.2 | 47.38 ± 7.0 | 42.43 ± 6.2 | 47.76 ± 9.3 | 48.95 ± 9.0 | 49.44 |
| Leaf width | 6.87 ± 1.2 | 6.69 ± 0.8 | 5.36 ± 1.2 | 6.59 ± 1.3 | 8.85 ± 1.8 | 6.95 |
| Length to max. leaf width | 19.20 ± 3.8 | 21.15 ± 4.2 | 22.29 ± 4.0 | 19.83 ± 4.2 | 17.50 ± 5.6 | 21.33 |
| Raceme internode length | 26.20 ± 5.4 | 24.69 ± 6.7 | 20.14 ± 5.8 | 22.66 ± 7.5 | 20.98 ± 6.0 | 23.00 |
| Petal length | 18.47 ± 2.2 | 21.46 ± 3.0 | 20.36 ± 2.3 | 20.03 ± 2.9 | 23.65 ± 2.8 | 21.27 |
| Petal width | 23.13 ± 3.1 | 26.62 ± 2.4 | 26.07 ± 3.3 | 22.07 ± 3.2 | 28.80 ± 4.1 | 24.89 |
| Hypanthium length | 8.33 ± 1.0 | 8.54 ± 1.1 | 7.64 ± 0.9 | 9.52 ± 1.2 | 8.93 ± 1.5 | 8.71 |
| Hypanthium width | 6.67 ± 1.0 | 6.23 ± 0.6 | 6.57 ± 0.5 | 6.67 ± 0.9 | 7.73 ± 0.9 | 6.75 |
| Style length | 18.93 ± 2.0 | 20.00 ± 2.4 | 18.86 ± 1.9 | 19.55 ± 1.9 | 21.40 ± 2.9 | 19.45 |
| Inflorescence length | 26.14 ± 12.3 | 30.30 ± 14.3 | 18.60 ± 9.6 | 24.32 ± 13.4 | 16.97 ± 6.4 | 26.20 |
| Bud number | 16.40 ± 6.0 | 19.54 ± 6.5 | 16.57 ± 5.5 | 17.38 ± 6.4 | 15.18 ± 6.0 | 18.28 |
| Capsule number | 15.20 ± 6.3 | 17.54 ± 5.4 | 15.86 ± 5.3 | 17.21 ± 6.2 | 14.38 ± 5.4 | 17.05 |

contrast to the neutral to purple color of these regions in polyantha (Table 2). The southern populations A through E (Fig. 1) showed no evidence of these nitens characters. The purple stigma characteristic of polyantha extends north as far as H but decreases in frequency abruptly at F where it is largely replaced by the yellow of nitens. The purple petal base of polyantha extends north through population I in high frequency. The neutral to purple color

characteristic of the polyantha style top extends north to population I but drops in frequency abruptly in population H where the nitens yellow begins and is predominant.

Analysis of all flower color characters produced two canonical variates which together account for 92% of the total dispersion (Table 4). The absolute value of the standardized coefficients for each character shows that the character petal base

TABLE 4. *Cumulative proportion of total dispersion of color characters accounted for by the first three canonical variates and the standardized coefficients of the 12 color characters (eight morphological regions) entered in the analysis.*

| Characters | Canonical variate | | |
|---|---|---|---|
| | 1(77%) | 2(92%) | 3(96%) |
| Stigma purple | .109 | .593 | −.140 |
| Stigma yellow | −.170 | −.308 | .308 |
| Style top purple | −.159 | .016 | −.123 |
| Style top yellow | −.626 | −.330 | −.566 |
| Style bottom purple | .109 | −.289 | −.108 |
| Inner hypanthium purple | .247 | .075 | −.441 |
| Inner hypanthium green | .087 | −.084 | .545 |
| Filament purple | .252 | −.345 | .125 |
| Petal base purple | −.083 | −.180 | −.149 |
| Petal base yellow | −1.064 | .271 | −.044 |
| Petal spot purple | −.199 | −.114 | −.139 |
| Petal top purple | −.094 | −.268 | −.045 |

yellow, for which nitens scores very high, contributed most heavily to the first canonical variate and was positively correlated with the next most important character, style top yellow. The two characters next in importance in the first canonical variate, inner hypanthium purple and filament purple, are shown to be positively correlated with each other and negatively correlated with the previous two characters. This first canonical variate strongly separates nitens from all other populations (Fig. 2A). The second canonical variate, accounting for 15% of the total dispersion, loaded stigma purple most heavily followed by filament purple, style top yellow and stigma yellow which all contributed about equally. The stigma purple character is negatively correlated with style top yellow and stigma yellow, whereas, filament purple is positively correlated with style top yellow and stigma yellow. The second canonical variate thus separates the four northern populations F through I from all more southern populations. These northern populations have purple filaments, as in polyantha, but yellow stigma and/or style top as in nitens (Fig. 1). Approximate 95% confidence circles (Seal, 1964) for the population centroids show that popula-

tions A, B, C, D and E are very similar to polyantha and do not vary toward nitens. On the first canonical axis, which functions to discriminate maximally between nitens and polyantha, the populations north of E generally become progressively more like nitens in flower color. The third canonical variate clusters populations C, D and E together with population I primarily on the basis of the purpleness of the inner hypanthium as distinct from the remaining populations which score high on inner hypanthium green. The remaining canonical variates do not appear to order the populations in any significant manner.

*Metric variation.*—A comparison of the polyantha reference population with the nitens reference population in terms of the metric variation of single characters shows the polyantha flowers to have shorter, narrower petals, shorter style and a narrower but equally long hypanthium (Table 3). The two populations have raceme internodes of similar length but polyantha racemes are longer and, therefore, produce more flower buds per raceme and perhaps also more capsules. The leaves of polyantha and nitens are about the same length but those of polyantha are narrower and, on the average, lanceolate rather than narrowly ovate as in nitens.

The metric data yielded four canonical variates, utilizing 10 of 12 available characters, which will account for 92% of the total dispersion of all 11 populations (Table 5). The first canonical variate, loading most heavily on leaf width, style length and raceme length, in that order, separated nitens clearly from all other populations (Fig. 2B). Populations B to E surround polyantha on this first axis while populations F to I form a cluster near polyantha but clearly in the direction of nitens. Population A is located almost midway between nitens and polyantha farther toward nitens than any other population. The most conspicuous aspect of the distribution of all populations on

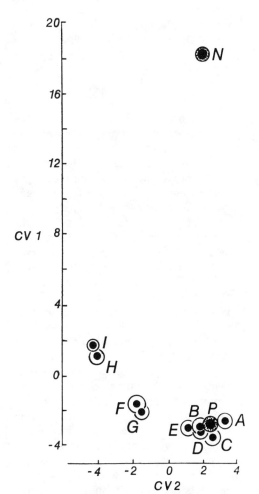

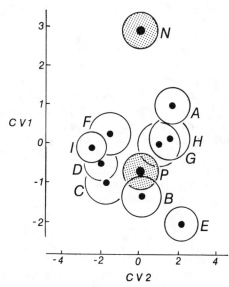

FIG. 2. Morphological relationships between polyantha, nitens and populations of hybrid origin as indicated by canonical variate analysis. A circle around a population mean represents the approximate 95% confidence circle. The first canonical variate (cv) is on the vertical axis and the second on the horizontal. A. Analysis of color variation of eight flower regions (see Table 2). B. Analysis of metric variation of 12 vegetative and reproductive characters (see Table 3). Compare with the geographical and chromosomal relationships of populations shown in Fig. 1.

TABLE 5. *Cumulative proportion of total dispersion of metric characters accounted for by the first four canonical variates and the standardized coefficients of the 10 metric characters (of 12 evaluated) entered in the analysis.*

| Character | Canonical variate | | | |
|---|---|---|---|---|
| | 1(57%) | 2(73%) | 3(85%) | 4(92%) |
| Leaf length | −.318 | .022 | .361 | −.676 |
| Leaf width | .710 | .444 | .018 | −.332 |
| Leaf length to maximum width | −.378 | −.352 | −.211 | .546 |
| Petal length | −.095 | −.380 | 1.141 | .494 |
| Petal width | .231 | −.748 | −.873 | −.460 |
| Hypanthium length | −.319 | .339 | .730 | .026 |
| Hypanthium width | .301 | .040 | .442 | .381 |
| Style length | .605 | .202 | −.900 | −.188 |
| Inflorescence length | −.566 | −.285 | −.102 | −.822 |
| Capsule number | .050 | .425 | −.009 | .711 |

metric canonical axes 2 through 6 is the consistent occurrence of polyantha and nitens together near zero, surrounded by the remaining populations. The characteristics distinguishing nitens from polyantha are virtually all included in the first canonical variate. Most of the remaining variation (43%) is among populations A to I, not the nitens and polyantha reference populations; and it requires five additional canonical variates to account for 98% of the total dispersion. Canonical variates 2 to 10 tend to place geographically adjacent populations near one another on a particular axis, but such groups are not constant from one axis to the next, for example, G and H (with B) are close together on metric axis 2 but are widely separated on axis 4.

The major phenotypic relationships between populations as revealed by the first canonical variate are very similar using metric or flower color data. In both cases the nitens and polyantha reference populations are clearly distinct from one another, populations B to E are clustered around polyantha rather than nitens, and populations F to I are grouped between nitens and polyantha though considerably closer to polyantha. The ordering of populations on metric canonical axes subsequent to the first shows little correlation with the ordering achieved using color data.

The one conspicuous discrepancy between the metric and flower color analyses is the placement of population A on the first axis of each. This population, geographically located at the northern edge of the distribution of the South chromosome arrangement, is extremely close to the polyantha reference population in flower color but scores midway between polyantha and nitens on the metric first canonical axis. Observations of natural populations subsequent to the common garden experiment indicate that the phenotype expressed in population A is common in polyantha throughout its distribution, particularly at somewhat higher elevations.

Thus, the differences between populations P and A in metric characters likely reflect the normal variation within polyantha rather than any influence from nitens.

## DISCUSSION

Morphological variation in Sierra Nevada populations of nitens and polyantha shows a conspicuous lack of correlation with chromosomal variation. The polyantha phenotype which characterizes the South chromosome race extends north through the chromosome boundary region and beyond it about 40 miles into the North chromosome race. For the next 60 miles northward the polyantha phenotype continues to dominate, though clear indications of the nitens phenotype are evident in some characters. The nitens phenotype becomes dominant in the region of the Kings River. Field observations subsequent to these common garden studies indicate the transition to nitens is relatively abrupt in this region. Chromosomal evidence suggests that the complex system of chromosome arrangements found in the North–South chromosome boundary region was evolved in response to hybridization between the two chromosome races (Bloom and Lewis, 1972). The structural relationships between the various boundary arrangements and the particular spatial distribution of each in this region appear to have eliminated a major barrier to gene exchange between the North and South races. It is likely that gene flow can occur stepwise from one population to the next across the chromosomal boundary with no significant fertility barrier.

The morphological evidence considered together with the chromosomal evidence strongly supports the hypothesis that the existing patterns of variation result from hybridization followed by extensive introgression—genetically and geographically. In contrast to the classical expectations for introgressive variation, the introgression from polyantha into nitens has been so extensive as to produce recipient populations which are virtually indistinguishable in external morphology from the donor populations and the introgressed populations in general do not show the expected increase in external morphological variation relative to the parental taxa.

Hybridization between polyantha, with the South chromosome arrangement, and nitens, with the North arrangement, apparently was initiated in the region currently marked by the North–South chromosomal boundary. At the time of initial contact and hybridization the two taxa were undoubtedly sufficiently distinct morphologically and reproductively to be considered different species. The initial contact between the two taxa was probably due to a northward movement by polyantha as suggested by the northward introgression of polyantha genes into nitens populations following hybridization. Further evidence that polyantha has moved northward is an early collection of polyantha from Tejon Pass (Lewis and Lewis, 1955), located at the southwest end of the Tehachapi Mountains considerably south of any known extant population of polyantha (cf. Fig. 1).

Three points are particularly important to a full understanding of the biological significance of this introgression. First, numerous genes are involved. Undoubtedly all of the metric characters by which nitens and polyantha differ are multigenic, and the lack of correlation between flower colors and metric characters in various individuals of populations F, G, H and I indicates that a number of different genes control these characters. For example, style color expression is independent of stigma color, flower color is independent of leaf shape, etc. In fact, the only two characters showing strong correlation in all populations were, not surprisingly, bud number and capsule number.

Second, the replacement of nitens genes by polyantha genes in the hybrid derivatives was not accomplished by a simple substitution of polyantha chromosomes for nitens chromosomes but rather a transfer of segments of polyantha chromosomes onto nitens centromeres by means of crossing over in hybrid individuals. This situation is clear because the introgressant populations still have the North chromosome arrangement of nitens, not the South arrangement of polyantha. The $F_1$ hybrid between nitens–North and polyantha–South is heterozygous for at least seven reciprocal translocations involving eight of the nine chromosomes of the haploid set. As a consequence, chromosome pairing normally produces a single ring of 16 chromosomes plus one bivalent. The highly irregular disjunction of the ring of 16 during meiosis results in a low ($<10\%$) proportion of gametes with the genetically normal complement of nine chromosomes (Bloom, 1974). Of these balanced gametes half will have the chromosome arrangement of nitens but with some genetic material of polyantha.

The actual number of polyantha alleles which are transferred onto the nitens chromosome complement by crossing over during meiosis depends on the distribution of chiasmata in the chromosomes. Most often one chiasma is formed near the end of each arm (all chromosomes are metacentric) but such distal localization is not rigorous and chiasmata are occasionally observed even in proximal regions of the euchromatic arms (Bloom, 1974). All or most of the alleles in the euchromatic portion of a particular chromosome arm could, therefore, be transferred from the polyantha to nitens chromosome arrangement as a block, but such a massive transfer probably is not required. Variation in the positioning of chiasmata along a particular chromosome arm in different meiotic cells would undoubtedly yield nitens chromosomes with sections of polyantha alleles of various lengths. Backcrossing of the hybrid to nitens would transfer the polyantha genes into the nitens gene pool to be subjected to selection independent of the low fertility associated with the chromosomal hybrid. Currently there appears to be no fertility barrier between the North and South races because of the sequence of boundary arrangements evolved there following hybridization. Whether this breakdown of the infertility barrier was crucial to the successful movement of polyantha genes into nitens is uncertain. The elimination of the barrier undoubtedly allows for an increase in the rate of gene flow and it is possible that had the rate not been increased incompatibility barriers to prevent hybridization could have been evolved. Selection for alleles preventing hybridization presumably would have been strong in both polyantha and nitens of the hybridizing populations. The evolution of such a premating barrier could have permitted the sympatric occurrence of nitens and polyantha, followed perhaps by character divergence to minimize competition. Whether the breakdown of the fertility barrier was important or not can be tested to some extent using crossing experiments to evaluate the ease of transferring polyantha genes to the nitens chromosome arrangement. Whether or not the breakdown of the barrier was important to the flow of genes, it is clear that once in the nitens population the polyantha genes would be subjected to continual intra- and interchromosomal recombination. Selection for polyantha alleles had to be continued to produce the populations which currently have the polyantha phenotype and the nitens chromosome arrangement. Although the completeness of the phenotypic replacement in this case is perhaps surprising, other evidence does exist indicating that selection certainly can maintain complexes of nonlinked coadapted alleles (Allard et al., 1972; Hamrick and Allard, 1972). The ease with which the polyantha phenotype can be transferred to the nitens chromosome arrangement via a nitens-polyantha hybrid and backcrossing will be tested experimentally.

Finally, it is likely that the introgression process involved primarily an interaction between nitens, hybrids and hybrid derivatives—not polyantha. Both nitens and polyantha are bee pollinated outcrossers

with very limited seed dispersal. It is likely that the polyantha genotype was introduced into the nitens population mainly via haploid pollen and that very little direct competition between nitens and parental polyantha plants has ever occurred. Even if polyantha seeds were dispersed to nitens populations their fate would be hybridization because of the numerical preponderance of nitens. Thus, the major interaction had to be largely between nitens and hybrids. In compensation for their infertility the hybrids clearly had to have a strong competitive advantage over nitens. It is possible that the competitive difference involved the ability to cope with water deficit stress. Natural populations of polyantha have been observed which suffered well over 75% mortality during flowering due to drought in a particular season (Bloom, unpubl.). With such severe conditions in the zone of hybridization a greater tolerance of water stress could easily compensate for the lowered fertility of the hybrids. The fact that selection ultimately produced populations (polyantha–North) which are morphologically indistinguishable from polyantha, rather than stabilized hybrid types, suggests that on the average the hybrids were less fit than polyantha, though superior to nitens.

The relative fitness of polyantha–South and polyantha–North, that is, the parental polyantha compared to the polyantha reconstructed on the nitens chromosome arrangement is a question of major interest. While the external phenotypes of these two are very similar this does not necessarily indicate that they are just as similar in all adaptively significant characteristics. It is conceivable that some nitens alleles have been retained in the polyantha–North populations which contribute significantly to their fitness such that these introgressant populations are superior to both polyantha–South and nitens. In fact, this could have been a basic cause of the introgression. The question is, therefore, whether the driving force for this extensive introgression was hybrid superiority or a superior fitness of the polyantha genotype as an integrated unit. Studies on this question have been initiated involving comparisons of parental and introgressed populations with regard to enzymatic and morphological characteristics and their relative fitness under different environmental regimes.

The thoroughness of the replacement of the nitens phenotype by the polyantha phenotype suggests great caution in the interpretation of hybrid zones. Without the chromosomal evidence it is most unlikely that polyantha–North would have been recognized as introgressant. In reviewing the known cases of geographically extensive introgression, Heiser (1973) suggested some possible reasons for its apparent rarity: (1) it does not occur, for genetic reasons or because gene flow is too limited or (2) it is undetected because of the small number of genes involved. The polyantha introgression suggests an additional possibility, that introgression may be so complete as to be normally undetected. There is no strong reason to suspect that the polyantha introgression is unique; other comparable cases probably exist but are unrecognized. The width of the hybrid zone and the morphological evidence of introgression, or lack of it, will be controlled by the relative fitness of the parental and hybrid genetic systems. If the donor taxon functions as a tightly integrated genetic unit with recombinants unfit relative to it then the immediate hybrid zone with its wide array of genetic recombinants would be expected to be geographically narrow. New selection pressures favoring the donor's coadapted gene complex in the hybrid zone could yield populations identical in essential features to the donor and morphologically indistinguishable from it. Thus, as in polyantha, the hybridizing boundary between species may shift extensively geographically and do so by means of introgression.

Two facts suggest that introgressive replacement may be quite common. First,

numerous cases of both plants and animals are known where two closely related, largely allopatric taxa hybridize abundantly in the zone of contact (Remington, 1968). Second, much evidence is available showing that biologically significant climatic changes have been frequent and abrupt (measured in decades and centuries—see Dort and Jones, 1970). Given such climatic changes, it appears that two major alternatives exist at the hybridizing interface; either character displacement (reproductive or ecological) will occur and eliminate the genetic interaction, or replacement will occur via introgression. The numerous currently hybridizing taxa clearly have not eliminated their genetic interaction and it seems unlikely that all (or even most) such pairs have come into contact for the first time only since the last climatic change. The possibility exists that many such interfaces have existed through numerous climatic changes which were sufficiently severe to cause geographic movement of the interface by means of introgressive replacement. It is important to recognize also that introgression in a particular direction which results in an extensive morphological replacement except for one or a few, perhaps neutral, characters would normally be interpreted as introgression of the neutral character in the reverse direction. For example, population I resembles polyantha in metric characters very strongly but shows definite similarity to nitens in flower color, the simplest but erroneous explanation for this would be introgression of flower color from nitens into polyantha. The interpretation of the history of hybrid zones would appear to be extremely difficult. As pointed out by Blackith and Reyment (1971), "A full study of the nature of hybridization and the selective forces operating at tension zones where two species meet is of greatest value in understanding the evolution of the species involved: the morphometric aspects of the work are then useful but where a cytogenetic analysis is possible the combination of the two approaches is greatly to be desired."

The effectiveness of gene flow between local populations in linking their genetic systems has been debated considerably. According to Sokal (1974), "It is not at all established that local populations of species are reproductively interconnected (persuasively argued by Ehrlich and Raven, 1969)." The rate of gene flow in polyantha is unknown but as concluded for *Clarkia rubicunda* (Bartholomew et al., 1973) it is probably low, certainly not above average for outcrossing flowering plants. Both nitens and polyantha are self-compatible but morphologically adapted to promote outcrossing which is primarily accomplished by solitary bees not likely to effect long distance movement of pollen (MacSwain et al., 1973). The seeds appear to be adapted for the colonial habitat characteristic of clarkias and show no special adaptation for dispersal. Though some seed dispersal by rodents and seed-eating birds is likely, the pollen would appear to have greater interpopulational mobility. Whatever the rate of gene flow the movement of genes from polyantha into nitens for a distance of about 100 miles seems clear. This movement of genes undoubtedly was not accomplished in opposition to selective pressures but rather in conjunction with them. It is clear, however, that without interpopulational gene flow the populations which are now polyantha–North would have remained nitens–North.

## SUMMARY

The hypothesis that introgression has occurred from *Clarkia s. polyantha* into *C. nitens* led to an examination of the morphological relationships between polyantha, nitens and nine geographically intermediate populations using canonical variate analysis of 12 metric characters and eight flower color characters. The geographically intermediate populations were found to be phenotypically very similar to the polyantha reference population. The morphological evidence considered together with chromosomal evidence from previous studies indicates that substantial introgression, both genetically and geographically, has occurred from polyantha into nitens. Populations from the initial zone of hybridization (identified chromosomally) north for about 40 air miles are extremely similar to polyantha in external morphology with only the nitens chromosome arrangement marking their nitens ancestry. Populations for an additional 60 air miles to the north are also very similar morphologically to polyantha but show clear variation toward nitens in both metric characters and flower color characters. The numerous phenotypic differences between nitens and polyantha appear to be due to numerous segregating genes. The extensive replacement of nitens alleles by polyantha alleles on the nitens chromosomes via introgression, as evidenced by the phenotypic replacement, argues that the alleles responsible for the polyantha phenotype are functioning as part of a coadapted gene complex. The existence of such coadapted complexes of specific alleles would seem to be required for such introgressive replacement.

There appears to be no reason to assume that the polyantha-nitens introgression is exceptional and the completeness of the phenotypic replacement urges great caution in interpreting the genetic history of populations in regions of hybridization. In many cases where suitable cytological and/or genetic markers are lacking the genetic history of populations around hybridization zones must remain unknown. However, given suitable markers additional comparable cases undoubtedly will be found and their study in detail should yield valuable information on the genetic nature of species—the extent to which coadapted allelic arrays define adaptive units at the local and regional level and the significance of gene flow between these units.

## Acknowledgments

I thank Peter Neely, Norm Slade, and especially John Pizzimenti and Larry Holden for their assistance with the data analysis, and also Don Duncan, director of the San Joaquin Experimental Range for his cooperation. This study was supported in part by University of Kansas General Research Fund grant, 3988-5038.

## Literature Cited

ALLARD, R. W., G. R. BABBEL, M. T. CLEGG, AND A. L. KAHLER. 1972. Evidence for coadaptation in *Avena barbata*. Proc. Nat. Acad. Sci. U.S. 69:3043–3048.

ANDERSON, E. 1949. Introgressive hybridization. John Wiley and Sons, Inc. New York. 109 p.

ANDERSON, E., AND L. HUBRICHT. 1938. Hybridization in *Tradescantia*. III. The evidence for introgressive hybridization. Amer. J. Bot. 25: 396–402.

BARTHOLOMEW, B., L. C. EATON, AND P. H. RAVEN. 1973. *Clarkia Rubicunda*: a model of plant evolution in semiarid regions. Evolution. 27: 505–517.

BLACKITH, R. E., AND R. A. REYMENT. 1971. Multivariate Morphometrics. Academic Press, London and New York. 412 p.

BLOOM, W. L. 1974. Origin of reciprocal translocations and their effect in *Clarkia speciosa*. Chromosoma (Berl.) 49:61–76.

BLOOM, W. L., AND H. LEWIS. 1972. Interchanges and interpopulational gene exchange in *Clarkia speciosa*, p. 268–284. *In* Darlington, C. D. and K. Lewis (eds.). Chromosomes Today III. Constable Ltd., Edinburgh.

DORT, W., JR., AND J. K. JONES, JR. (eds.). 1970. Pleistocene and recent environments of the Central Great Plains. University of Kansas Press. Lawrence 433 p.

HAMRICK, J. L., AND R. W. ALLARD. 1972. Microgeographical variation in allozyme frequencies in *Avena barbata*. Proc. Nat. Acad. Sci. U.S. 69:2100–2104.

HEISER, C. B., JR. 1973. Introgression reexamined. Bot. Rev. 39:347–366.

LEWIS, H., AND M. E. LEWIS. 1955. The genus *Clarkia*. Univ. of Calif. Publ. Bot. 20:251–392.

MACSWAIN, J. W., P. H. RAVEN, AND R. W. THORP. 1973. Comparative behavior of bees and Onagraceae. IV. *Clarkia* bees of the western United States. Univ. Calif. Publ. Entom. 70:1–80.

MOSQUIN, T., AND H. LEWIS. 1959. (Abstract). Variation in relation to reciprocal translocations in a diploid species of *Clarkia* (Onagraceae). Proc. IX Int. Bot. Congr. 2:272.

REMINGTON, C. L. 1968. Suture-zones of hybrid interaction between recently joined biotas. Evolutionary Biol. 2:321–428.

SEAL, H. 1964. Multivariate Statistical Analysis for Biologists. Methuen, London. 207 p.

SOKAL, R. R. 1974. The species problem reconsidered. Syst. Zool. 22:360–374.

WAGNER, W. H., JR. 1970. Biosystematics and evolutionary noise. Taxon 19:146–151.

WATT, W. B. 1972. Intragenic recombination as a source of population genetic variability. Amer. Natur. 106:738–753.

Part VI

# REPRODUCTIVE CHARACTER DISPLACEMENT

# Editor's Comments
# on Papers 19 Through 22

19  **BROWN and WILSON**
    *Character Displacement*

20  **KNIGHT, ROBERTSON, and WADDINGTON**
    *Selection for Sexual Isolation Within a Species*

21  **EHRMAN**
    *Direct Observation of Sexual Isolation Between Allopatric and between Sympatric Strains of the Different* Drosophila Paulistorum *Races*

22  **LEVIN and KERSTER**
    *Natural Selection for Reproductive Isolation in* Phlox

Brown and Wilson (Paper 19) refer to differences of characters in areas of species contact as a consequence of reinforcement of reproductive isolating mechanisms or ecological divergence as character displacement. Their paper synthesizes a body of data on character displacement that shows its widespread occurrence among animals, and attempts to explain its significance for evolutionary theory. It also shows that areas of species contact may be one of divergence as opposed to hybridization and coalescence. Moreover, ecological character displacement may incidentally reinforce existing isolating mechanisms. P. R. Grant (1972) notes that the detection of character displacement involves two questions: (1) what was the precontact state, and (2) is the difference between pre- and post contact character states due to selection arising from the presence of a related species, or to some other cause? He redefines character displacement as "the process by which a morphological character state of a species changes under natural selection arising from the presence, in the same environment, of one or more species similar to it ecologically and/or reproductively." In addition to defining character displacement as a process as opposed to a product, Grant argues for a reassessment of the putative cases of character displacement, because of the weakness of most data sets vis à vis the questions posed above.

Of fundamental importance is the demonstration that selection can indeed reinforce pre-existing isolating mechanisms. *Drosophila* has been the focal point for this demonstration. Koopman (1950) mixed populations *D. pseudoobscura* and *D. persimilis* and showed that the level of ethological isolation between the species increased rapidly. After only five generations of selection, the proportion of hybrids among the offspring fell to a fraction of the former value. Working with the same species, Kessler (1966) successfully selected for both weaker and stronger ethological isolation in both species after eighteen generations.

In contrast to experiments involving two species, Knight, Robertson, and Waddington (Paper 20) initiated ethological isolation between two populations of *D. melanogaster* that previously showed no mating discrimination and that were marked by different recessive genes. A significant preference for homogamic mating developed after several generations of selection. Crossley (1974) repeated the experiment of Knight et al. and demonstrated some specific changes in mating behavior that contribute to ethological isolation. Dobzhansky, Pavlovsky, and Powell (1976) selected for the erection of ethological isolation between Llanos and Orinocan strains of *D. paulistorum*. Artificial selection favoring the offspring of homogamic mating was applied for 131 generations. By that time, the strains showed a pronounced preference for homogamic matings, but complete ethological isolation was not obtained.

Evidence that selection reinforces or builds isolating mechanisms in nature is of necessity indirect. Ehrman's observations (Paper 21) on allopatric and sympatric strains of *Drosophila paulistorum* are among the most elegant demonstrations that premating isolation between population systems can be enhanced in the geographic areas where hybridization is most likely to occur. There is less ethological isolation among allopatric than among sympatric strains of the same species pair. Subsequent studies on this species complex (Ehrman and Probber, 1978) have shown that aged females' sexual selection was similar to that of young females, and that previous heterogamic copulatory experience did not consistently change the level of sexual isolation. However, females with homogamic copulatory behavior displayed a significantly higher preference for homogamic males. This behavior can act as a barrier to gene exchange between population systems, and can lead to more rapid speciation. Wasserman and Koepfer (1977) have recently provided evidence for character displacement for ethological isolation between *D. mojavensis* and *D. arizonensis* in their region of sympatry. The increased level of homogamic mating is due to the behavior of the sympatric *D. mojavensis* flies, which are more isolated from all strains of *D. arizonensis* than are the allopatric *D. mojavensis*.

Although *Drosophila* has been the focal point thus far because both laboratory and field data are available, character displacement in animals also has been described in other invertebrates, fish, amphibians, birds, and mammals. Several references are given in the papers cited above, and the journals *Evolution* and *Systematic Zoology* have published several papers on the subject in the past decade. The most detailed analysis of character-displacement in birds involves the genus *Sitta* (P. R. Grant, 1975). The symposium on character displacement published in 1974 in *American Zoologist* also is a prime source.

Character displacement in plants has received much less fanfare than it has in animals. Yet there are a few instances where is may have occurred. Levin and Kerster (Paper 22) present evidence indicating natural selection for ethological isolation in *Phlox*. *Phlox pilosa* and *P. glaberrima* are both pink-flowered, and sympatric in Illinois and Indiana. In several areas of close contact, the pink corolla phase of *P. pilosa* has been replaced by a white one. Pollen flow from *P. glaberrima* to *P. pilosa* is nearly five times greater to the pink than to the white phase. The shift from pink to white reduces hybrid seed from 4 percent to 2 percent (Levin and Schaal, 1970).

A parallel to character displacement in *Phlox* is seen in *Clarkia* (Lewis and Lewis, 1955). In *Clarkia dudleyana*, flower color varies from the typical pale lavender to nearly white on the one hand, and bright pink on the other. In areas of contact with *C. biloba*, which has lavender to pink flowers, a pale white-streaked variant of *C. dudleyana* prevails. As hybrids between these species are most difficult to obtain, we may infer that the shift is not consequence of selection against hybridization, but for more reliable pollinator service.

Breedlove (1969) has described an apparent case of character displacement in *Fuchsia* involving flower color, flower dimensions, and pollinators. Where allopatric, *F. parviflora* and *F. encliandra* display red sepals and hypanthium, the latter being of similar dimensions in both species. The petals of *F. parviflora* are red, whereas those of *F. encliandra* are white or pink. In areas of sympatry, *F. parviflora* displays white sepals and petals, and pink to pale red hypanthium. The hypanthium is shorter and broader than the norm, and flowers appear almost totally white because the sepals are reflexed and the petals are spreading. On the other hand, the flowers of *F. encliandra* have sepals, petals, and a hypanthium that are longer and narrower than in allopatric populations. When not in contact, both species are pollinated by hummingbirds and bees. Character displacement has reinforced an ethological barrier and introduced a mechanical one. The color change of *F. encliandra* has made it more attractive to hummingbirds, while the lengthening and narrowing of the hypanthium precludes effective feeding by bees. The color change of *F. parviflora* has made it less attractive to hummingbirds, while the shortening and broadening of the hypanthium facilitates bee pollination. The stimulus for character displacement and pollinator specialization presumably has been competition for pollinators, because the species are not known to hybridize.

Whalen (1977) recently described an apparent case of character displacement in *Solanum. Solanum lumholtzianum* and *S. grayi* have medium-sized flowers throughout most of their respective ranges, and they are effectively pollinated by large bees such as species of *Bombus, Xylocopa,* and *Protoxea.* In the area of sympatry, the flowers of *S. grayi* are much smaller than those in conspecific populations outside the area of contact and smaller than those of *S. lumholtzianum.* The smaller flowers of *S. grayi* are pollinated by small bees, including species of *Nomia* and *Exomalopsis.* These pollinators are ineffective on the flowers of the congeneric species as well as the flowers of *S. grayi* beyond the area of sympatry. The shift in flower size in *S. grayi* has resulted in the emergence of an effective mechanical barrier to interspecific pollen exchange. The two species are cross-compatible, but the hybrid seed is much smaller than that of the parents. Competition for pollinators probably was the stimulus for divergence.

There is one strongly suggestive case of character displacement in the breeding system (Hinton, 1976). *Calyptridium monospermum* is an outcrosser over most of its range; it has rose-colored petals and faint fragrance, and it is pollinated primarily by bees. *Calyptridium umbellatum* is a facultative inbreeder. This species has white petals and a strong floral fragrance and is pollinated by butterflies, flies, and beetles. In a sympatry with *C. umbellatum,* some populations of *C. monospermum* have adopted a complex of floral modifications that promotes selfing similar to that of *C. umbellatum.* The former has white petals and strong fragrance, and many of the same pollinators as *C. umbellatum.* The two species hybridize extensively in some areas. Perhaps *C. monospermum* has shifted its breeding posture in the areas of sympatry because the novel breeding posture is adaptive in its own right vis-à-vis a pollinator and has nothing to do with retarding hybridization. On the other hand, facultative inbreeding would reduce the proportion of seeds sacrificed to hybridization or backcrossing, and the amount of pollen carried outside the species.

V. Grant (1966a) studied cross-compatibility among species of the leafy-stemmed gilias and among the cobwebby gilias. He showed the presence of an inverse correlation between geographical proximity and cross-compatibility, and concluded that this pattern is the result of selection against hybridization in areas of sympatry. He also reported the presence of polymorphism for interspecific cross-compatibility, which is a prerequisite for the selective alteration of crossability.

Selection for reproductive isolation may be stimulated by the loss of reproductive potential associated with heterospecific mating or by competition for a resource upon which mating success rests. Levin (1970) proposed that such selection in response to heterospecific mating is more likely to occur in animals (insects and vertebrates) than in plants (angiosperms) because animals generally experience a greater reproductive handicap per breeding episode. It has also been proposed that selection for reproductive isolation in response to competition is more likely to occur in animal-pollinated plants than in animals, because plants, unlike animals, depend on a specific resource (the pollinators). These relationships suggest that selection for reproductive isolation in plants will usually be the product of competition, whereas in animals it will usually be the product of heterospecific mating.

# 19

Reprinted from *Syst. Zool.* **5**:49–64 (1956)

# Character Displacement

## W. L. BROWN, JR. and E. O. WILSON

IT IS the purpose of the present paper to discuss a seldom-recognized and poorly known speciation phenomenon that we consider to be of potential major significance in animal systematics. This condition, which we have come to call "character displacement," may be roughly described as follows. Two closely related species have overlapping ranges. In the parts of the ranges where one species occurs alone, the populations of that species are similar to the other species and may even be very difficult to distinguish from it. In the area of overlap, where the two species occur together, the populations are more divergent and easily distinguished, i.e., they "displace" one another in one or more characters. The characters involved can be morphological, ecological, behavioral, or physiological; they are assumed to be genetically based.

The same pattern may be stated equally well in the opposite way, as follows. Two closely related species are distinct where they occur together, but where one member of the pair occurs alone it converges toward the second, even to the extent of being nearly identical with it in some characters. Experience has shown that it is from this latter point of view that character displacement is most easily detected in routine taxonomic analysis.

By stating the situation in two ways, we have called attention to the dual nature of the pattern: species populations show displacement where they occur together, and convergence where they do not. Character displacement just might in some cases represent no more than a peculiar and in a limited sense a fortuitous pattern of variation. But in our opinion it is generally much more than this; we believe that it is a common aspect of geographical speciation, arising most often as a product of the genetic and ecological in-

teraction of two (or more) newly evolved, cognate species during their period of first contact. This thesis will be discussed in more detail in a later section.

Character displacement is not a new concept. A number of authors have described it more or less in detail, and a few have commented on its evolutionary significance. We should like in the present paper to bring some of this material together, to illustrate the various aspects the pattern may assume in nature, and to discuss the possible consequences in taxonomic theory and practice which may follow from a wider appreciation of the phenomenon.

## Two Illustrations

An example of character displacement outstanding for its simplicity and clarity has been reviewed most recently by Vaurie (1950, 1951). This involves the closely related rock nuthatches *Sitta neumayer* Michahelles and *S. tephronota* Sharpe. *S. neumayer* ranges from the Balkans eastward through the western half of Iran, while *S. tephronota* extends from the Tien Shan in Turkestan westward to Armenia. Thus, the two species come to overlap very broadly in several sectors of Iran (Fig. 1). Outside the zone of overlap, the two species are extremely similar, and at best can be told apart only after careful examination by a taxonomist with some experience in the complex (Vaurie, personal communication). Both species show some geographical variation, and it seems clear from Vaurie's account (1950, Table 5, pp. 25–26) that such races as bear names have been raised for character discordances in various combinations. It therefore appears safe to ignore the subspecies analysis as such and to concentrate on the variation of the independent characters themselves.

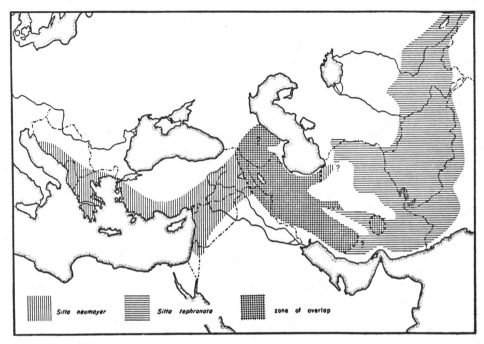

FIG. 1. Distribution of *Sitta neumayer* and *S. tephronota*. (After Vaurie.)

These show quite remarkable displacement phenomena in the Iranian region of overlap between the species, where the two species apparently usually occur in more or less equal numbers (see Fig. 2). In this region, *S. neumayer* shows distinct reductions in overall size and bill length, as well as in width, size, and distinctness of the facial stripe. *S. tephronota*, on the other hand, shows striking positive augmentation of all the same characters in the overlap zone, so that it is distinguishable from sympatric *neumayer* at a glance. Vaurie concludes, we think quite correctly, that the differences within the zone of overlap constitute one basis upon which the two species can avoid competition where they are sympatric. The case of these two nuthatches has already received considerable attention both in the literature and elsewhere, and it bids fair to become the classic illustration of character displacement.

A more complicated case involving multiple character displacement is seen in the ant genus *Lasius* (Wilson, 1955). Where they occur together, in forested eastern North America, the related species *L. flavus* (Fabr.) and *L. nearcticus* Wheeler show differences in the following seven characters: antennal length, ommatidium number, head shape, degree of worker polymorphism, relative lengths of palpal segments, cephalic pubescence, and queen size. In western North America and the Palaearctic Region, where *nearcticus* is absent, *flavus* is convergent to it in all seven characters. In this shift, each character behaves in an independent fashion; e.g., scape length becomes exactly intermediate between that of the two eastern populations, ommatidium number increases in variability and overlaps the range of the two, and queen size changes to that of *nearcticus*. In North Dakota, at the western fringe of the *nearcticus* distribution, the *flavus* population is at an intermediate level of convergence (Fig. 3).

There is some evidence that this dual displacement-convergence pattern is associated with competition and ecological

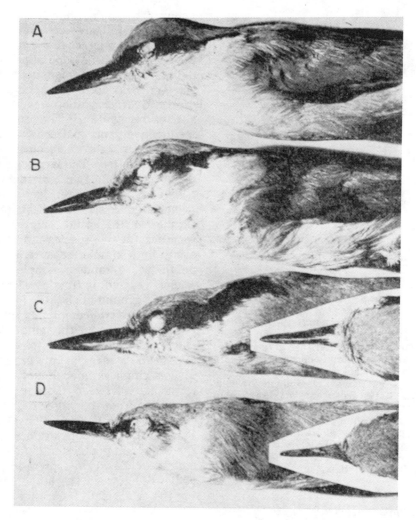

FIG. 2. Size and shape of the bill and facial stripe in *Sitta neumayer* and *S. tephronota*: *A, S. neumayer* from Dalmatia; *B, S. tephronota* from Ferghana; *C, S. tephronota* and *D, S. neumayer*, both from Durud, Luristan, in western Iran. (After Vaurie.)

displacement between the two species. So far as is known, they have similar food requirements. But in eastern North America, where they occur together, *flavus* is mainly limited to open, dry forest with moderate to thin leaf-litter, while *nearcticus* is found primarily in moist, dense forest with thick leaf-litter. There is little information available on the western North American and Asian *flavus* populations, but in northern Europe this species is known to be highly adaptable; preferring open situations, but also occurring commonly in moist forests.

### Some Additional Examples

In the following paragraphs we wish to present a number of cases selected from the literature (with two additional unpublished examples) which we have interpreted as showing character displacement. In so doing we are trying to document the thesis that character displacement occurs widely in many groups of animals and in a range of particular patterns. But at the same time we are obliged to give warning, perhaps unnecessarily for the critical reader, that most of these cases in-

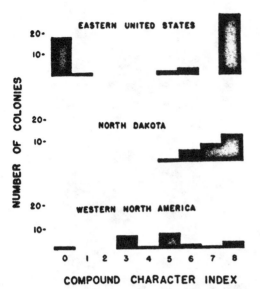

FIG. 3. Frequency histograms of the compound character index of the ants *Lasius nearcticus* (0–1) and *L. flavus* (3–8) in three broad geographic samples. For each colony typical *nearcticus* characters are given a score of 0, typical eastern *flavus* characters a score of 2, and intermediate characters a score of 1. The four characters most clearcut in the eastern United States are used: maxillary palp proportions, antennal scape index, compound eye ommatidium number and head shape. Thus, completely typical *nearcticus* colonies score a total of 0 and completely typical eastern *flavus* 8, with the various ranks of intermediates falling in between (after Wilson, 1955).

volve discontinuously distributed populations, that as a result the species status of these populations with respect to one another has not been ascertained with complete certainty, and that explanations alternative to character displacement are therefore assuredly possible. We ask only that the reader bear through and consider our interpretation in each case.

*Birds of the Genus* Geospiza. A striking case of character displacement has been described by David Lack in his classic, *Darwin's Finches* (1947). Lack has shown that in the Galapagos certain species of *Geospiza* are often absent on smaller islands, in which case their food niche is filled by other species of the

genus. The populations of the latter tend to converge in body size and beak form to the absent species, so much so as to make placement of these populations to species difficult. Lack has demonstrated that body size and beak form are generally important in *Geospiza* in both food getting and species recognition. The dual displacement-convergence pattern we are interested in occurs, at least once, in the following situation. The larger ground-finch *Geospiza fortis* Gould and the smaller *G. fuliginosa* Gould differ from each other principally in size and beak proportion. On most of the islands, where they occur together, the two species can be separated easily by a simple measurement of beak depth, i.e., a random sample of ground-finches (excluding from consideration the largest ground-finch *G. magnirostris* Gould) gives two completely separate distribution curves in this single character. But on the small islands of Daphne and Crossman a sample of ground-finches gives a single unimodal curve exactly intermediate between those of *fortis* and *fuliginosa* from the larger islands. Analysis of beak-wing proportions has shown that the Daphne population is *fortis* and the Crossman population is *fuliginosa;* according to Lack's interpretation each has converged toward the other species, filling the ecological vacuum its absence has created.

*Birds of the Genus* Myzantha. Among the Australian honey-eaters of the genus *Myzantha,* a light-colored species, *M. flavigula,* occupies the greater part of the arid inland. Toward the wet southwestern corner of the continent, *flavigula* blends gradually into a darker population, usually referred to as "subspecies *obscura.*" In southeastern Australia, in higher-rainfall country, *flavigula* is replaced by two forms—*M. melanocephala,* mostly in the wettest districts, and *M. melanotis* of the subarid Victorian-South Australian mallee district. The southwestern (*obscura*) and one of the southeastern populations (*melanotis*) are extremely similar, differing by what are described as trifling characters of plumage shading, so that some authors consider them conspecific.

The members of an ornithological camp-out in the Victorian mallee, however, have found that *melanotis* there nests sympatrically with both *melanocephala* and *flavigula*, and that at this place the three behave as distinct species without intergradation. Thus we find the two morphologically very similar forms, *obscura* and *melanotis*, flanking the much more widely distributed and differently colored species, *flavigula*, but showing exactly opposite interbreeding reactions with *flavigula*. *Obscura* appears to represent merely the terminus of a cline for melanism produced by *flavigula* in the southwest, where, it may be noted, there is no other competing dark form of the same species group (Fig. 4).

Judging by the findings of the mallee observers, *melanotis* is clearly to be regarded as a species distinct from *flavigula*, including the southwestern *obscura* population. In this we follow Condon (1951), and not Serventy (1953), though

the latter has furnished the most comprehensive analysis of the situation.

Serventy's dilemma is keyed by his statement that ". . . it would be unreal to treat *melanotis*, obviously so akin to south-western *obscura*, as a separate species from it. . . ." Here one plainly sees the conflict between two species criteria: one based on morphological similarity, and one on interbreeding reaction in the zone of sympatry.

From the data presented, we interpret the *Myzantha* situation as a case of character displacement. *M. flavigula* tends to produce, in the less arid extremities of its range, populations with darker plumage. In the southwest, it has done just this; presumably, melanism is connected adaptively in some way, directly or indirectly, with increased moisture ("Gloger's Rule"), or plant cover, or both. In the southeastern mallee, however, the melanistic tendencies presumed to be latent or potential in *flavigula* toward the wetter extremes of its range are suppressed in the presence of the darker species *melanotis* (and possibly also *melanocephala*). It would be interesting to know more about

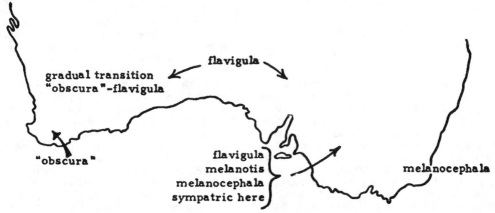

FIG. 4. Map showing the geographical relationships of three species of the bird genus *Myzantha* in southern Australia, based on the discussion of Serventy, 1953. *M. flavigula*, the light-colored bird of arid central Australia, grades into a darker population ("race *obscura*") in southwestern higher-rainfall districts. In southeastern Australia, in the Victorian mallee belt, transitional and mixed ecological conditions allow three non-intergrading species to breed side by side: *M. flavigula; M. melanotis*, a species characteristic of the mallee scrub; and *M. melanocephala*, a southeastern bird of the higher-rainfall districts. *M. melanotis* and the "*obscura*" population are extremely similar, and have been considered synonymous or at least conspecific in the past.

the ecological distribution, food, and habits of the three *Myzantha* species within the region where they occur together.

*Parrots of the Genus* Platycercus. Serventy (1953) also reviews, among other cases that may involve character displacement, the situation in the rosellas of southeastern Australia (Fig. 5). The crimson rosella (*Platycercus elegans*) is a species of the wooded eastern areas—mostly those with higher rainfall nearest the coast. On Kangaroo Island, off the coast of South Australia, occurs a crimson population that appears to be *elegans* from a strictly morphological viewpoint. Beginning on the mainland opposite Kangaroo Island is a cline connecting the crimson form to an inland, arid-country yellow form (*P. flaveolus*) inhabiting the red gums of the rivers and dry creeks in

FIG. 5. Map showing the approximate distribution of color forms of the rosellas (parrots) of the *Platycercus elegans* complex in southeastern Australia. The heavy pecked line indicates roughly the inland margin of the southeastern highlands and the higher-rainfall districts, and also the inland limit of the range of the crimson-trimmed *P. elegans*. Inside this line, along the upper reaches of the Murray-Darling river systems, the closely related *P. flaveolus*, a yellow-trimmed form, approaches and may even meet the range of *P. elegans* at some points without producing intergrades. Downstream, *P. flaveolus* grades through a series of intermediately-colored populations culminating in the crimson-trimmed flocks of Kangaroo Island, which are apparently outwardly indistinguishable from those of the true eastern *elegans*. (Adapted from Cain, 1955.)

the Murray-Darling Basins. However, in the Albury district of the upper Murray River and elsewhere up the other rivers, *flaveolus* overlaps or closely approaches the true southeastern *elegans* along a wide front without interbreeding (for a recent detailed account, see Cain, 1955).

It is interesting to note that the cline from yellow to crimson in South Australia follows broadly the regional increase in moisture and luxuriance of forest vegetation; both rise to peaks in the ravines at the western end of Kangaroo Island. We suggest that the South Australian clinal population on the mainland, and probably even the crimson populations of Kangaroo Island, are referable to *flaveolus*, which can here produce a wet-adapted crimson form free of displacement pressure from *elegans*.

*Birds of the Cape Verde Islands.* Bourne (1955) in his review of the birds of the Cape Verde Islands, has presented several cases of character displacement so concisely and pointedly that we can quote him directly:

The two shearwaters [breeding in the Cape Verde Islands], Cory's shearwater *Procellaria diomedea* and the Little Shearwater *Procellaria baroli*, take similar foods (fish and cephalopods) differing only in size; competition for food between the two species is reduced by the development of different breeding seasons. Elsewhere in its range *Procellaria diomedea* breeds at the same stations as the medium Manx Shearwater *Procellaria puffinus*, which takes similar foods but breeds slightly earlier. There is a dramatic difference in size, and particularly the size of bill, between those races of *Procellaria diomedea* which breed with *Procellaria puffinus* and the form [*P. diomedea*] *edwardsi* which breeds alone at the Cape Verde Islands, the latter having a bill exactly intermediate in size between that of the northern races and that of *Procellaria puffinus*. It seems likely that *edwardsi* takes the food that is divided between both species elsewhere. It may be remarked that one race of *Procellaria puffinus*, *mauretanicus* of the Balearic Islands, avoids competition with *Procellaria diomedea* by breeding unusually early and leaving the area when the larger species prepares to nest; it is significant that this is the only race of the species which has a large bill resembling that of *P. d. edwardsi.* It would appear that the bill-size and the breeding seasons of these shearwaters vary with the amount of competition occurring between different species breeding at the same site. . . .

Where the two kites *Milvus milvus* and *Milvus migrans* occur together the latter is the species which commonly feeds over water. The race of *Milvus milvus* found in the Cape Verde Islands closely resembles *Milvus migrans* in the field, and very commonly feeds along the shore and over the sea. It may replace *Milvus migrans,* but it seems likely that with the Raven *Corvus corax,* which also abounds along the shore, it replaces the gulls *Larus* spp. which usually scavenge along the shore elsewhere but have failed to colonize the barren coast of the islands.

Bourne's opinion concerning which species are replaced is a little confusing in this case, since elsewhere *Milvus,* notably *M. migrans* in India, often tends to replace or at least dominate the gulls in scavenger-feeding situations around seaports (Brown, personal observation). The absence of *migrans* seems to us the probable chief reason for the convergence characteristics in the Cape Verde Islands populations of *milvus.*

Bourne cites one additional case:

The Cane Warbler *Acrocephalus brevipennis* [a species precinctive to the Cape Verde Islands] is closely related to large and small sibling species *Acrocephalus rufescens* and *A. gracilirostris* which occur together in the same habitats on the [African] mainland. Where the ranges of these two species overlap they are sharply distinct in size and voice; where they occur apart these distinctions are less marked (Chapin, 1949). *A. brevipennis* is probably related to the larger species, *A. rufescens,* but in the absence of the smaller species it is exactly intermediate in all its characters except the bill, which is large, resembling that of *A. rufescens.* The large bill may be part of the general trend seen on islands, or a consequence of competition for food with the smaller *Sylvia* warblers.

*Birds of the Genus* Monarcha. Mayr (1955 and personal communication) has described a case of displacement in the monarch flycatchers of the Bismarck Archipelago. *Monarcha alecto* and *M. hebetior eichhorni* occur together through the main chain of the Bismarcks, from New Britain north onto New Hanover, but beyond, on isolated St. Matthias, *M. hebetior hebetior* occurs alone; this last is an ambiguous variant combining several features of *alecto* and *eichhorni.* Mayr suggests the following evolutionary scheme: *hebetior* differentiated from *alecto* as an isolate on St. Matthias and later reinvaded the range of *alecto* on New Britain and New Ireland, where it diverged further under displacement pressure from the latter until it became the present *eichhorni.*

It seems to us that this situation can be more simply explained by assuming that the Bismarcks were first populated by a stock which evolved within the Archipelago and became the species *hebetior.* The later entry of *alecto* into the chain was followed by the displacement of *hebetior* as far as the sympatry extended, leaving the St. Matthias isolate to represent the undisplaced relict of the original *hebetior.*

*Fishes of the Genus* Micropterus. The two basses *Micropterus punctulatus* and *M. dolomieu* have ranges which include a large part of the eastern United States and are mostly coextensive (Hubbs and Bailey, 1940). Of the two, however, only *punctulatus* is known to occur in Kansas, western Oklahoma, and the Gulf States south of the Tennessee River drainage system. In the Wichita Mountains of western Oklahoma there is a population, described as *M. punctulatus wichitae,* which is intermediate between typical *punctulatus* and *dolomieu.* Its affinity to *punctulatus* is shown by the fact that in a number of characters it grades without a break into *punctulatus,* so that some specimens are indistinguishable from typical *punctulatus,* and in its agreement with *punctulatus* in the critical character of scale-row counts. Hubbs and Bailey seem to favor the theory of a hybrid origin for *wichitae,* but they consider this "no more plausible than the view that the similarities between *wichitae* and *dolomieu* are caused by parallel development, or the view that *wichitae* is a relict of a generally extinct transitional stage between *punctulatus* and *dolomieu*." We, of course, are inclined to favor parallel development, resulting specifically from the absence of the displacing influence of *dolomieu*, as the simplest and most plausible explanation.

Away to the south, many of the Texas populations of *punctulatus* are peculiar in showing converging trends toward *dolomieu*, but less strongly, so that Hubbs and Bailey consider them as possible intermediates between *punctulatus* and *wichitae*. In northern Alabama and Georgia there is a form described as a distinct species (*M. coosae*), which combines some of the characters of *punctulatus* and *dolomieu*, besides showing some peculiar to itself. *Coosae* is completely allopatric to *dolomieu*, and there is some evidence that it may hybridize extensively with the sympatric *punctulatus*. We should like to suggest the possibility here that *coosae* is conspecific with *punctulatus* and represents a section of the *punctulatus* population tending to converge toward *dolomieu* where that species is absent.

In summary, it appears to us likely that *wichitae*, the Texas populations, and possibly even *coosae*, each of which shows intermediate characters, are not products of introgressive hybridization, but may instead represent true *punctulatus* stocks that have tended to converge toward *dolomieu* in the absence of displacing influence from that species.

*Frogs of the Genus* Microhyla. W. F. Blair (1955) concludes from his study of two North American frogs of the genus *Microhyla*:

The evidence now available shows that there are geographic gradients in body size in both *Microhyla olivacea* and *M. carolinensis*. The former species shows a west to east decrease in body length, while the latter shows an east to west increase. The clines are such, therefore, that the largest *carolinensis* and the smallest *olivacea*, on the average, occur in the overlap zone of the two species. This pattern of geographic variation in body size parallels the pattern of geographic variation in mating call reported by W. F. Blair (1955) [in press] in which the greatest call differences in frequency and in length occur in the overlap zone. One of these call characteristics, frequency, probably is directly related to body size, for smaller anurans of any given group tend to have a higher pitched call than larger ones of the same group. The other, length of call, appears unrelated to size.

The differences in body size, like those in mating call, belong to a complex of isolation mechanisms (W. F. Blair, 1955) which tends to restrict interspecific mating in the overlap zone of the two species. The existence of the greatest size differences as well as the greatest call differences where the two species are exposed to possible hybridization supports the argument (*op. cit.*) that these potential isolation mechanisms are being reinforced through natural selection.

*Frogs of the Genus* Crinia. A most interesting case in the Australian genus *Crinia* has recently been called to our attention by A. R. Main (*in litt.*). Where they occur together in Western Australia, as around Perth, the two species *C. glauerti* and *C. insignifera* have markedly different calls. *C. glauerti* has a rattling call resembling "a pea falling into a can and bouncing"; oscilloscope analysis shows this to consist of evenly spaced single impulses at the rate of about 16 per second. *C. insignifera* produces a call "similar to a wet finger being drawn over an inflated rubber balloon . . . we refer to this call as a 'squelch.'" Oscilloscope analysis shows the squelch to have a duration of about 0.25 second and to consist of impulses crowded together. Around Perth and in other localities where it is sympatric with *insignifera*, *glauerti* individuals are occasionally heard to produce the beginnings of the "squelch" by running 12–15 impulses together, but this occurrence is extremely rare. Along the south coast of Western Australia, however, where *glauerti* occurs alone, the call is commonly modified by running 30 or more single impulses together to produce a squelch almost identical to the ear with that of *insignifera*. Thus, in effect, where this species occurs alone it has extended the variability of its call to include the sounds typical of both species. According to Main, the two species show color differences in the breeding males and different ecological preferences; laboratory crosses show reduced $F_1$ viability. It seems evident to us (Brown and Wilson) that displacement in this case is associated with the reinforcement of reproductive barriers, the breakdown of which would result in inferior hybrids. This aspect will be discussed more fully in a later section.

*Ants of the Genus* Rhytidoponera. The ants of the Australian *Rhytidoponera metallica* group (revised by Brown, ms.) are widespread and often among the dominant insects of given localities. The common greenhead (*R. metallica*) is the most successful species—a metallescent green or purple ant adapted to a variety of habitats ranging from desert to warm, open woodland, and the only species of the group at all abundant across the dry interior of Australia. In the southeastern and southwestern ("Bassian") corners of the continent, where the rainfall is higher and luxuriant forests occur, *metallica* is replaced by similar species of the same group that nearly or quite completely lack metallic coloration (Fig. 6).

In the east, two such species make the replacement, *R. tasmaniensis* Emery and *R. victoriae* André. *R. tasmaniensis* is the larger of the two, has the fine gastric sculpture of *metallica*, and is usually reddish brown, with bronzy-brown gaster. It is virtually identical with *metallica*, except for color. *R. victoriae* is smaller, more blackish, and has relatively coarser gastric striation. *R. tasmaniensis* is found in a variety of woodland situations, but apparently is excluded from the very wettest forests, which are occupied by *victoriae*. Nevertheless, the two species exist in abundance side by side over large parts of southeastern Australia without a sign of interbreeding. At some points, such as on the moist temperate grasslands west of Melbourne, both species occur together with *metallica*, but maintain their distinctness.

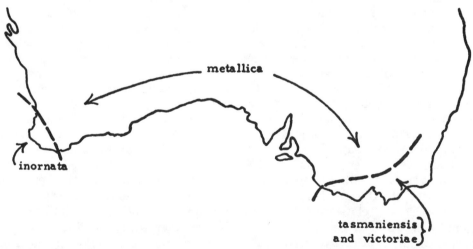

FIG. 6. Map showing the approximate distribution in southern Australia of four closely related common species of ants of the *Rhytidoponera metallica* group. *R. metallica* is nearly or quite the only representative of its group in the more arid central regions, and occurs in open situations in the southeast and southwest as well. In the moister forests of the southeast, *metallica* is replaced by the small, dark *R. victoriae* and the larger, more reddish *R. tasmaniensis*, which frequently occur side by side in the same localities. In the mesic wooded areas of the extreme southwestern corner of Australia, *metallica* is replaced by *R. inornata*, a distinct species which in size and color resembles closely, and broadly overlaps in variation, the two eastern forest species.

In the southwestern corner of Australia, *metallica* is replaced in the wetter parts of the region by non-metallescent *R. inornata* Crawley (though the two species overlap in the Darling Range and undoubtedly elsewhere). The interesting feature here is that *R. inornata* varies in size and color so as to cover the variation in these attributes of both southeastern non-metallic species, *tasmaniensis* and *victoriae*. In fact, one might speak of the two southeastern forms as mutually-displacing equivalents of the southwestern *inornata*, the latter being nearest the generalized type of the group because it has never suffered close competitive pressure and the character displacement that helps to relieve that pressure. This example illustrates the existence of a dual character-displacement pattern where the convergent population is clearly at, or above, the species level.

*Slave-making* Formica *Ants.* A simpler case in the ants involves the famous Holarctic slavemakers of the *Formica sanguinea* group. In a recent revision (Wilson and Brown, 1955) only three really distinct species are recognized in the group: *F. sanguinea* Latreille, widely distributed through temperate and northern Eurasia, where it is the only species; *F. subnuda* Emery, of boreal and subboreal North America; and *F. subintegra* Emery, ranging through temperate North America and overlapping the range of *subnuda* in the northern United States and along the Rocky Mountain chain.

The two most different forms are *subnuda* and *subintegra*, which can be separated on several external characters. *F. sanguinea* is closely related to *subnuda* in form and habits and is treated as a separate species only arbitrarily, on the basis of slight morphological discontinuities. At the same time, *sanguinea* has pilosity intermediate between that of the two American species, and its clypeal notch, a second important diagnostic character, is more like that of *subintegra* than like that of *subnuda*. We have interpreted this

pattern to represent a displacement of *subnuda* away from *subintegra* where these two species meet and interact, while the Palaearctic equivalent of *subnuda* (i.e., *sanguinea*) has tended to converge toward *subintegra* as a consequence of its filling the "adaptive vacuum" which a companion species might otherwise occupy. Of course in this case, as in all others under present consideration, there is no way of determining how much "displacement" has occurred as a process in the sympatric populations as opposed to "convergence" in the unispecific one. The final pattern observed may in fact be the result of one of these two processes alone.

*American Scarabaeid Beetles.* Howden (1955, p. 207) discusses the status of two geotrupine beetles considered by him to represent subspecies of the species *Eucanthus lazarus* (Fabricius). His *E. l. lazarus* is stated to range widely over the United States, but records from the Gulf States, excepting Florida, are scanty. *E. l. subtropicus* Howden, on the other hand, is restricted to the southeastern states, and is best represented in Florida, Georgia, Alabama and neighboring states.

Howden is puzzled by the apparent fact that "intermediates" between the two forms came from areas "not bordering the Gulf of Mexico," despite the circumstance that it is in this region that the main overlap falls. Intergrades came from areas "on the East Coast," and from Miami, Florida, and, "Occasional northern specimens appear to exhibit most of the characters of *subtropicus*." However, in particular limited localities, presumably near or in the zone of overlap, Howden was able to name the populations one way or the other with little difficulty.

Although the situation in the Florida Peninsula is not clear from Howden's account, the "intermediate" and more typical-appearing *E. l. lazarus* occurring together with *subtropicus* in the Miami area may really represent undisplaced populations of *subtropicus*. If this is the case, then we would favor Howden's alternative interpretation, and consider *lazarus* and *subtropicus* as closely related but distinct species.

*Crabs of the Genus* Uca. Jocelyn Crane (in Allee *et al.*, 1950, p. 620) notes that in fiddler crabs of the genus *Uca* differentiation in behavior and often in the coloration of the male is greater if the species are found together than if they are found in different habitats or regions.

## The Evolution of Character Displacement

Divergence between two species where they occur together, coupled with convergence where they do not, is a pattern that strongly suggests some form of interaction in the evolutionary history of the pair. The usual case may be one in which the members of the pair are cognate (derived from the same immediate parental population) and have recently made secondary contact following the geographical isolation that has mediated their divergence to species level. In such cases, the "terminal" populations, to which overlap does not yet extend, are not affected by the contact and remain closely similar to each other. But where contact has been made, there are two important ways in which the sympatric populations can interact to augment their initial divergence.

The first type of interaction might best be termed *reinforcement* [1] of the reproductive barriers. It may happen that the species continue to interbreed to some extent, and either the resulting inseminations are ineffectual, or the hybrids produced are inviable or sterile, resulting in what geneticists have termed "gamete wastage." Consequently, any further ethological or genetic divergence reducing this wastage will be strongly favored by natural selection (Dobzhansky, 1951; Koopman, 1950; Kawamura, 1953).

Of conceivably equal or greater importance is the process of *ecological displacement*. It seems clear from an *a priori*

---

[1] *Reinforcement* is a familiar term in psychology that has been applied to speciation processes (Blair, 1955).

basis that any further ecological divergence lessening competition between the overlapping populations will be favored by natural selection if it has a genetic basis (Mayr, 1949). That such a process actually occurs is suggested by abundant indirect evidence from ornithology (Lack, 1944), as well as the cases already cited above.

It seems unnecessary to go into a detailed discussion of these previously elaborated concepts, except to point out that secondary divergence of this nature inevitably entails phenotypic "characters" of the type employed in ordinary taxonomic work. Character displacement therefore may be considered as merely the aspects of such divergence that are recognizable to the taxonomist and some other favored organisms. It is interesting to note that the tendency toward displacement of characters is opposed by the pressure for mimicry. One can imagine some elaborate interactions between the two tendencies, particularly in the evolutionarily fertile tropics.

## Competition

The concept of competition has been the focus of much important disagreement among ecologists and other biologists, and it deserves close and persistent investigation. However, were it not that Andrewartha and Birch (1954) criticize the use of the concept by Lack and others to explain distribution and variation of birds and other animals, we might well have avoided discussing it here altogether. Andrewartha and Birch (p. 25) seem to consider that competition is an idea of lesser, perhaps even negligible, importance in biology. They think that the tendency for closely related species to inhabit different areas or exploit different ecological niches (as reported, for instance, by Lack) may conceivably have originated from causes "quite different" from competition. They do not offer alternatives that seem to us anything like as satisfactory as Lack's hypotheses.

Andrewartha and Birch make a point

when they ask for more direct evidence for the action of competition, but it is clear that they have failed to appreciate the amount of evidence that does exist in the literature. However, interspecific competition of the direct, conspicuous, unequivocal kind is apparently a relatively evanescent stage in the relationship of animal individuals or species, and therefore it is difficult to catch and record (just as is the often parallel crisis in the rise of reproductive barriers between two newly diverging species). What we usually see is the result of an actually or potentially competitive contact, in which one competitor has been suppressed or is being forced by some form of aggressive behavior to take second choice, or in which an equilibrium has been established when the potential competitors are specialized to split up the exploitable requisites in their environment. A third possible result is the dispersion of potential competitors in space (Lack, 1954). Surely the cases of character displacement we have considered above, especially those for which we have some ecological data, are pertinent examples of correlation between sympatry (with the possibility of competition) and genetic fixation of specializations resulting in the avoidance of competition. The respective convergent unispecific populations outside the sympatric zones are the "controls" for these observations.

The case in which Lack (1944) cites the distribution of the chaffinches (*Fringilla*) in the Canary Islands is held up to special criticism by Andrewartha and Birch. Lack demonstrates that *F. teydea,* endemic to the islands of Gran Canaria and Tenerife, occupies only the coniferous forests at middle altitudes. On the same islands there also occurs a form of the widespread *F. coelebs,* presumably a relatively recent arrival from the Palaearctic mainland, but this bird occurs only in the tree-heath zone above, and in the broadleaf forests below, the coniferous belt. On the island of Palma, however, *F. teydea* is absent, and there a form of *F.*

*coelebs* occupies the coniferous forest as well as the broadleaf zone. Andrewartha and Birch conclude that, "So far as the case is stated, there is no direct evidence that the two species could not live together if they were put together." It is obvious from this that Lack's critics are not going to be satisfied by any ordinary kind of evidence.

What emerges starkly from contemporary discussion of "competition" is the great variation in the meanings with which different authors freight the word. Andrewartha and Birch, while differing with Nicholson (1954) on most important points, do manage to agree with him that the correct kernel of meaning of competition is contained in the expression "together seek." We would adopt the part of their definition that deals with the common striving for some life requisite, such as food, space or shelter, by two or more individuals, populations or species, etc. This seems to us to be close to the definitions preferred by the larger dictionaries we have consulted.

But Andrewartha and Birch, following many other writers, allow their competition concept to include another idea—that expressing direct interference of one animal or species with the life processes of another, as by fighting. On the surface, this inclusion of aggression as an element of competition may seem to some familiar and reasonable, but we wonder whether the concept of competition could not be more useful in biology if it were more strictly limited to "seeking, or endeavoring to gain, what another is endeavoring to gain at the same time," the first meaning given in *Webster's New International Dictionary, Second Edition, Unabridged.* It is noteworthy that competition as defined by this dictionary fails to include the idea of aggression in any direct and unequivocal way.

It may therefore be more logical in the long run to regard the various kinds of aggression between potential competitors (the outcome of which is so often predictable) as another method, parallel with character displacement and dispersion—and genetically conditioned in a similar fashion—by which organisms seek to lessen or avoid competition. Surely it is significant that aggressive behavior often seems most highly developed in cases where a conspecific, or closely related, potential competitor occurs with the aggressor, yet shows little or no displacement in behavior or form. In contrast are the many cases of complete mutual tolerance shown by closely related organisms that live side by side and are differentially specialized in behavior or form.

## Character Displacement versus Hybridization

Since both divergent and "intermediate" populations are involved in the displacement patterns we have been describing, it is clear that the convergent populations might easily be mistaken as representing products of interspecific hybridization between the two species displacing each other. This is especially true if the convergent populations are small and isolated, or if only a single one is developed. Lack, for instance, in an early paper (1940) interpreted the Daphne and Crossman populations of *Geospiza* as being of hybrid origin, changing his mind only after he had begun to consider more fully the influence of competition on speciation (in *Darwin's Finches*, 1947).

To take another possible example, Miller (1955) describes what he calls a "hybrid" between the woodpeckers *Dendrocopos scalaris* and *D. villosus*. This specimen, a female, was shot in the Sierra del Carmen, Coahuila, Mexico, at about 7000 feet altitude, near the lower limits of the coniferous belt capping the Sierra. Up to, or near, this altitude, Miller found the Sierra to support a population of *scalaris*, but despite intensive collecting, he found no sign of occupancy by the other putative parent species, *villosus*. *D. scalaris* reaches a higher point in these mountains than it usually does in the neighboring regions of desert scrub and bottomland—its habitat wherever it has been studied—

in Mexico, Arizona, New Mexico and parts of Texas. In general, the *villosus* populations of this part of North America are restricted to the higher coniferous belts, but *villosus* and *scalaris* are in contact at some stations where pinyon-oak-juniper meets coniferous forest. Presumably *scalaris* extends farther vertically in the Sierra del Carmen because *villosus* is not present to limit its upward expansion. According to Miller, *villosus* probably does not occur within 200 miles of the Sierra at the present time.

The specimen, thoroughly described and figured by Miller, is indeed intermediate in many respects between the *scalaris* and *villosus* of northern Mexico. However, there seems to be nothing in the information presented to prevent one's interpreting this as a large, unusually dark specimen of *scalaris*, instead of as a hybrid. There is no good reason to deny the possibility that *scalaris* can produce somewhat *villosus*-like variants at the upper limits of its range when *villosus* is absent.

Other examples we have already cited in the present paper show the difficulty in deciding between displacement and hybridization where the species involved are incompletely known. This situation adds considerable complication to the analysis of interspecific hybridization in nature, for it is clear that the alternative explanation of displacement should at least be taken into account.

One thing seems certain; the "hybrid index," better called "compound character index," can by itself be no sound indication that the situation plotted really involves hybridization. This leads us to ask whether even such elaborate and beautifully documented studies of "hybrid" situations as that made by Sibley (1950, 1954) on the towhees of southern Mexico (*Pipilo erythrophthalmus s. lat.* and *P. ocai*) are not really just illustrations of character displacement. In some of the higher mountains of the southeast (Orizaba, Oaxaca), the two very differently colored forms (species) meet but remain distinct. Farther west are found various populations that apparently grade between the extreme *erythrophthalmus* form and the *ocai* form to various degrees of intermediacy, as expressed by Sibley in his "hybrid index."

Some of the *ocai*-form populations at the western end of the range (*P. ocai alticola*) are stated to be distinct from the other races of *ocai* by a characteristic melanization of the head region, which Sibley thinks is due to introgression from *erythrophthalmus* populations found to the north in the Sierra Madre Occidental. Despite this indication of introgression, the western populations at the *ocai* end of the gradients studied are indexed at, or extremely close to, zero, the figure indicating a population of "pure" *ocai*. Aside from what seems to be a variation in "purity" standards for *ocai* here, it is interesting to note that the western populations and those others among the apparent intermediates of the southern Plateau Region can all conceivably, on present evidence, be interpreted as *erythrophthalmus* that have converged toward *ocai* in the absence of the "true" *ocai* form represented by the upland, sympatric southeastern samples.

It seems possible that some strong selective pressure may be acting in the southern Plateau region to produce an *ocai* coloration-type in finchlike birds, and that *erythrophthalmus* may yield to this pressure wherever the true *ocai* is absent in this area. A very *ocai*-like bird of a related genus, *Atlapetes brunneinucha*, reaches the northern limit of its range in the southern Plateau area, and it is possible that the striking similarity marks some adaptive relationship to which both it and the *Pipilo* stock respond. It might even be that mimicry is involved between the sympatric *Atlapetes* and *Pipilo* stocks, although this is nothing more than the sheerest speculation in view of our very incomplete knowledge of the relative distribution of the two forms and other aspects of their biology and their environment, including their predators. At any rate, character displacement must for the time being be considered a reasonable alternative explanation of the variation of southern Mexican *Pipilo* in this group.

It may perhaps be argued that the "hybrid" populations of *Pipilo* are more variable than the presumed parental populations, and that this in itself is a strong indication of hybridization. We do not believe, however, that the case should be decided on this kind of evidence. To start with, tailspot length, the one character used in Sibley's study that has also been analyzed at length in other populations of *P. erythrophthalmus*, shows very considerable variation in areas far removed from the likely influence of *ocai*. According to the data of Dickinson (1953), the Florida population ("race *alleni*") has a coefficient of variation in this character of about 22 in the male; the range of variation is from 6.1 to 27.5 mm. The northeastern (nominate) race shows a corresponding coefficient of about 12, with a range of variation of from 24.0 to 55.0 mm. Furthermore, the chestnut-tinted pileum characteristic of *ocai-erythrophthalmus* "hybrids" occasionally crops up in the eastern North American samples of *erythrophthalmus*. But even if it were true that variation in the direction of *ocai* could be demonstrated only in the *ocai* "area of influence," this could not be taken as proof of hybridization, because an increase in variation is also a common quality of the "convergent" populations in character displacement patterns.

### Character Displacement and Taxonomic Judgment of Allopatric Populations

Foremost among the problems of taxonomic theory today is the tantalizing conundrum concerning the status of the allopatric (isolated) population. Few authors hesitate to assign such populations either subspecific or specific rank, and most, it is hoped, appreciate the fact that their decisions are essentially arbitrary. As Mayr (1942) says, "The decision as to whether to call such forms species or subspecies is often entirely arbitrary and subjective. This is only natural, since we cannot accurately measure to what extent reproductive isolation has already evolved." There does not seem to be any definable threshold between polytypic species composed of such subspecific "units" and the superspecies composed of allopatric sister species. However, it is entirely possible that by the time an isolated population attains an ascertainable level of character concordance, it has already passed the species line; i.e., the more sharply defined an isolated subspecific population is by conventional standards, the less likely it is to be infraspecific in reality.

The phenomenon of character displacement should be borne heavily in mind in considering this matter of allopatric populations. If the present conception is correct, related sympatric species will generally show more morphological differences than similarly related allopatric ones. Hence the degree of observed difference between sympatric species cannot be considered a reliable yardstick for measuring the real status of related allopatric populations, nor can the differences among the latter be taken too seriously as indications of their relationships. In fact, the morphological standards set for determining which completely allopatric populations have reached species level may be much too strict in current practice. Despite impressions that might be gained from recent literature, many systematists have realized that in different allopatric populations (of the same species-group or genus), the degree of morphological divergence may be poorly correlated with the amount of reproductive isolation holding between them (Moore, 1954; Kawamura, 1953). In other words, where there is any question whatsoever about the objective species status of two closely related but geographically separated populations, morphology alone cannot be expected to answer it definitely.

Unfortunately, allopatric species or "subspecies" designated as such on a purely morphological basis frequently enter into theoretical discussions as though they were objectively established realities, when in fact they are usually no more than arbitrary units drawn for curatorial convenience.

## Summary

*Character displacement* is the situation in which, when two species of animals overlap geographically, the differences between them are accentuated in the zone of sympatry and weakened or lost entirely in the parts of their ranges outside this zone. The characters involved in this dual divergence-convergence pattern may be morphological, ecological, behavioral, or physiological. Character displacement probably results most commonly from the first post-isolation contact of two newly evolved cognate species. Upon meeting, the two populations interact through genetic reinforcement of species barriers and/or ecological displacement in such a way as to diverge further from one another where they occur together. Examples of the phenomenon, both verified and probable, are cited for diverse animal groups, illustrating the various aspects that may be assumed by the pattern.

Character displacement is easily confused with a different phenomenon: interspecific hybridization. It is likely that many situations thought to involve hybridization are really only character displacement examples, and in cases of suspected hybridization, this alternative should always be considered. Displacement must also be taken into account in judging the status (specific *vs.* infraspecific) of completely allopatric populations. It is clear that, in the case where the species are closely related, sympatric species will tend to be more different from one another than allopatric ones. Thus, degrees of difference among related sympatric populations cannot be used as trustworthy yardsticks to decide the status of apparently close, allopatric populations.

## Acknowledgements

We are grateful for information, advice and other aid received from numerous colleagues in the course of preparing this contribution. Especially to be thanked are J. C. Bequaert, W. J. Bock, W. J. Clench, P. J. Darlington, A. Loveridge, A. R. Main, E. Mayr, A. J. Meyerriecks, K. C. Parkes, R. A. Paynter, and E. E. Williams. Dr. C. Vaurie kindly offered the use of his figures to illustrate the *Sitta* case and gave us the benefit of some unpublished observations. Our acknowledgement is not meant to imply that any of those listed necessarily support the arguments we advance.

## REFERENCES

ALLEE, W. C., EMERSON, A. E. and others. 1950. Principles of animal ecology. W. B. Saunders Co.

ANDREWARTHA, H. G., and BIRCH, L. C. 1954. The distribution and abundance of animals. Univ. Chicago Press.

BLAIR, W. F. 1955. Size differences as a possible isolating mechanism in *Microhyla*. *Amer. Naturalist*, 89:297–301.

BOURNE, W. R. P. 1955. The birds of the Cape Verde Islands. *Ibis*, 97:508–556, cf. 520–524.

CAIN, A. J. 1955. A revision of *Trichoglossus haematodus* and of the Australian platycercine parrots. *Ibis*, 97:432–479, cf. 457–461, 479.

CONDON, H. T. 1951. Notes on the birds of South Australia: occurrence, distribution and taxonomy. *S. Aust. Ornith.*, 20:26–68.

DICKINSON, J. C. 1952. Geographical variation in the red-eyed towhee of the eastern United States. *Bull. Mus. Comp. Zool. Harv.*, 107:273–352.

DOBZHANSKY, TH. 1951. Genetics and the origin of species. 3rd Ed. Columbia Univ. Press.

HOWDEN, H. F. 1955. Biology and taxonomy of the North American beetles of the subfamily Geotrupinae . . . *Proc. U. S. Nat. Mus.*, 104:159–319, 18 pls.

HUBBS, C. L., and BAILEY, R. M. 1940. A revision of the black basses (*Micropterus* and *Huro*) with descriptions of four new forms. *Misc. Publ. Zool. Univ. Mich.*, No. 48, 51 pp.

KAWAMURA, T. 1953. Studies on hybridization in amphibians. V. Physiological isolation among four *Hynobius* species. *J. Sci. Hiroshima Univ.* (*B, 1*) *14*:73–116.

KOOPMAN, K. F. 1950. Natural selection for reproductive isolation between *Drosophila pseudoobscura* and *Drosophila persimilis*. *Evolution*, 4:135–148.

LACK, D. 1940. Evolution of the Galapagos finches. *Nature, 146*:324–327.

———— 1944. Ecological aspects of species formation in passerine birds. *Ibis*, 86:260–286.

———— 1947. Darwin's finches. Cambridge Univ. Press.

———— 1954. The natural regulation of animal numbers. Oxford Univ. Press.

MAYR, E. 1942. Systematics and the origin of species. Columbia Univ. Press.

———— 1949. Speciation and selection. *Proc. Amer. Phil. Soc.*, 93:514–519.

———— 1955. Notes on the birds of northern Melanesia. *Amer. Mus. Novitates*, No. 1707: 1–46, cf. p. 29.

MOORE, J. A. 1954. Geographic and genetic isolation in Australian amphibia. *Amer. Naturalist*, 88:65–74.

MILLER, A. H. 1955. A hybrid woodpecker and its significance in speciation in the genus *Dendrocopos*. *Evolution*, 9:317–321.

NICHOLSON, A. J. 1954. An outline of the dynamics of animal populations. *Australian J. Zool.*, 2:9–65.

SERVENTY, D. L. 1953. Some speciation problems in Australian birds . . . *Emu*, 53:131–145, with further references.

SIBLEY, C. G. 1950. Species formation in the red-eyed towhees of Mexico. *Univ. Calif. Publ. Zool.*, 50:109–194.

———— 1954. Hybridization in the red-eyed towhees of Mexico. *Evolution*, 8:252–290.

VAURIE, C. 1950. Notes on Asiatic nuthatches and creepers. *Amer. Mus. Novitates*, No. 1472:1–39.

———— 1951. Adaptive differences between two sympatric species of nuthatches. *Proc. Xth Internat. Ornith. Congr., Uppsala, June 1950*:163–166, 3 figs.

WILSON, E. O. 1955. A monographic revision of the ant genus *Lasius*. *Bull. Mus. Comp. Zool. Harv.*, 113:1–205, ill.

WILSON, E. O., and BROWN, W. L., JR. 1955. Revisionary notes on the *sanguinea* and *neogagates* groups of the ant genus *Formica*. *Psyche*, 62:108–129.

WILLIAM L. BROWN, JR. is Associate Curator of Insects at the Museum of Comparative Zoology, Harvard University. EDWARD O. WILSON is a Junior Fellow of the Society of Fellows of Harvard University.

## 20

Reprinted from *Evolution* **10**:14–22 (1956)

# SELECTION FOR SEXUAL ISOLATION WITHIN A SPECIES

G. R. Knight, Alan Robertson and C. H. Waddington

*Institute of Animal Genetics, Edinburgh*

Received March 17, 1955

Two mechanisms have been advanced for the origin of reproductive isolation between species. Muller (1939), dealing in the main with barriers to crossing in the later stages of species divergence, such as hybrid inviability and infertility, suggests that these arise almost by chance as a product of change in the genetic background either by genetic drift or as adaptation to different biological situations. This would lead to accelerating divergences as the process continues, or, as Muller puts it, "ever more pronounced immiscibility as an inevitable consequence of non-mixing." Dobzhansky's suggestion (1937), which is perhaps complementary rather than antagonistic to Muller's, is that when sufficient divergence between two species has arisen so that the hybrids are less well adapted for any available habitat than either parental type, there will be selection for sexual isolation. That is to say—if mating can take place and if the resulting hybrids are inviable or infertile, then natural selection will operate to reduce the chance that mating will occur, either by reducing the chance of encounter or the chance of mating with members of the other species when they are encountered.

Some writers, in discussing the mechanism proposed by Dobzhansky have suggested that "natural selection will favour any mechanism which prevents the wastage of gametes involved in unsuccessful hybridisation." This seems to be unduly teleological. Natural selection will only tend to suppress crossbreeding if those individuals which hybridise will in consequence pass on fewer gametes in the form of pure-bred offspring. It would seem probable that this would be more often the case in females than in males. In *Drosophila melanogaster*, for instance,

females seem reluctant to mate again for a period of two or three days after an effective mating. If the first mating has been heterogamic, this will reduce the number of purebred offspring that she will produce in her lifetime. Gestation in mammals will have a similar effect. But the male, who must on the average have the same number of effective matings in his life as the female, is usually capable of many more if willing females are available. It follows then that willingness to cross-breed, which may merely be a sign of greater general sexual activity, will not necessarily reduce the number of purebred progeny that a male will leave. If Dobzhansky's mechanism for the establishment of sexual isolation is correct, it follows that it should be in the main a matter of female preference. Merrell (1954) has recently presented evidence that it is the female which exercises discrimination in matings between *D. pseudobscura* and *D. persimilis*.

Koopman (1950) has shown that selection leads to an intensification of the sexual isolation between these two species. Using marked stocks of the two species, he selected continually for purebred flies—the progeny of parents that had mated homogamically. He showed that the proportion of hybrids emerging declined dramatically after a few generations of selection. More recently, Wallace (1950) and King (private communication) attempted to demonstrate the production of sexual isolation by selection within a species. They used two stocks of *D. melanogaster*, from widely separated localities, which had each been marked by a different recessive gene. After 12 generations, when the experiment was first reported, little change in the proportion of wild-type flies emerging had been observed, but in subsequent generations the proportion declined significantly, showing that sexual isolation had been to some extent established. This was confirmed by observation of individual matings.

Our own experiment on very similar lines was started before we were aware of Wallace's work, and as our work was slightly different in conception, we decided to proceed with it. In Wallace's experiment, the mutants were used solely as markers, the stocks because of their origin presumably differing in many genes. As it happens, we had used in our work stocks marked with the autosomal recessive genes, ebony and vestigial, which has been extracted from a population in which the two had been segregating for many generations. The original stocks making up this population were actually those used by Rendel (1951) in his work on the effect of light on the mating of these mutants. Our two foundation stocks, both of which contained a considerable amount of genetic variability, were thus probably genetically very similar except for the marker genes. These genes were chosen because of the ease of scoring but they do react differently to light and, as Rendel has shown, ebony males mate more frequently in the dark than in the light.

In the first experiment of this type that we carried out, there appeared in the seventh generation some flies that were both ebony and vestigial, indicating that in previous generations either a non-virgin female or else a wild-type heterozygote had been used as a parent with the result that each mutant stock was contaminated with the other gene. Theoretical consideration of the effect of this showed that the proportion of double recessive flies should increase by a factor of four each generation until they reached a level of 11% of all flies emerging. At that point, the proportion of flies in each mutant stock that were heterozygous for the other gene would be $\frac{2}{3}$. There would then be a continual inter-

change of genes between the two stocks. In addition, one-third of the apparently pure mutants used as parents would be derived from heterogamic matings, thereby reducing the selection for sexual isolation. We therefore discarded the line and started afresh with stringent precautions against non-virginity, parents being collected over a 7 hour period. In the two experiments presented here in detail, no double recessive flies were ever observed.

## DESCRIPTION OF EXPERIMENTS

### Box Experiment

Two mutant strains of *D. melanogaster* homozygous for the genes ebony and vestigial respectively were used. They had been extracted from the same population, after segregation for many generations. At the start, 54 males and 54 virgin females from each of the stocks were put together into a breeding box (size $18'' \times 18'' \times 7''$) which contained 10 unstoppered $\frac{1}{4}$ pint bottles of maize meal–molasses–yeast–agar medium. Flies were etherised for counting, but were not put into the box until three hours after complete recovery from anaesthesia. The box was then placed in a constant temperature room at 25° C. All phases of the experiment were done at this temperature. The box was always put in the same part of the room, where, due to the direction of the light, two sides of the box near the edge were in slight shadow. The ebony flies, immediately the box was positioned, migrated towards the light source, that is, towards the shaded edge. After some time, the majority of them moved more freely about the cage.

After six days of mating, the ten food bottles were removed, cleared of any flies which remained inside, and stoppered. The parents were discarded. The count of the next generation was started five days afterwards, i.e., on the eleventh day after the parents were put into the box. Three types of flies emerged; hybrids from heterogamic matings, and the two mutants ebony and vestigial from homogamic fertilizations. For 3½ days every fly which emerged was counted. The culture bottles were completely cleared at 10 A.M. Flies which emerged by 5 P.M. on the same day were segregated and mutants were kept in separate vials to be used as parents for the next generation when 1 to 4 days old. When insufficient virgins were obtained, those collected were bred with their own kind, and the experiment carried on from their progeny. In the box experiment, this was done three times in 38 generations.

In order to ensure that any changes in external conditions had not affected the course of the experiment, controls were done on the box experiment in the later generations. Parent virgin flies were obtained from the original stocks and put into a box of identical proportions to the experimental one. The control box and the experimental one received exactly the same treatment throughout. This was done seven times between the 25th and 35th generations.

### Jar Experiment

An experiment on similar lines was run in conjunction with the cage one. A 2 lb. glass jar containing approximately

1″ of food was used as the breeding chamber. The number of parent flies employed in this case was between 20 and 30 of each sex of the mutants. Again, it was sometimes necessary to mate the virgins with their own kind to produce sufficient numbers for the next preferential mating. This was done three times in 33 generations. From generations 1 to 12 the parent flies were still under ether when put into the jar, as it was thought that they might otherwise escape. This was found, however, to be unsatisfactory. So from the 13th generation onwards the parents were introduced into the jar three hours after recovery from the ether. The jar was put into the same constant temperature room and at the same time as the box. Thereafter, all operations, such as clearing parents from the jar, counting and segregating flies of each generation, etc., were carried out at the same time and in an exactly similar manner to the box experiment.

Because of the small capacity of the jar compared with that of the box and the fact that there was little or no variation in the light within the jar, it was assumed that any tendency towards an eco-

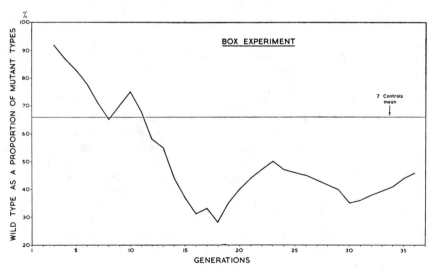

FIG. 1. Results of box experiment. Number of hybrids expressed as a percentage of sum of ebony and vestigial emergences.

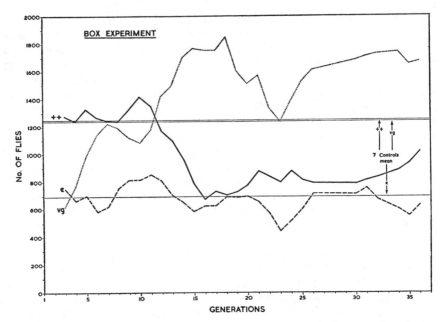

FIG. 2. Results of box experiment. Numbers of 3 types of flies emerging shown separately.

logical isolation between the ebony and vestigial flies would be eliminated.

### RESULTS

One of the first impressions at the start of the experiments was of the great fluctuation in results from generation to generation. The jar experiment was in fact started to try to remove this by having all flies developing in one food mass. Our criterion of isolation has been the ratio of wild-type flies, produced by heterogamic matings, to the total number of mutants produced by homogamic matings. The standard deviation of this ratio due to chance fluctuation, determined from the mean square difference between successive generations, was 0.16 for the box experiment and 0.24 for the jar. This fluctuation is equal to that produced by random sampling of 160 units and 70 units respectively from a population made up of two types of objects with equal frequency. The total count was actually of the order of 2,000 flies in both cases. But the number of female parents was 108 and 50 in the box and jar respectively.

The observed fluctuations suggest that the effective units are the initial inseminations of the individual females. In this respect, it is of interest that of the individual platings of females taken from the box after six days of mating, 660 gave offspring all of the same type and only 75 had mixed offspring. However, whatever the reason, it is still true that too many flies were counted each generation and a sample of a quarter of the size that we took would have been quite adequate.

#### Box Experiment

The results of the box experiment are set out graphically in figures 1 and 2. The graphs are moving averages over 5 generations to smooth out fluctuations. In figure 1, the number of hybrids is expressed as a percentage of the sum of the ebony and vestigial emergences. Figure 2 shows separately the numbers of the three types of flies emerging. From the first to the eighteenth generation a more or less steady decline in the percentage of hybrids is noted. The lowest percentage of hybrids in any individual gen-

278

eration was 10.3% at the eighteenth, with emergences of ++ 246, e 736, vg 1640. Only once afterwards, at the 23rd generation, does the vestigial line graph fall as low as the control mean for this mutant. Thereafter the values remain high for vestigial emergences. The hybrid figure drops, and is lowest between the 16th and 18th generations, only rising a little and slowly towards the end of the experiment. During the whole 38 generations the emergence values for ebony alternate slightly above and below the figure for the control mean. This suggests that the sexual isolation, after the 18th generation, is due mainly to the increase in the number of homogamic matings of the vestigial flies.

The average values for the seven control generations are also shown in figures 1 and 2. The proportion of wild-type flies to mutants averages 0.66, compared to the proportion in the selected population at the same period of 0.38. The figures for the individual mutants show that the change is due to a decrease of wild-type flies and an increase of vestigial.

It has been shown by Rendel (1951) that ebony reacts to light intensity in its mating behaviour. It seemed possible that the sexual isolation was due to an accentuation of this response. Towards the end of the experiment, therefore, duplicates of the selection box were made up from parents from the selected stock but were kept instead in complete darkness. The ratio of wildtype to mutant offspring was 0.48 compared to 0.46 for

### TABLE 1

| Controls ♀ | Inseminated by | |
|---|---|---|
| | e | vg |
| e | 71 | 69 |
| vg | 41 | 63 |
| Selected stocks | | |
| e | 151 | 108 |
| vg | 77 | 142 |

the three contemporary generations in the light. It seems therefore that the demonstrated sexual isolation is not concerned with phototropic response. However, there were many more ebony flies in the dark boxes—in fact the average of the three tests (1102 flies) had only once been exceeded by a single generation in the light, and the average in the last few generations of the latter was about 650. There was correspondingly a shortage of vg flies, but the proportional effect was not so great. This agrees with Rendel's observation that ebony males show greater sexual activity in the dark.

Between the 20th and 30th generations, the females were placed in individual vials after they had been removed from the box, and their progeny were examined on emergence. This was done with 6 generations of the selected stocks and with three of the controls. The results in terms of effective matings are given in table 1.

There is a slight tendency to homo-

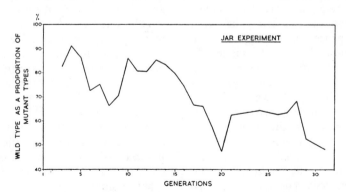

FIG. 3. Results of jar experiment. Compare figure 1.

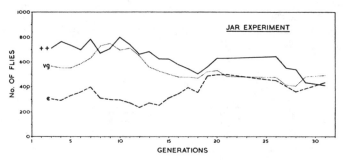

FIG. 4. Results of jar experiment. Compare figure 2.

gamic mating in the controls but the heterogeneity $\chi^2$ is only 2.15. In the selected stocks the tendency is much more marked and the $\chi^2$ value is 27.14. This is confirmatory evidence that some degree of sexual isolation has been obtained in the selected population.

It might have happened that this type of selection, picking out always the mutant flies, would have affected the segregation ratio by selecting those genes favouring the survival from egg to adult of the mutant types. However, a check based on several thousand flies at the end of the experiment showed no differences between control and selected stock in the segregation ratio for either mutant.

## Jar Experiment

The results for the jar experiment are given in figures 3 and 4. In the jar, the light intensity was much more uniform than in the box and in addition the flies were more confined—the volume of the jar being of the order of $\frac{1}{50}$ of that of the box. Here again there is a decline in the proportion of wild-type flies as the experiment proceeds, although the proportion at the end is a trifle higher than in the box experiment. However, the ratio of wild-type to mutant flies has declined from 80% to between 50% and 60%. As was noted above, there was a change in method in the middle of this experiment. Up to generation 12, flies were put into the jar etherized but afterwards they were put in an active state. This change does not appear to have affected the sexual isolation. In the five

generations before the change, the average ratio of wild-type to mutants was 0.86 and in the five after the change it averaged 0.80. It may however have affected the separate types. The numbers of vg and wild-type flies decline by about one-third as a result of the change, whereas the ebony count is unchanged. The subsequent change in the wild-type/mutant ratio appears in this case to be due to an increase in the number of ebony flies. Control experiments were not carried out on the jar population, as the latter was subsidiary to the main experiment in the box.

## Sexual Preferences in Inbred Lines and Closed Populations

It is convenient to present here a small amount of data on sexual preferences between lines and populations chosen at random, with a bearing on the "chance" occurrence of sexual isolation. These experiments were carried out by the usual "male choice" method in which males are given equal numbers of two types of female, one of which is recognisable—in our case by a spot of silver paint. The females are then examined for the presence of sperm in the seminal tract. In the first case, two wild-type inbred lines of completely different origin were used, given in our stock list the symbols W20 and K7. The results are given in table 2, which shows the proportion of females inseminated. In all cases, the ♂♂ were equal in number to the ♀♀ of each of the separate lines.

A similar experiment was then done

using lines which, although of common origin, had been selected in different directions without inbreeding for 20 generations for number of chaetae on the 4th and 5th abdominal sternites. There was no overlap in chaetae count between the high and low lines used, so that this character could be used for identifications. The results are given in table 3, for matings between one high line, H1 and two low lines, L4 and L5.

In these rather meagre results, there is little suggestion of sexual isolation having developed by chance in either the inbred experiment or in that with the selected lines. In the latter, it is of interest that the selection for the quantitative character has caused a differentiation in mating ability. The H1 ♂♂ are poorer than those from the two low lines but on the other hand the high ♀♀ seem to be better. But this seems to be a general change in sexual drive, not specifically adapted to the other sex of the same line. Wallace (1955 in press, and personal communication) has also tested whether the mere isolation of two populations is sufficient to cause sexual isolation to arise between them. His populations had been separated for 80–100 generations, and differed in certain morphological characters (primarily abdominal pigmentation). In an extensive series of tests, no tendency towards preferentially mating could be detected nor does there seem to have been any evidence of

TABLE 3

| ♂ | H1 ♀ | L4 ♀ | Duration of mating |
|---|---|---|---|
| H1 | 3/10 | 1/9 | 30 mins. |
| L4 | 9/14 | 3/14 | 30 mins. |
|  | H1 ♀ | L5 ♀ |  |
| H1 | 11/20 | 5/19 | 60 mins. |
| L5 | 20/24 | 17/25 | 60 mins. |

differences in the intensity of general sexual drive.

## DISCUSSION

Our results may be summarised in the statement that some sexual isolation developed when we selected for a tendency towards homogamic matings, but that none was found to have arisen by chance in a few lines which had been selected for abdominal chaeta number or inbred. Laboratory experiments on evolutionary mechanisms can, of course, only be indicative and not demonstrative—they can show what might happen in wild populations, rather than what has happened. As far as they go, our experiments lend support to the mechanism suggested by Dobzhansky rather than that discussed by Muller. But when attempting to apply these results to occurrences in nature, one must bear in mind the ways in which artificial populations may fail to imitate conditions in the wild.

It is perhaps misleading to put Muller's hypothesis of the chance origin of sexual isolation in antithesis to that of Dobzhansky, which attributes it to the action of selection. In all probability, both mechanisms have operated in the wild in different cases. Some evidence supporting Dobzhansky's hypothesis comes from Dobzhansky and Koller (1938) who found, in an analysis of crosses between *Drosophila pseudoobscura* and *D. miranda*, that the isolation was greatest between races close to each other in their range. King (1947) has similar evidence from the *guarani* group. However, even between races of the two species widely

TABLE 2

| ♀s marked | ♂ | W20 ♀ | K7 ♀ | Duration of mating |
|---|---|---|---|---|
| W20 | W20 | 22/23 | 11/25 | 40 min s. |
|  | K7 | 17/23 | 9/25 | 40 min s. |
| W20 | W20 | 17/19 | 16/20 | 70 mins. |
|  | K7 | 4/20 | 1/19 | 70 mins. |
| K7 | W20 | 9/10 | 5/10 | 30 mins. |
|  | K7 | 8/14 | 9/14 | 60 mins. |
| K7 | W20 | 10/20 | 2/19 | 30 mins. |
|  | K7 | 10/20 | 5/20 | 45 mins. |

separated in origin, the isolation was considerable. One has the suspicion that sexual isolation is common between species which have never had the opportunity to crossbreed, though the evidence is rarely conclusive since seldom if ever do we know the full evolutionary history of the populations.

It is perhaps not surprising that differences in sexual behaviour arose in the experiments involving selection, but were not found in the comparison of populations which has originated independently. If they had occurred in the latter, they could only have appeared by chance, or as a correlated response. It might be expected that random changes in sexual behavior would be slow, even though they arose as a secondary response to an adaptive change in the population. Mating involves the cooperation of the two sexes and it seems unlikely that a genetic shift in the population causing a change of sexual behaviour in the female, perhaps by a modification of the pattern by which a male recognises an animal of his own species, would also change male behaviour in a compensating manner (although an exception to this might be in habitat preference). An individual with aberrant sexual behaviour is not likely to leave many progeny. A population gradually changing its genetic situation could only change its pattern of mating by the selection of males capable of responding to the altered female behaviour. This must constitute a brake on the change of mating behaviour either by chance changes in the genetic situation or even as a correlated response to an adaptive change. This will be particularly true of inbred lines in which selection between potential mates is small or non-existent. Reproductive behaviour, excepting perhaps choice of habitat, would therefore be more stable than other physiological systems to genetic changes.

Both the hypotheses that we have discussed demand the development of a previous geographic isolation before sexual isolation can be established. In this sense, the selection hypothesis is perhaps clumsy, since in the formation of a new species showing sexual isolation with the parent species it demands first a geographic isolation and then an overlapping of the species range so that members of the two species can be selected for refusal to crossbreed. It seems to us that sexual isolation instead of being a consequence of geographic isolation, may be a contributory factor in its establishment. The spread of a population into new territory will often involve the occurrence of genotypes with new hereditary habitat preferences. The existence of such preferences amongst Drosophila stocks has recently been shown by Waddington, Woolf and Perry (1954). In organisms such as birds, in which rather sudden changes in geographical or ecological range are well-known, learning may play a part, but this may also have an important genetic component. In the genetic constitution of a sub-population which has broken out of the original species boundaries and is spreading into new territory, one must expect to find that a number of adjustments are occurring simultaneously. There is most likely to be, in the first place, an evolution of a new system of habitat-preferences and/or of general activity; in the second, the adaptive characters and general fitness of the migrating group will be attuned to the new circumstances which it has to meet. Both these necessary modifications of the gene pool will be made more easily if the genetic constitution of the sub-population is prevented from continual intermingling with that of the original stay-at-home group. Thus any tendency for preferential mating within the migrating group, and sexual isolation between it and the main population, will acquire selective value. It seems rather probable that a species may be able to spread into new territory even if no sexual isolation develops between the main population and the migrating one; but if the increase in species-range demands considerable adjustment of the genotype to fit the new environment, the evolution of some degree of mating-barrier will undoubtedly be of considerable advantage. Our experiments show that the necessary genetic variability is likely to be present in a population; and the fact that the change of environment involves alterations to the behaviour pattern of the migrating animals makes it more likely that their preferences for sexual partners as well as for habitats, will exhibit evolutionary flexibility. Thus species-spread and sexual isolation will tend to act synergistically.

## Summary

1. Partial sexual isolation (between two stocks of *D. melanogaster* differing only in marker genes) has been established by selection of the offspring of flies mating with their own type. This has been demonstrated by a reduction in the number of cross-bred offspring found and also by examination of the progeny of individual females.

2. In a small series of tests, no tendencies towards preferential mating were found to have arisen by chance in a sample of lines which had been inbred or selected for number of abdominal chaetae, although there were differences in intensity of sexual drive.

3. Changes in reproductive behaviour brought about by selection are more likely to affect female than male behaviour. Willingness to cross breed, in a male, will not materially reduce the number of his pure-bred offspring but in a female it usually will.

4. It is argued that selection pressure against cross-breeding of two partially separated populations, although probably effective when it occurs, is not likely to be the only mechanism by which sexual isolation between taxonomic groups develops in nature. It is suggested that an important part in the origin of such isolation may be played by the factors (e.g. changes in hereditarily controlled behaviour patterns) which bring about the spread of an initial panmictic population into new geographical or ecological situations.

### Literature Cited

Dobzhansky, T. 1937. Genetics and the Origin of Species. Columbia Univ. Press, New York.

Dobzhansky, T., and P. C. Koller. 1938. An experimental study of sexual isolation in Drosophila. Biol. Zentral., **58**: 589–607.

King, J. C. 1947. Interspecific relationships within the guarani group of Drosophila. Evolution, **1**: 143–153.

Koopman, K. F. 1950. Natural selection for reproductive isolation between Drosophila pseudoobscura and Drosophila persimilis. Evolution, **4**: 135–148.

Merrell, D. J. 1954. Sexual isolation between Drosophila persimilis and Drosophila pseudoobscura. Am. Nat., **88**: 93–100.

Muller, H. J. 1939. Reversibility in evolution considered from the standpoint of genetics. Biol. Rev., **14**: 261–280.

Rendel, J. M. 1951. Mating of ebony, vestigial and wild-type Drosophila melanogaster in light and dark. Evolution, **5**: 226–230.

Wallace, B. 1950. An experiment on sexual isolation. D.I.S., **24**: 94–96.

Waddington, C. H., B. Woolf, and M. Perry. 1954. Environment selection by Drosophila mutants. Evolution, **8**: 89–96.

## 21

Reprinted from *Evolution* 19:459–464 (1965)

# DIRECT OBSERVATION OF SEXUAL ISOLATION BETWEEN ALLOPATRIC AND BETWEEN SYMPATRIC STRAINS OF THE DIFFERENT *DROSOPHILA PAULISTORUM* RACES

LEE EHRMAN[1]

*The Rockefeller University, New York*

Accepted June 11, 1965

Dobzhansky and Spassky (1959) first suggested that *Drosophila paulistorum* was a cluster of species *in statu nascendi*, a borderline case of uncompleted speciation. Their suggestion has been borne out by much laboratory work since then. It is indeed a superspecies composed of six races or subspecies or incipient species. This situation is of interest precisely because these six may be considered about equally legitimately as very distinct races or as very closely related species. Each race inhabits a geographic area different from the others, but the distributions of some of the races do overlap. Where two or more races share a common territory (four is the maximum number of races occurring sympatrically), they seem not to interbreed, and thus they behave like full-fledged species (Dobzhansky *et al.*, 1964).

Sexual (or ethological or behavioral) isolation wherein potential mates meet but do not mate, is a most efficient isolating mechanism; it does away with the wastage of gametes, food and space for developing hybrids, etc. This sexual isolation which makes matings between the females and males of different *D. paulistorum* races much less likely to occur than matings within the races, is due to polygenes scattered in every one of the three pairs of chromosomes which the species possesses. These polygenes efficiently control the sexual preferences of their carriers. Their effects seem to be simply cumulative (Ehrman, 1961). A female of hybrid origin which carries a majority of the chromosomal material derived from a given race

is most likely to accept a male of that race. And conversely, a hybrid male is most likely to be accepted by females whose chromosomal constitution is closest to his. This clearly is not at all comparable with the genetic basis of the hybrid male sterility, where the properties of an egg are determined, as far as the sterility of the backcross males is concerned, by the genotype present in it before meiosis, and not by that formed following fertilization. Thus, while the hybrid sterility depends solely on the genetic constitution and the source of the cytoplasm contributed by the mother, the sexual isolation certainly does not.

The experiments reported here were undertaken to test the degree of sexual isolation for its strength between *sympatric* and between *allopatric* strains of the *same pairs of D. paulistorum races* since different populations are exposed to the risk of hybridization when they share the same territory. There is no doubt that hybridization, in this particular case, will result in the production of reproductively inferior individuals so that those individuals which do not engage in interracial matings will contribute more to the gene pool of subsequent generations than those that do engage in interracial matings. In this way, natural selection favors genotypes which hinder or prevent hybridization altogether.

## MATERIAL

There are six *Drosophila paulistorum* races, and 15 interracial crosses can be made. Since each cross might be tested with sympatric and with allopatric strains, 30 crosses are conceivable. However, only the following races occur *both* sympatri-

---

[1] The work reported here has been carried out under Contract No. AT-(30-1)-3096, U. S. Atomic Energy Commission.

cally and allopatrically and obviously only they could be used in this analysis. The strains employed and the numbers given them follow the names of the races (see Dobzhansky *et al.*, 1964, and Dobzhansky and Spassky, 1959, for further identification):

*Centro-American* × *Amazonian*

Allopatric

Boquete, Panama (27) × Georgetown, British Guiana N (77)

Cerro Campana C, Panama (32) × Georgetown, British Guiana N (77)

Goofy Lake C, Panama (81) × Simla D, Trinidad (58)

Cerro Campana C, Panama (32) × Simla D, Trinidad (58)

Lancetilla, Honduras (2) × Apoteri C, British Guiana (68)

Tikal, Guatemala (1) × Belem A, Brazil (11)

Sympatric

Goofy Lake C, Panama (81) × Goofy Lake A, Panama (80)

Boquete, Panama (27) × Barro Colorado A, Panama (33)

Boquete, Panama (27) × Goofy Lake A, Panama (80)

Cerro Campana C, Panama (32) × Cerro Campana A, Panama (31)

Boquete, Panama (27) × Cerro Campana A, Panama (31)

*Amazonian* × *Orinocan*

Allopatric

Bucaramanga A, Colombia (26) × Georgetown, British Guiana C (42)

Georgetown, British Guiana X (78) × Tuy, Venezuela (41)

Simla D, Trinidad (58) × Bucaramanga 8, Colombia (28)

Piña, Panama (29) × Georgetown, British Guiana C (42)

Goofy Lake A, Panama (80) × Georgetown, British Guiana C (42)

Belem A, Brazil (11) × Apoteri Y, British Guiana (74)

Barro Colorado A, Panama (33) × Georgetown, British Guiana (42)

Sympatric

Caripe G, Venezuela (38G) × Caripe L, Venezuela (38L)

Georgetown, British Guiana N (77) × Georgetown, British Guiana C (42)

*Amazonian* × *Andean*

Allopatric

Elena N, Venezuela (48) × Apoteri O, British Guiana (72)

Georgetown, British Guiana N (77) × Elena A, Venezuela (49)

Goofy Lake A, Panama (80) × Simla M, Trinidad (54)

Goofy Lake A, Panama (80) × Elena A, Venezuela (49)

Apoteri C, British Guiana (68) × Cucuta C, Colombia (90)

Belem A, Brazil (11) × Elena A, Venezuela (49)

Belem A, Brazil (11) × Cucuta C, Colombia (90)

Sympatric

Simla D, Trinidad (58) × Simla M, Trinidad (54)

Cucuta A, Colombia (88) × Cucuta C, Colombia (90)

Apoteri C, British Guiana (68) × Apoteri N, British Guiana (71)

*Amazonian* × *Guianan*

Allopatric

Georgetown, British Guiana (77) × Apoteri X, British Guiana (73)

Barro Colorado A, Panama (33) × Apoteri X, British Guiana (73)

Simla D, Trinidad (58) × Apoteri X, British Guiana (73)

Belem A, Brazil (11) × Georgetown, British Guiana B (10B)

Belem A, Brazil (11) × Apoteri X, British Guiana (73)

Apoteri C, British Guiana (68) × Georgetown, British Guiana B (10B)

Sympatric

Georgetown, British Guiana N (77) × Georgetown, British Guiana B (10B)

Apoteri A, British Guiana (66) × Apoteri X, British Guiana (73)

Apoteri C, British Guiana (68) × Apoteri X, British Guiana (73)

*Orinocan* × *Guianan*

Allopatric

Georgetown, British Guiana C (42) × Apoteri X, British Guiana (73)

Bucaramanga 8, Colombia (28) × Georgetown, British Guiana B (10B)

Sympatric

Apoteri Y, British Guiana (74) × Apoteri X, British Guiana (73)

Georgetown, British Guiana C (42) × Georgetown, British Guiana B (10B)

*Orinocan* × *Andean*

Allopatric

Georgetown, British Guiana C (42) × Apoteri O, British Guiana (72)

Apoteri Y, British Guiana (74) × Elena A, Venezuela (49)

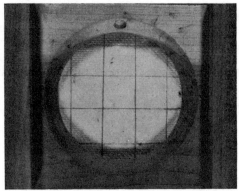

FIGS. 1 and 2. Two views of the Elens-Wattiaux chamber used for the direct observation of mating behavior; see the Method section for an explanation of its operation.

Caripe L, Venezuela (38L) × Simla M, Trinidad (54)

Bucaramanga 8, Colombia (28) × Cucuta C, Colombia (90)

Apoteri Y, British Guiana (74) × Cucuta C, Colombia (90)

Sympatric

Apoteri Y, British Guiana (74) × Apoteri O, British Guiana (72)

Apoteri Y, British Guiana (74) × Apoteri N, British Guiana (71)

*Andean × Guianan*

Allopatric

Simla M, Trinidad (54) × Georgetown, British Guiana B (10B)

Elena A, Venezuela (49) × Georgetown, British Guiana B (10B)

Cucuta C, Colombia (90) × Georgetown, British Guiana B (10B)

Cucuta C, Colombia (90) × Apoteri X, British Guiana (73)

Sympatric

Apoteri O, British Guiana (72) × Apoteri X, British Guiana (73)

Apoteri M, British Guiana (70) × Apoteri X, British Guiana (73)

Apoteri N, British Guiana (71) × Apoteri X, British Guiana (73)

*Centro-American × Orinocan*

Allopatric

Goofy Lake C, Panama (81) × Georgetown, British Guiana C (42)

Cerro Campana C, Panama (32) × Georgetown, British Guiana C (42)

Lancetilla, Honduras (2) × Apoteri Y, British Guiana (74)

Tikal, Guatemala (1) × Bucaramanga 8, Colombia (28)

Tikal, Guatemala (1) × Apoteri Y, British Guiana (74)

Lancetilla, Honduras (2) × Bucaramanga 8, Colombia (28)

Sympatric

Cerro Campana C, Panama (32) × Madden Forest, Panama (86)

The number of strains involved differ in the crosses between the races simply because different numbers of strains were or were not available; almost always, the number of sympatric strains tested was less than the number of allopatric strains tested —fewer do occur sympatrically in nature.

## THE METHOD

Elens (1958) first described an apparatus for measuring the sexual isolation between different mutants of *Drosophila melanogaster* via direct observation of the individuals involved; Elens and Wattiaux (1964) extended the use of these observation chambers to other species (*e.g.*, Ehrman *et al.*, 1965, successfully used it with different inversion types in *D. pseudoobscura*) and described the apparatus more fully. It is to this later paper that the reader is referred for a thorough description of what is essentially a glass and wood sandwich, pictured in Figs. 1 and 2.

This method is superior to the older "male choice" technique for several reasons, the most important of which is that it easily permits the observation of four types of matings, *i.e.*, A♀ × A♂, A♀ × B♂, B♀ × A♂, and B♀ × B♂ in any single experiment. Briefly, the "male choice" method (see Merrell, 1960) involves confining equal numbers of two kinds of virgin females with males of only one kind. After a certain lapse of time, the sperm receptacles of the females are dissected out and examined for the presence of spermatozoa (or, more time consuming, the inseminated females are permitted to produce offspring). Thus, only the matings A♂ × A♀ or B♀ can be recorded at one time, while B♂ × A♀ or B♀ must be experimentally analyzed separately.

In the observation chambers employed here, as soon as a mating occurs five facts about it are recorded: (1) the time it takes place, (2) its sequence among the other copulae which occur, (3) where in the chamber the couple is located (for this purpose, a grid forms the floor, see Fig. 1 and 2), (4) the type of female involved, and, (5) the type of male involved. It is also a simple matter to note the duration of each mount though this was not considered necessary here. No flies need be sacrificed for dissection with this procedure, and they may subsequently be used to breed additional generations.

A possible weakness of this technique is that it in no way prevents the scoring of a single male more than once, but then neither does the "male choice" technique; however, if a single male is successful more than once in a single experimental run, perhaps it should be recorded twice. Furthermore, data accumulate more slowly with the direct observation method. This is so because each individual mating is directly observed and a limited number of chambers can be run in one day. For the crosses tested here, four chambers per day was the maximum.

These observations were begun on October 4, 1963, and concluded on September 8, 1964; there is at least one advantage in stretching such experiments out over a rather long time (actually, it is unavoidable in the case of the direct observation

technique), since seasonal variations in behavior, etc. would be expected to occur even in the laboratory. The flies introduced into the chambers were aged while isolated from individuals of the opposite sex for about eight days and alternating strains and sexes had the margin of one wing clipped for ease of identification with a 4× hand lens while in the chambers.

## Results and Discussion

Table 1 tells the entire story; for it, the new joint isolation coefficient and its standard error were calculated according to the formula used by Malogolowkin-Cohen *et al.* (1965). Briefly, if $P_{11}$, $P_{22}$, $q_{12}$ and $q_{21}$ equal the proportion of matings observed between males and females of race 1, males and females of race 2, males of race 1 and females of race 2, and males of race 2 and females of race 1 respectively, then the coefficient equals:

$$P_{11} + P_{22} - q_{12} - q_{21}$$

and the variance is $\sqrt{4pq/N}$ where $p =$ the proportion of homogamic matings, $q =$ the proportion of heterogamic matings, and $V =$ the total number of matings observed. Random mating results in an isolation index of zero and complete isolation produces a coefficient $= 1.00$.

In all but one case, that entered in line 7 of Table 1, Centro-American × Amazonian, the coefficients are higher for the sympatric than for the allopatric strains, and in this one case, the difference is not significant so that the two measures of isolation may be considered about the same. Furthermore, the average sympatric isolation coefficient is 0.85 while the average allopatric one is 0.67. It seems reasonable to conclude that any isolating mechanism which decreases the chances of the production of inferior hybrid offspring may be expected to be stronger between sympatric than between allopatric populations of the same pairs of subspecies or incipient species. This is certainly borne out by the data reported here. These data also concur with those obtained by Dobzhansky *et*

TABLE 1. *Numbers of matings observed and isolation coefficient calculated for sympatric and for allopatric crosses; the total number of matings observed was 1,695.*

| Races | Origin | | Matings | Coefficient |
|-------|--------|---|---------|-------------|
| 1. Amazonian × Andean | { | Sym | 108 | 0.86 ± 0.049 |
| | { | Allo | 100 | 0.66 ± 0.074 |
| 2. Amazonian × Guianan | { | Sym | 104 | 0.94 ± 0.033 |
| | { | Allo | 109 | 0.76 ± 0.061 |
| 3. Amazonian × Orinocan | { | Sym | 106 | 0.75 ± 0.065 |
| | { | Allo | 124 | 0.61 ± 0.070 |
| 4. Andean × Guianan | { | Sym | 109 | 0.96 ± 0.026 |
| | { | Allo | 102 | 0.74 ± 0.066 |
| 5. Orinocan × Andean | { | Sym | 100 | 0.94 ± 0.033 |
| | { | Allo | 111 | 0.46 ± 0.084 |
| 6. Orinocan × Guianan | { | Sym | 104 | 0.85 ± 0.053 |
| | { | Allo | 100 | 0.72 ± 0.069 |
| 7. Centro-American × Amazonian | { | Sym | 102 | 0.68 ± 0.072 |
| | { | Allo | 103 | 0.71 ± 0.070 |
| 8. Centro-American × Orinocan | { | Sym | 110 | 0.85 ± 0.052 |
| | { | Allo | 103 | 0.73 ± 0.069 |

Average (sympatric) = 0.85
Average (allopatric) = 0.67

*al.*, (1964) using the traditional "male choice" method, discussed above, as applied to similar material. Malogolowkin-Cohen *et al.* (1965) found that this relationship held even when allopatric strains of *Drosophila paulistorum* that were closer or further apart in nature were compared for the strength of their behavioral isolation from one another. Indeed, one may hypothesize that the peculiar sterility of the hybrids between these races (genically induced modification of the egg cytoplasm, Ehrman, 1962, pp. 287–291), came first in the history of this species-complex, as a by-product of adaptation of allopatric races to environments of their respective distribution regions. Sexual isolation which is more effective came later, and its biological function is to limit or prevent the appearance of hybrids with reduced reproductive fitness. As a result, six races which can be considered incipient species have appeared. Alexander (1964) carefully reviewed this sort of situation (among others), *i.e.*, variation in the efficiency of reproductive isolation between sympatric and between allopatric populations, in sound-producing insects where acoustical communication often proves to be a most dramatic isolating device. He accurately noted that: "In closely related sympatric species, very little is known about which differences developed when the species were allopatric and which developed subsequently."

Finally, we should consider time and any variation in the occurrence of interracial or intraracial matings which might be correlated with it; this was done, also with *Drosophila paulistorum*, in careful detail by Malogolowkin-Cohen *et al.* (1965) who note, ". . . regardless of whether the isolation is low or high, the isolation index is about the same immediately after the flies are introduced as when most of the available females have already mated." The data reported here certainly do confirm their conclusion. For instance, as is recorded in Table 1, a total of 1,695 matings were observed. The observation chambers were run a total of 185 separate times so that an average of 9.2 matings took place in each run (10 represents the optimal 50 per cent). 12.86 per cent or

218 of these matings were heterogamic; 103 occurred during the first half of individual runs with the observation chambers (this is, of course, just one of many possible ways of examining the time data) whereas 115 took place during the second half. The difference is not statistically significant since $x_1^2 = 0.66$.

To conclude, according to Grant (1963): "The relation between reproductive isolation and ecological coexistence is a reciprocal one. Not only does the formation of reproductive isolating mechanisms permit species to coexist more or less closely, but the attainment of sympatry promotes the further development of breeding barriers between species. Reproductive isolation is a cause of sympatry, and sympatry in turn can be a cause of reproductive isolation."

## SUMMARY

Those races of the *Drosophila paulistorum* species-complex of six known races which occur sympatrically and retain their separate identities in nature, exhibit a greater sexual isolation than do strains of the same races which occur allopatrically. This behavioral isolation was tested by a recently developed direct observation technique and bespeaks the correctness of considering these races as incipient species.

## ACKNOWLEDGMENTS

This is to acknowledge a profitable correspondence with Dr. J. M. Wattiaux about the method employed here. Professors Th. Dobzhansky and H. Levene helped by criticizing the manuscript and by offering advice about the statistics, respectively.

## LITERATURE CITED

ALEXANDER, R. D. 1964. The evolution of mating behavior in arthropods. *In* Insect Reproduction, K. C. Highnam (Ed.). Royal Ent. Soc. Symposium no. 2, London.

DOBZHANSKY, TH., L. EHRMAN, O. PAVLOVSKY, AND B. SPASSKY. 1964. The superspecies *Drosophila paulistorum*. Proc. Nat. Acad. Sci., **54**: 3–9.

DOBZHANSKY, TH., AND B. SPASSKY. 1959. *Drosophila paulistorum*, a cluster of species *in statu nascendi*. Proc. Nat. Acad. Sci., **45**: 419–428.

EHRMAN. L. 1961. The genetics of sexual isolation in *Drosophila paulistorum*. Genetics, **46**: 1025–1038.

——. 1962. Hybrid sterility as an isolating mechanism in the genus Drosophila. Quart. Rev. Biol., **37**: 279–302.

EHRMAN, L., B. SPASSKY, O. PAVLOVSKY, AND TH. DOBZHANSKY. 1965. Sexual selection, geotaxis, and chromosomal polymorphism in experimental populations of *Drosophila pseudoobscura*. Evolution, **19**: 337–346.

ELENS, A. A. 1958. Le rôle de l'heterosis dans la compétition entre ebony et son allèle normal. Experientia, **14**: 274–278.

ELENS, A. A., AND J. M. WATTIAUX. 1964. Direct observation of sexual isolation. Drosophila Information Service, **39**: 118–119.

GRANT, V. 1963. The origin of adaptations. Columbia University Press, New York, pp. 514–515.

MALOGOLOWKIN-COHEN, CH., A. S. SIMMONS, AND H. LEVENE. 1965. A study of sexual isolation between certain strains of *Drosophila paulistorum*. Evolution, **19**: 95–103.

MERRELL, D. J. 1960. Mating preferences in Drosophila. Evolution, **14**: 525–526.

## 22

Reprinted from *Evolution* 21:679–687 (1967)

## NATURAL SELECTION FOR REPRODUCTIVE ISOLATION IN *PHLOX*

Donald A. Levin and Harold W. Kerster

*Department of Biological Sciences, University of Illinois at Chicago Circle, Chicago, Illinois*

Divergence in the genic and chromosomal architecture of local populations from the population system norm has been documented in numerous plant and animal species (cf. Grant, 1963, 1964; Mayr, 1963; Ford, 1965). The stimulus for such divergence may be provided by one of an array of environmental challenges. The immediate presence of a potentially interbreeding congener, or one which competes for biotic or physical resources of the environment may constitute such a challenge. A shift in the adaptive mode of one of the congeners could intensify existing external and internal barriers to gene exchange, or could permit the two species more efficiently to exploit their habitats. Population divergence in areas of congener sympatry, which may involve morphological, physiological, ecological, or ethological traits, has been termed "character divergence" (Darwin, 1859) and "character displacement" (Brown and Wilson, 1956; Wilson, 1965). The term "Wallace effect" has been applied to shifts in population structure whose primary adaptations are to strengthen reproductive isolating barriers (Grant, 1966).

### Description of the Study

The eastern alliance of the plant genus *Phlox* (Polemoneaceae) is a favorable object for investigating character displacement in perennial or long-lived organisms. The alliance displays extensive inter- and intrapopulation heterogeneity, broad zones of polyspecies sympatry, weak to moderate ecological differentiation, overlapping flowering periods, and various degrees of reproductive isolation. Populations may exhibit polymorphisms in vegetative and reproductive characteristics, as well as in crossing behavior and karyology.

Inter- and intrapopulation variation in corolla pigmentation is particularly striking. This variation is expressed in a multitude of color phases ranging from deep lavender or red-pink to white. Most populations are comprised almost exclusively of pigmented forms in which the color is manifested in an array of nuances and intensities. The frequency of dimorphic populations, i.e., those containing pigmented and non-pigmented color phases, ranges from less than 1% to 10%, depending on the taxon. White-flowered plants maintain their notable character through successive seasons in the experimental garden and greenhouse; the hues displayed by pigmented corollas are subject to environmental modification. An analysis of pigment inheritance in a series of annual *Phlox* cultivars suggests that pigment synthesis is under complex polygenic control (Kelly, 1920).

Exceptions to typical color structure may be found in populations of several species, especially *P. pilosa* L. This *Phlox* inhabits much of the eastern United States and is composed almost entirely of populations in which the white corolla phase is rare or absent. There are, however, discrete populations or populations systems in which the white phase prevails. These populations, which contain tens to hundreds of plants, are comparable to the norm in all respects except corolla pigmentation. Although very sporadic in occurrence, such colonies often occur in habitats supporting congeners with similar ecological requirements, typical corolla form and pigmentation, and overlapping flowering periods. We do not mean to imply, however, that all biotically sympatric *P. pilosa* populations deviate from their biotically allopatric allies.

The congener of particular interest in this discussion is *P. glaberrima* subsp. *interior* Wherry. *Phlox glaberrima* L. and *P. pilosa* are conspicuous elements of the late spring and early summer prairie floras of Illinois, Indiana, and adjacent states. The species have similar ecological tolerances as evidenced by the frequent formation of contiguous or confluent populations. In spite of these spatial associations, opportunities for interspecific gene exchange are precluded by seasonal isolation, *P. pilosa* being in fruit before *P. glaberrima* commences flowering. If some unusual climatic event afforded an opportunity for interbreeding, hybridization could occur as the species are obligate outbreeders and interfertile (Levin, 1963, 1966; Levin and Kerster, 1967). The incompatibility barriers, however, are formidable.

*Phlox pilosa* is composed of several morphologically, ecologically, and chromosomally distinct phylads (Wherry, 1955; Levin, 1966; Smith and Levin, 1967). A late-flowering race of *P. pilosa* subsp. *pilosa* [= *P. argillacea* Clute and Ferris] is of special significance with respect to hybridization with *P. glaberrima*. This race, indigenous to the sand prairies of northeastern Illinois and northwestern Indiana, consistently flowers 3 weeks later than the norm. Thus the flowering period of this *Phlox* overlaps that of *P. glaberrima* for as long at 2 weeks. The opportunities for hybridization are multiple as populations of these phloxes are often juxtaposed or intermixed, and are pollinated by the same pollinators (Grant and Grant, 1965; Levin and Kerster, 1967). The general floral mechanisms of *P. pilosa* and *P. glaberrima* render them especially suited for Lepidoptera pollination, butterflies being the most important group (op. cit.).

The corolla color structure of populations of the late-flowering *P. pilosa* typically is similar to that of standard populations, red-pink being the predominant or sole phase. The color structure is often reversed, however, in populations where this *Phlox* is in contact with *P. glaberrima*.

The frequency of the white phase appears to be related to the proximity of the congener. As populations of the two taxa become more distant, the frequency of the white phase decreases abruptly. These observations suggest that there is a compelling relationship between white phase predominance and the presence of *P. glaberrima*.

The investigations reported here were undertaken to determine whether *P. glaberrima* was the stimulus for the local displacement of the pigmented phase of *P. pilosa* by the white phase, and to ascertain the adaptive significance of the white phase.

EXPERIMENTAL PROCEDURES

Lepidoptera pollinators are capable of differentiating between species pairs on the basis of flower color and form, and scent (cf. Grant, 1949, 1963; Dronamraju, 1960; Faegri and van der Pilj, 1966). Of particular interest is their ability to distinguish between a number of colors including red-pink and white. The similarity of corolla pigmentation of *P. glaberrima* and typical *P. pilosa*, and the pollinator-discernible difference between *P. glaberrima* and the white phase of *P. pilosa* suggested that flower color displacement in *P. pilosa* might be an adaptation to aid pollinator discrimination and flower-constancy.

The merits of this hypothesis were tested using a pollen "tracer" technique which permits an objective and quantitative appraisal of interspecific pollen flow. The pollen grains of *P. glaberrima* and *P. pilosa* are well-marked by virtue of large, discontinuous, differences in pollen diameter, the average of the former being 55 microns as compared to 30 microns in the latter (Levin and Kerster, 1967). By examining pollen on the stigma of the species in populations comprised of the two phases of *P. pilosa*, and *P. glaberrima*, the flow of pollen between the species and the role of flower color differences in regulating that flow can be ascertained with facility.

The population chosen for pollen flow investigations, located in a sand prairie in southern Cook County, Illinois, contained over 150 plants of each species. The population was 70 meters long and 20 meters wide. The species were biotically sympatric in approximately 50% of the area under investigation, *P. pilosa* inhabiting the more xeric sites in the area of allopatry. *Phlox pilosa* was represented almost exclusively by the white phase so that it was necessary to introduce a series of red-pink transplants in order to achieve some measure of color balance. The transplants were introduced in a random fashion in the microhabitats supporting the white phase. The red-pink phase comprised 25% of the modified *P. pilosa* population.

Collections for pollen analysis were performed on June 21, 1966, a time of maximum overlap in species flowering periods. In all, 238 flowers of the white phase and 238 flowers of the red-pink phase of *P. pilosa* were collected from 50 plants of each phase, the plants being located in various sectors of the study area. The stigma of each flower was removed and stained in a solution of aniline-blue in lactophenol for 24 hours. Each stigmata containing over 25 pollen grains was examined for foreign grains. Over 300 stigma of *P. glaberrima* were treated in a similar fashion.

Consideration was given to the problem of whether pollinator discrimination also might occur in dimorphic populations of *P. pilosa*. The pollen of the red-pink and white phases is comparable in size, color and sculpture. Pollinator favor or disfavor, however, might be detected by differential seed production of the two phases. Reproductively mature inflorescences were collected from red-pink and white-flowered plants in 13 populations with typical color structure, and in 3 populations in which white-flowered plants were the predominant form. Nearly 3500 flowers were collected for capsule and seed analyses.

### RESULTS

An examination of stigma of red-pink and white-flowered *P. pilosa* revealed a striking pollinator preference for the red-

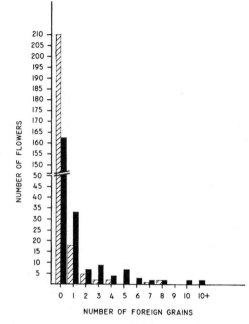

FIG. 1. The numerical distribution of *P. glaberrima* pollen grains on stigma of the red-pink phase (solid bars) and white phase (crosshatched bars) of *P. pilosa*.

pink form when making an interspecific move from *P. glaberrima*. This preference was manifested in the percentage of stigma with *P. glaberrima* grains and by the total alien pollen load. Thirty per cent of the stigma of the red-pink form carried contraspecific pollen in contrast to 12% of the white form. The distribution of alien pollen loads on the stigma of the two phases is depicted in Figure 1. It is significant to note that 4.8 times as many grains were transported from *P. glaberrima* to the red-pink phase as to equal numbers of the white phase. Foreign pollen loads on *P. pilosa* were distributed in a manner which suggests that the usual number of *P. glaberrima* grains deposited by a single pollination is one (Fig. 1; Levin and Kerster, 1967). Accordingly, we may infer that pollinator moves from *P. glaberrima* to the red-pink phase of *P. pilosa* were nearly five times as frequent as moves to the white phase.

Pollen flow to *P. glaberrima* greatly exceeded that to *P. pilosa*, color notwithstanding. Thirty-seven per cent of the stigma examined bore contraspecific pollen with an average of 14.7 grains per "contaminated" stigma. The number of contraspecific grains on *P. glaberrima* stigma varied from 1 to 38. These data, as well as those presented above, indicate that interspecific pollen flow in the study area was more extensive in 1966 than in 1965 (Levin and Kerster, 1967). The overall balance with respect to direction of alien pollen transport, however, did not change significantly.

Seed-set data suggest that pollinators strongly favor the red-pink phase of *P. pilosa* over the white phase in populations where the typical color structure prevailed. The average seed-set per capsule in the red-pink phase was 2.0 as compared to 1.4 in the white phase. Even in colonies where white predominated, its seed production per capsule lagged behind that of red-pink. The average seed-set for the white phase, as compiled from a series of populations containing from 85% to 97% "white" plants, was 1.8 per capsule. The red-pink phase continued to average 2.0 seeds per capsule in spite of its low frequency. The difference in seed-set between the two forms may not entirely be attributable to pollinator color preferences, although such preferences ostensibly are very important. Butterflies have well-developed visual sense, and favor vividly colored blossoms to those which are white or faintly colored (Faegri and van der Pilj, 1966).

### Discussion and Conclusions

The eastern *Phlox* alliance is composed of 18 species whose ranges generally are extensive, thereby creating broad zones of polyspecies sympatry. Although the potential for hybridization between many species combinations is rather great, a series of ecological, seasonal, ethological, incompatibility and sterility barriers severely limits this event (Erbe and Turner,

1962; Levin, 1963, 1966, 1967*a*, *b*, unpubl.; Levin and Smith, 1965, 1966; Levin and Kerster, 1967). Ecological and seasonal barriers normally preclude pollen flow between sympatric species. In localized areas where edaphic or climatic conditions afford opportunities for such flow, one may find a correspondingly local divergence in the floral pigmentation of one of the species pair. The divergence is particularly conspicuous in the case of interfertile species pairs, but also occurs among intersterile species.

The often compelling relationship between congener proximity and species divergence suggests that certain patterns of variation in flower pigmentation are a consequence of character displacement. In view of the trait involved, one might surmise that local divergence in corolla pigmentation strengthens ethological barriers and thus conserves the reproductive potential of the species. The merits of this hypothesis were validated by data which disclosed that interspecific pollen flow between congeners with comparable corolla pigmentation far exceeded flow between the same congeners with contrasting pigmentation. The reader will recall that nearly five times as much alien pollen was carried to the red-pink phase of *P. pilosa* as to the white phase. In view of the magnitude of the pollen load differential and pollinator preference, we may infer that pollen flow to *P. glaberrima* also is related to corolla pigmentation, and that the bulk of alien pollen which it receives comes from the red-pink phase of *P. pilosa*.

We may now ask whether character displacement was stimulated by a need to impede introgression, or was stimulated by a need to reduce pollen wastage. Stringent incompatibility barriers all but eliminate the potential for hybridization. Interspecific crosses rarely are effectual, and seeds are produced only when *P. pilosa* serves as the maternal parent (Levin, 1966). Since the bulk of the pollen in the population in question is towards *P. glaberrima*, relatively few alien grains will be deposited on stigma of *P. pilosa*, and only a small percentage of these grains will be able to surmount the incompatibility barriers. Nevertheless, there is evidence that genes have trickled from *P. glaberrima* to *P. pilosa* (Levin, 1964, unpubl.). Characteristics of *P. glaberrima* such as glabrous corolla tube and imbricate corolla lobes were found in populations of *P. pilosa*, although examination of bispecific populations from 1962 through 1966 failed to reveal the existence of $F_1$ hybrids or recent backcross derivatives. Not all mixed populations, however, showed evidence of past hybridization. It seems unlikely that divergence in flower color arose solely in response to a need for stronger reproductive barriers. However, we cannot discount the possibility that displacement is related to gene exchange and low hybrid fitness (Wilson, 1965). Wilson (op. cit.) contends that species with imperfect internal isolating barriers will either fuse or displace reproductively in areas of species contact depending on the level of gene flow and fitness of hybrids.

Interspecific pollen flow data suggest that the pollinating mechanism of *Phlox* is somewhat inefficient (Levin and Kerster, 1967). The foreign pollen load on stigma of *P. glaberrima* indicates that typically one or two *P. pilosa* grains are transferred per pollinator move. We may assume that the pollen load carried on intraspecific moves is comparable to that carried on interspecific moves. The relatively large size of *Phlox* pollen and the organ of dispersal, i.e., a smooth butterfly proboscis (Grant and Grant, 1965), contribute to the diminutive pollen load. The butterfly inserts its proboscis into several flowers on a single plant, but most or all of the pollen carried from the previously visited plant is deposited on the first flower (Levin and Kerster, 1967). The species is self-incompatible, so that most pollinations are ineffectual as they are self-pollinations.

The paucity of intraspecific pollen flow is portrayed in the percentage of flowers which form capsules. Collections made during 1965 and 1966 revealed that only 60% of the flowers formed capsules and that seed-set per capsule averaged 2.0, three being the maximum. If seed-set per flower is considered, it becomes apparent that seed production is only 40% of the maximum. Can *P. pilosa* afford to lose potentially effectual grains to another species?

The variability in flower-color genes necessary for a character displacement response is usually available. Maintenance of the white phase of *P. pilosa* at low levels is quite common; one or a few of such plants usually can be found in any colony composed of a few hundred plants. Seed-set data indicate that "white" plants, however, are at a distinct selective disadvantage in typical populations. The average seed-set per capsule for the red-pink phase was 2.0 as compared to 1.4 in the white phase. Accordingly, the relative fitness value of the white phase is .70 for this stage of the life cycle; for every 100 seeds produced by the red-pink flowers, the same number of white flowers produce 70 seeds. The retention of the white phase in *P. pilosa* populations might be attributed to the recessive nature of white corolla color, relatively high mutation rates in the polygenes controlling pigment synthesis, or to a selective advantage accorded during the vegetative stage of the life cycle.

The white phase of *P. pilosa* is more highly adapted to co-exist with *P. glaberrima* than is its red-pink ally. We may assume that pollen wastage by the red-pink phase is nearly five times as great as that by the white phase. This supposition is based on the inference that pollinator movements between *P. glaberrima* and red-pink phase are nearly five times greater than between the former and the white phase. Presumably at the time of initial postglacial contact between the species, populations of *P. pilosa* "carried" the white phase in very low frequency. As a consequence of this contact, local colonies of *P. pilosa* experienced differential pollen wastage which resulted in a gradual displacement of the red-pink phase by the white.

Interesting color-sympatry correlations exist between other species of eastern *Phlox*. *Phlox pilosa* and *P. amoena* Sims, a close ally, are sympatric throughout much of the southeastern United States, but generally remain spatially isolated because of different ecological requirements. Close spatial associations and hybridization are observed in southern Tennessee and northern Alabama, where diverse habitats often are in juxtaposition (Levin and Smith, 1966). Many populations of *P. pilosa* in this region contain a high proportion of the white phase, in contrast to the typical red-pink phase.

The color-sympatry correlation may involve entire population systems rather than single populations as illustrated by *P. bifida* Beck and *P. pilosa*. These species typically are seasonally isolated and differ markedly in ecological tolerances, *P. bifida* inhabiting rocky woodland slopes and ledges in dry soils and *P. pilosa* inhabiting more mesic sites in prairies and woodland borders. Deeply pigmented forms typically comprise approximately 99% of the population of both species. There is, however, a late-flowering race of *P. bifida* whose populations are often near or in contact with those of *P. pilosa* in the sand prairies of Illinois and western Indiana. Populations of this race of *P. bifida* are composed almost exclusively of a white-flowered phase. Pollen "tracer" studies show that when contiguous or confluent, pollen is transferred from one species to another (Levin, unpubl.).

An examination of corolla pigmentation in regions of sympatry and allopatry of *P. cuspidata* Scheele and *P. drummondii* Hook. reveals that a color-sympatry association may occur at the subspecies level of divergence. *Phlox drummondii* is composed of six subspecies, five of which are completely or largely allopatric with respect to *P. cuspidata*. The sixth, subspecies *drummondii*, is almost completely sympatric. It is particularly interesting to note that subsp. *drummondii* displays a red corolla while its allies all have pink corollas as does *P. cuspidata*. Natural hybridization between *P. cuspidata* and *P. drummondii* subsp. *drummondii* occurs infrequently. Hybrids are relatively few in number and are sterile (Levin, 1967).

Ethological barriers are the most efficient pre-mating barriers as related to the conservation of reproductive potential (Grant, 1966). The strengthening of ethological barriers would be particularly advantageous to species whose hybrids are adaptively inferior, or to species which can ill afford the loss of gametes involved in actual or potential hybridization (Grant, 1963). Perhaps the latter is the case in *Phlox*.

As we have seen, corolla color displacement may be a highly localized phenomenon or may involve population systems of various magnitudes. Displacement may involve single populations, races or subspecies. Displacement may occur between species which hybridize and exchange genes readily (*P. pilosa* and *P. amoena*), which interbreed but form sterile hybrids (*P. cuspidata* and *P. drummondii*), which rarely hybridize but exchange some genes (*P. pilosa* and *P. glaberrima*), and between species which never hybridize (*P. pilosa* and *P. bifida*). It is significant to note that flower color displacement has occurred irrespective of species' abilities to exchange genes. We must recognize that whether or not hybridization ensues, pollen will flow between intermixed species pairs and the flow may be substantial. We therefore maintain that *Phlox* can ill afford the loss of pollen typically incurred in confluent or adjacent colonies, and that color displacement is the product of selection for ethological isolation, the latter enhancing pollinator discrimination and thereby reducing pollen wastage. The presence of a congener which competes with a species for domestic pollen ostensibly has served as the stimulus for color divergence.

The reinforcement of ethological barriers in areas of sympatry has been described in *Microhyla* (Blair, 1955), *Pseudacris*

(Mecham, 1961), *Hyla* (Littlejohn, 1965), and *Drosophila* (Dobzhansky and Koller, 1938; Dobzhansky, 1951; Dobzhansky et al., 1957; Dobzhansky et al., 1964; Ehrman, 1965), *Erebia* (Lorković, 1958), and *Peromyscus* (McCarley, 1964). Artificial selection for reproductive isolation leading to a strengthening of interstrain or interspecific ethological barriers also has been demonstrated in *Drosophila* (Koopman, 1950; Wallace, 1954; Knight et al., 1956).

Grant (1966) has designated the secondary and supplementary process of selection for reinforcement of existing reproductive isolating mechanisms as the "Wallace effect." The term refers only to those mechanisms which are operative in the parental generation. The "Wallace effect" may manifest itself in the form of enhanced ethological or incompatibility barriers. Selection for reproductive isolation is most effective in short-lived organisms such as annual plants or ephemeral animals where a loss of reproductive potential could have grave consequences (Dobzhansky, 1958; Stebbins, 1958). Evidence suggesting the occurrence of the "Wallace effect" has been obtained exclusively from short-lived organisms (see Grant, 1966, for a discussion and critique of the evidence for the "Wallace effect").

The role of corolla color displacement is reducing pollen wastage, and the selective advantage of this reduction have been clearly demonstrated in *Phlox*. The evidence presented above indicates that the "Wallace effect" indeed occurs in perennial plants. Therefore, the suggestion that selection for reproductive isolation would be relatively ineffective in perennial plants (Dobzhansky, 1958; Stebbins, 1958) is open to conjecture. If effectual pollen was at a premium in a perennial, a shift in the adaptive mode of that plant which would conserve its reproductive potential indeed would be at a selective advantage. Color displacement in *Phlox* must be interpreted as selection for reproductive isolation.

## SUMMARY

Populations of the eastern species of *Phlox* typically are composed of organisms with pigmented corollas. A small percentage of these populations show color dimorphism, i.e., they contain pigmented and non-pigmented or white forms. The white phase often comprises less than 1% of such populations. Exceptions to the normal color structure occur in a number of species, especially *P. pilosa*. Discrete populations or population systems are known to occur in which the white phase is the predominate or sole form. Although such populations are infrequent, they often grow in habitats supporting a congener with similar ecological requirements and flowering period. This congener often is *P. glaberrima*.

The relationship of color displacement and congener presence was studied in an assemblage comprised of *P. glaberrima* and the two phases of *P. pilosa*. The pollen of both species is well-marked so that an objective, quantitative appraisal of interspecific pollen flow can be made with facility. An examination of the stigma *P. glaberrima* and *P. pilosa* disclosed that there was considerable interspecific pollen flow. It was most enlightening to find that while 30% of the pigmented phase of *P. pilosa* bore alien pollen, only 12% of the white phase bore such pollen. Moreover, 4.8 times as many grains were transported from *P. glaberrima* to flowers of the pigmented phase as to equal numbers of flowers of the white phase. Thus it appears that color divergence aids pollinator discrimination and conserves reproductive potential.

Character displacement in the eastern *Phlox* alliance occurs irrespective of the species' abilities to exchange genes. Displacement occurs between species which are incapable of hybridization, but are capable of exchanging pollen.

We conclude that *Phlox* can ill afford the loss of pollen typically incurred in bispecific populations, and that color displacement is the product of selection for enhanced ethological isolation. The presence of a congener which competes for domestic pollen ostensibly has served as the stimulus for color divergence.

ACKNOWLEDGMENTS

The work reported here was aided by National Science Foundation grant GB-4326 to the first author.

LITERATURE CITED

BLAIR, W. F. 1955. Mating call and stage of speciation in the *Microhyla olivacea-M. carolinensis* complex. Evolution **9:** 469–480.

BROWN, W. L., JR., AND E. O. WILSON. 1956. Character displacement. Systematic Zool. **5:** 49–64.

DARWIN, C. 1859. On the origin of species by means of natural selection. First ed. John Murray, London.

DOBZHANSKY, TH. 1951. Genetics and origin of species. Third ed. Columbia Univ. Press, New York.

——. 1958. Species after Darwin. *In* S. A. Barnett [ed.], A century of Darwin. Heinemann, London.

DOBZHANSKY, TH., L. EHRMAN, O. PAVLOVSKY, AND B. SPASSKY. 1964. The superspecies *Drosophila paulistorum*. Proc. Nat. Acad. Sci. **51:** 3–9.

DOBZHANSKY, TH., AND P. C. KOLLER. 1938. An experimental study of sexual isolation in *Drosophila*. Biol. Zentralbl. **58:** 589–607.

DOBZHANSKY, TH., L. EHRMAN, AND O. PAVLOVSKY. 1957. *Drosophila insularis*, a new sibling species of the wilistoni group. Univ. Texas Publ. **9:** 39–47.

DRONAMRAJU, K. R. 1960. Selective visits of butterflies to flowers: a possible factor in sympatric speciation. Nature **186:** 178.

EHRMAN, L. 1965. Direct observation of sexual isolation between allopatric and between sympatric strains of the different *Drosophila paulistorum* races. Evolution **19:** 459–464.

ERBE, L., AND B. L. TURNER. 1962. A biosystematic study of the *Phlox cuspidata-P. drummondii* complex. Amer. Midl. Natur. **67:** 257–281.

FAEGRI, K., AND L. VAN DER PIJL. 1966. The principles of pollination ecology. Pergamon Press. New York.

FORD, E. B. 1965. Ecological genetics. Second ed. John Wiley and Sons, New York.

GRANT, V. 1949. Pollination systems as isolating mechanisms in flowering plants. Evolution **3:** 82–97.

——. 1963. The origin of adaptations. Columbia Univ. Press, New York.

——. 1964. The architecture of the germplasm. John Wiley and Sons, New York.

——. 1966. The selective origin of incompatibility barriers in the plant genus *Gilia*. Amer. Natur. **100:** 99–118.

GRANT, V., AND K. GRANT. 1965. Flower pollination in the *Phlox* family. Columbia Univ. Press, New York.

KELLY, J. P. 1920. A genetical study of flower form and color in *Phlox drummondii*. Genetics **5:** 189–248.

KNIGHT, G. R., A. ROBERTSON, AND C. H. WADDINGTON. 1956. Selection for sexual isolation within a species. Evolution **14:** 14–22.

KOOPMAN, K. F. 1950. Natural selection for reproductive isolation between *Drosophila pseudoobscura* and *Drosophila persimilis*. Evolution **4:** 135–148.

LEVIN, D. A. 1963. Natural hybridization between *Phlox maculata* and *P. glaberrima* and its evolutionary significance. Amer. J. Botany **50:** 714–720.

——. 1966. The *Phlox pilosa* complex: crossing and chromosome relationships. Brittonia **18:** 142–162.

——. 1967a. Variation in *Phlox divaricata*. Evolution **21:** 92–108.

——. 1967b. Hybridization between annual species of *Phlox*: population structure. Amer. J. Botany **54:** 1122–1130.

LEVIN, D. A., AND H. W. KERSTER. 1967. Interspecific pollen exchange in *Phlox*. Amer. Natur. **101:** in press.

LEVIN, D. A., AND D. M. SMITH. 1965. An enigmatic *Phlox* from Illinois. Brittonia **17:** 254–266.

—— AND ——. 1966. Hybridization and evolution in the *Phlox pilosa* complex. Amer. Natur. **100:** 289–302.

LITTLEJOHN, M. J. 1965. Premating isolation in the *Hyla ewingi* complex (Anura: Hylidae). Evolution **19:** 234–243.

LORKOVIĆ, Z. 1958. Some peculiarities of spatially and sexually restricted gene exchange in the *Erebia tyndarus* group. Cold Spr. Harb. Symp. Quant. Biol. **23:** 319–325.

MAYR, E. 1963. Animal species and evolution. Harvard Univ. Press, Cambridge, Massachusetts.

MCCARLEY, H. 1964. Ethological isolation in the coenospecies *Peromyscus leucopus*. Evolution **18:** 331–332.

MECHAM, J. S. 1961. Isolating mechanisms in anuran amphibians. *In* W. F. Blair [ed.], Vertebrate speciation. Univ. Texas Press, Austin, Texas.

SMITH, D. M., AND D. A. LEVIN. 1967. Karyotypes of eastern North American *Phlox*. Amer. J. Botany **54:** 324–334.

STEBBINS, G. L. 1958. The inviability, weakness

and sterility of interspecific hybrids. Advances in Genet. **9:** 147–215.

WALLACE, B. 1954. Genetic divergence of isolated populations of *Drosophila melanogaster.* Proc. Ninth Internat. Congr. Genet., Caryologia, suppl. vol., p. 761–764.

WHERRY, E. T. 1931. The eastern short-styled phloxes. Bartonia **12:** 24–53.

——. 1955. The genus Phlox. Morris Arbor. Mono., III, Philadelphia.

WILSON, E. O. 1965. The challenge from related species. *In* H. G. Baker and G. C. Stebbins [eds.], The genetics of colonizing species. Academic Press, New York.

# A FINAL NOTE

Various aspects of hybridization have been reviewed with the hope of understanding and assessing its role in evolution. Perusal of the papers and commentary, and of the literature in general, make it apparent that in spite of the vast numbers of genera studied and the application of new techniques, we still know relatively little about the process and implications of hybridization in natural populations. With the exception of relatively few long-term experimental and field studies, we are attempting to understand a dynamic process from single points in time. How are we doing this? By inferring, surmising, speculating, and hoping from the products of hybridization. It seems that major advances in understanding the process will come from experimental analyses (such as those highlighted in Part II), and manipulations of natural hybrid swarms or hybrid zones, and a very thorough analysis of the biology of first-generation, advanced-generation, and backcross hybrids relative to the parental taxa under a range of environments. They will not come from further documentation of hybridization in the fashion of our time. In the case of plants, advances also will come from a better understanding of pollen and seed dispersal in the hybridizing taxa. If pollen and seed dispersal are highly restricted, should we be surprised to find that hybrids are located close to the place of parental contact? Would we be tempted to assume from this distribution that hybrids had a narrow niche width or were inferior to their parents in all but the most intermediate habitats? A similar set of questions could be asked of animals.

In the epilogue to his *Introgressive Hybridization*, Anderson (1949) made the following statement:

> How important is introgressive hybridization? I do not know. One point seems fairly certain: its importance is paradoxical. The more imperceptible introgression becomes, the greater is its biological significance. It may be of the greatest fundamental importance when by our present crude methods we

can do little more than to demonstrate its existence. When, on the other hand, it leads to bizarre hybrid swarms, apparent even to the casual passer-by, it may be of little general significance . . . . Only by the exact comparisons of populations can we demonstrate the phenomenon . . . . The wider spread of a few genes (if it exists) might well be imperceptible even from a study of population averages, but it would be of tremendous biological import . . . .Hence our paradox. Introgression is of the greater biological significance, the less is the impact apparent to casual inspection.

More than two decades later, we still await the answer to Anderson's question, and to the overall significance of hybridization. In a review of the literature, Heiser (1973) writes that "Although not all cases reported as introgression may actually be such, it seems fairly evident that introgression does occur. However, most introgression appears to be highly localized, extending only a short distance from the area of active hybridization. Dispersed or widespread introgression . . . appears to be extremely rare . . . ." He suggests that the rarity of widespread introgression may be due to the selective elimination of alien genes, or gene flow restriction, but he acknowledges that introgression may only appear to be rare. The lack of adequate sophistication in technology may preclude our detection of genes that are incorporated by the recipient. Gottlieb (1972) argues that technology notwithstanding, our approach to hybridization is too superficial and that many conclusions that have been drawn are tenuous. More criteria should be employed and more positive answers obtained before we conclude hybridization or hybrid origin.

There still seems to be a lingering notion that hybridization is bad: that gametes are wasted, hybrids are inferior, and the infiltration of alien genes will pollute an otherwise "pure" gene pool that must be on some adaptive peak already. Wagner (1970) refers to hybridization as an evolutionary mode producing something abnormal. However, Epling (1947) suggested that hybridization and gene exchange may actually be advantageous in some groups, and others (see Part 1) are in accord. Does the abundance of hybrids or hybrid swarms indicate the advantage that may accrue from hybridization, or does it speak to the matter of weak barriers to hybridization? In some groups it may provide an indication; in others, it will not. Will we be able to identify those groups accurately? Raven's (1978) recent review on hybridization in plants suggests that the answer can often be yes.

Finally, there is the matter of hybridization coupled with polyploidy. It is generally accepted that polyploidy is a mechanism to stabilize and "sexualize" sterile or partially sterile $F_1$ or advanced-generation hybrids. Correlatively, it is generally accepted that the success of polyploids derives from their hybrid nature. As we have treaded softly when it comes to the role of hybridization at the diploid level, we have stampeded to embrace hybridization as the key to the success of polyploids. Yet we are only beginning to appreciate the effects of chromosome doubling per se, and they may be manifold. In many polyploid complexes, it would be well to re-examine the bases for the divergence from the parental modes. It seems unlikely that the abrupt and major discontinuities between diploids and their polyploid derivatives is simply a product of hybridization.

# REFERENCES

Adams, R. P., and Turner, B. L. (1970) Chemosystematic and numerical studies of natural populations of *Juniperus ashei* bush. *Taxon* **19**:728–751.

Alexamder, R. D., and Moore, T. E. (1962) The evolutionary relationship of 17-year and 13-year cicadas, and three new species. *Misc. Publ. Mus. Zool. Univ. Mich.* **121**:1–57.

Alston, R. E. (1967) Biochemical systematics. *Evol. Biol.* **1**:197–305.

Anderson, E. (1936) Hybridization in American tradescantias. *Ann. Miss. Bot. Gard.* **23**:511–525.

Anderson, E. (1939) Recombination in species crosses. *Genetics* **24**:668–698.

Anderson, E. (1949) *Introgressive Hybridization.* New York: John Wiley & Sons, Inc.

Anderson, E. (1953) Introgressive hybridization. *Biol. Rev.* **28**:280–307.

Anderson, E., and Brown, W. L. (1952) Origin of Corn Belt maize and its genetic significance. In Gowen, J. W. (ed.), *Heterosis.* New York: Hafner Press.

Anderson, E., and Hubricht, L. (1938) Hybridization in *Tradescantia.* III. The evidence for introgressive hybridization. *Am. J. Bot.* **25**:396–402.

Ayala, F. J. (1976) Molecular genetics and evolution. In Ayala, F. J. (ed.), *Molecular Evolution.* Sunderland, Mass.: Sinauer Associates, Inc.

Baker, H. G. (1953) Race formation and reproductive method in flowering plants. *Symp. Soc. Exp. Biol.* **7**:114–143.

Barrett, D. H. P., and Flavell, R. B. (1977) Mitochondrial complementation and grain yield in hybrid wheat. *Ann. Bot.* **41**:1333–1343.

Bauer, E. (1932) Artungrenzung und Arbildung in der Gattung *Antirrhinum,* Sektion *Antirrhinum astrum. Z. Ind. Abstamm., Verebungsl.* **63**:256–302.

Bayer, M. H., and Ahuja, M. R. (1968) Tumor formation in *Nicotiana*: auxin levels and auxin inhibitors in normal and tumor-prone genotypes. *Planta* **79**:292–298.

Berger, L. (1973) Systematics and hybridization in European green frogs of *Rana exculenta* complex. *J. Herpetol.* **7**:1–10.

Bicudo, H. E. M. C., and Richardson, R. H. (1977) Gene regulation in *Drosophila mulleri, D. arizonensis,* and their hybrids: the nucleolar organizer. *Proc. Natl. Acad. Sci. USA* **74**:3498–3502.

Blair, A. P. (1941) Variation, isolating mechanisms and hybridization in certain toads. *Genetics* **26**:398–417.

Blair, W. F. (1956) Comparative survival of hybrid toads (*B. woodhousii* × *B. valliceps*) in nature. *Copeia* **1956**:259–260.

Bogorad, L. (1975) Evolution of organelles and eukarotypic genomes. *Science* **188**:891–898.

Breedlove D. E. (1969) The systematics of *Fuchsia* section *Encliandra (Onagraceae)*. *Univ. Calif. Publ. Bot.* **53**:1–69.

Brown, W. L., and Wilson, E. O. (1956) Character displacement. *Syst. Zool.* **5**:49–64.

Carlquist, S. (1974) *Island Life*. New York: Columbia University. Press.

Chapin, J. P. (1948) Variation and hybridization among the Paradise Flycatchers of Africa. *Evolution* **2**:111–126.

Charles, D. R., and Goodwin, R. H. (1943) An estimate of the minimum numbers of genes differentiating two species of goldenrod with respect to their morphological characters. *Am. Nat.* **77**:53–69.

Chase, V. C., and Raven, P. H. (1975) Evolutionary and ecological relationships between *Aquilegia formosa* and *A. pubescens (Ranunculaceae)*, two perennial plants. *Evolution* **29**:474–486.

Clausen, J. (1951) *Stages in the Evolution of Plant Species*. Ithaca, N. Y.: Cornell University Press.

Clausen, J., and Hiesey, W. M. (1958) Experimental studies on the nature of species. IV. Genetic structure of ecological races. *Carnegie Inst. Wash. Publ.* 615.

Clausen, J., and Hiesey, W. (1960) The balance between coherence and variation in evolution. *Proc. Natl. Acad. Sci. USA* **46**:494–506.

Clausen, J.; Keck, D.; and Hiesey, W. (1940) Experimental studies on the nature of species. I. Effect of varied environments on western American plants. *Publ. Carnegie Inst. Wash.* No. 520.

Crenshaw, J. W. (1965) Serum protein variation in an interspecific hybrid swarm of turtles of the genus *Pseudomys*. *Evolution* **19**:1–15.

Critchfield, W. B. (1967) Crossability and relationships of the closed-cone pines. *Silvae Genetica* **16**:89–97.

Crosby, J. L. (1966) Reproductive capacity in the study of evolutionary processes. In Hawkes, J. G. (ed.), *Reproductive Biology and Taxonomy of Vascular Plants*. New York: Pergamon Press.

Crossley, S. A. (1974) Changes in mating behavior produced by selection for ethological isolation between ebony and vestigial mutants of *Drosophila melanogaster*. *Evolution* **28**:631–647.

Darwin, C. (1859) *On the Origin of Species by Means of Natural Selection*, 1st ed. London: John Murray, Ltd.

Davidson, E. H. (1976) *Gene Activity in Early Development*, 2nd ed. New York: Academic Press, Inc.

Davidson, R. L. (1974) Gene regulation in somatic cell hybrids. *Ann. Rev. Genet.* **8**:195–218.

Dessauer, H. C.; Fox, W.; and Pough, F. H. (1962) Starch-gel electrophoresis of transferrins, esterases, and other plasma proteins of hybrids between two subspecies of whiptail lizards (genus *Cnemidophorus*). *Copeia* **1962**:767–774.

Dobzhansky, T. (1937, 1941, 1951) *Genetics and the Origin of Species*, 1st, 2nd, and 3rd eds. New York: Columbia University Press.

Dobzhansky, T. (1970) *Genetics of the Evolutionary Process*. New York: Columbia University Press.

Dobzhansky, T. (1973) Is there gene exchange between *Drosophila pseudo-obscura* and *Drosophila persimilis* in their natural habitats? *Am. Nat.* **107**:312–314.

Dobzhansky, T.; Pavlovsky, O.; and Powell, J. R. (1976) Partially successful attempt to enhance reproductive isolation between semispecies of *Drosophila paulistorum*. *Evolution* **30**:201–212.

Doerschug, E. B.; Miksche, J. P.; and Stern, J. (1976) DNA variation and ribosomal-DNA constancy in two *Crepis* species and the interspecific hybrid exhibiting nucleolar-organizer suppression. *Heredity* **37**:441–450.

Ehrendorfer, F. (1959) Differentiation hybridization cycles and polyploidy in *Achillea*. *Cold Spring Harbor Symp. Quant. Biol.* **24**:141–152.

Ehrendorfer, F. (1968) Geographical and ecological aspects of intraspecific differentiation. In Heywood, V. H. (ed.), *Modern Methods in Plant Taxonomy*. New York: Academic Press, Inc.

Ehrman, L., and Probber, J. (1978) Rare *Drosophila* males: the mysterious matter of choice. *Am. Sci.* **66**:216–222.

Epling, C. (1947) Actual and potential gene flow in natural populations. *Am. Nat.* **81**:104–113.

Epstein, C. J.; Weston, J. A.; Whitten, W. K.; and Russell, E. S. (1972) The expression of the isocitrate dehydrogenase locus (Idh) during mouse embryogenesis. *Dev. Biol.* **27**:430–433.

Etkin, L. D. (1977) Preferential expression of the maternal allele for alcohol dehydrogenase (ADH) in the amphibian bird *Ambystoma mexicanum* Axolotl) × *Ambystoma texanum*. *Dev. Biol.* **60**:93–100.

Fisher, R. A. (1930). *The Genetical Theory of Natural Selection*. Oxford: Clarendon Press.

Flake, R. H.; von Rudloff, E.; and Turner, B. L. (1973) Confirmation of a clinal variation in *Juniperus virginiana* using terpenoid data. *Proc. Natl. Acad. Sci. USA* **64**:487–494.

Flake, R. H.; von Rudloff, E.;and Turner, B. L. (1973) Confirmation of a clinal pattern of chemical differentiation in *Juniperus virginiana* from terpenoid data obtained in successive years. *Recent Adv. Phytochem.* **6**:215–228.

Flake, R. H.; Urbatsch, L.; and Turner, B. L. (1978) Chemical documentation of allopatric introgression in *Juniperus scopulorum* and *J. virginiana*. *Syst. Bot.*, in press.

Focke, W. O. (1881) *Die Pflanzen-Mischlinge*. Berlin: Borntraeger.

Fox, W.; Dessauer, H. C.; and Maumus, L. T. (1961) Electrophoretic studies of blood proteins of two species of toads and their natural hybrid. *Comp. Biochem. Physiol.* **3**:52–63.

Gartner, C. (1849) *Versuche und Beobachtunge uber die Bastarderzeugung im Planzenreich*.

Gerassimova. H. (1939) Chromosome alterations as a factor of divergence of forms. I. New experimentally produced strains of *C. tectorum* which are physiologically isolated from the original forms owing to reciprocal translocation. *Compt. Rend. Acad. Sci. USSR* **25**:148–154.

Gerhold, H. D., and Plank, G. H. (1970) Monoterpene variations in vapors from white pines and hybrids. *Phytochem.* **9**:1393–1398.

Gillard, E. T. (1959) The ecology of hybridization in New Guinea honey-eaters (Aves). *Am. Mus. Novitates.* **1937**:1–26.

Gillett, G. W. (1972) The role of hybridization in the evolution of the Hawaiian flora. In Valentine, D. H. (ed.), *Taxonomy, Phytogeography, and Evolution*. New York: Academic Press, Inc.

Gottlieb, L. D. (1972) Levels of confidence in the analysis of hybridization in plants. *Ann. Miss. Bot. Gard.* **59**:435–446.

Grant, P. R. (1972) Convergent and divergent character displacement. *Biol. J. Linn. Soc.* **4**:39–68.

Grant, P. R. (1975) The classical case of character displacement. *Evol. Biol.* **8**:237–337.

Grant, V. (1950) Genetic and taxonomic studies in *Gilia*. I. *Gilia capitata*. *Aliso* **2**:239–316.

Grant, V. (1954) Genetic and taxonomic studies in *Gilia*. IV. *Gilia achilleaefolia*. *Aliso* **3**:1–18.

Grant, V. (1956) The influence of the breeding habit on the outcome of natural hybridization in plants. *Am. Nat.* **90**:319–322.

Grant, V. (1966a) Selection for vigor and fertility in the progeny of a highly sterile species hybrid in *Gilia*. *Genetics* **53**:757–775.

Grant, V. (1966b) Block inheritance of viability genes in plant species. *Am. Nat.* **100**:591–601.

Grant, V. (1967) Linkage between morphology an viability in plant species. *Am. Nat.* **101**:125–139.

Grant, V. (1971) *Plant Speciation*. New York: Columbia University Press.

Grant, V., and Grant, K. (1971) Natural hybridization between the cholla cactus species *Opuntia spinosior* and *Opuntia versicolor*. *Proc. Nat. Acad. Sci. USA* **68**:1993–1995.

Hagen, D. W. (1967) Isolating mechanisms in three-spine sticklebacks (*Gasterosteus*). *J. Fish. Res. Board Can.* **24**:1637–1692.

Hall, W. P., and Selander, R. K. (1973) Hybridization in karyotypically differentiated populations of the *Sceloporus grammicus* complex (Iguanidae). *Evolution* **27**:226–242.

Harlan, J. R., and de Wet, J. M. J. (1963) The compilospecies concept. *Evolution* **17**:497–501.

Harland, S. C. (1936) The genetical conception of species. *Biol. Rev.* **11**:83–112.

Heiser, C. B. (1947) Hybridization between the sunflower species *Helianthus annuus* and *H. petiolaris*. *Evolution* **1**:249–262.

Heiser, C. B. (1949a) Natural hybridization with particular reference to introgression. *Bot. Rev.* **15**:645–687.

Heiser, C. B. (1949b) Study in the evolution of the sunflower species *Helianthus annuus* and *H. bolanderi*. *Univ. Calif. Publ. Bot.* **23**:157–208.

Heiser, C. B., (1951a) Hybridization in the annual sunflowers: *Helianthus annuus* × *H. debilis* var. *cucumerifolius*. *Evolution* **5**:42–51.

Heiser, C. B. (1951b) Hybridization in the annual sunflowers: *Helianthus annuus* × *H. argophyllus*. *Am. Nat.* **85**:65–72.

Heiser, C. B. (1954) Variation and subspeciation in the common sunflower, *Helianthus annuus*. *Am. Midl. Nat.* **51**:287–305.

Heiser, C. B. (1973) Introgression re-examined. *Bot. Rev.* **39**:347–366.

Herbert, W. (1837) *Amaryllidaceae: Preceded by an Attempt to Arrange the Monocotyledonous Orders and Followed by a Treatise on Cross-bred Vegetables and Supplement*. London:

Heslop-Harrison, J. (1964) Forty years of genecology. *Adv. Ecol. Res.* **2**:159–247.

Heslop-Harrison, Y. (1953) *Nuphar intermedia* Ledeb., a presumed relict hybrid, in Britain. *Watsonia* 3:7–25.

Hiesey, W. M.; Nobs, M. A.; and Bjorkman, O. (1971) Experimental studies in the nature of species. V. Biosystematics, genetics, and physiological ecology of Erythranthe section of *Mimulus. Carnegie Inst. Wash. Publ.* 628.

Hinton, W. F. (1976) Introgression and the evolution of selfing in *Calyptridium monospermum* (Portulacaceae). *Syst. Bot.* 1:85–90.

Hitzeroth, H.; Klose, J.; Ohno, S.; and Wolf, U. (1968) Asynchronous activation of paternal alleles at the tissue specific gene loci observed on hybrid trout during early development. *Biochem. Genet.* 1:287–300.

Honjo, T., and Reeder, R. H. (1973) Preferential transcription of *X. laevis* ribosomal RNA in interspecies hybrids between *X. laevis* and *X. mulleri. J. Mol. Biol.* 80:217.

Hovanitz, W. (1948) Ecological segregation of interfertile species of *Colias. Ecology* 29:461–469.

Hunt, R. S., and von Rudloff, E. (1974) Chemosystematic studies in the genus *Abies*. I. Leaf and twig analysis of alpine and balsam firs. *Can. J. Bot.* 52:477–487.

Irving, R. S., and Adams, R. P. (1973) Genetic and biosynthetic relationships of monoterpenes. *Recent Adv. Phytochem.* 6:187–214.

Jeffrey, E. C. (1915) Some fundamental morphological objections to the mutation theory of De Vries. *Am. Nat.* 49:5–21.

Johnson, K. E., and Chapman, V. M. (1971) Expression of paternal genes during embryogenesis in the viable interspecific hybrid amphibian embryo *Rana pipiens* × *Rana palustris*. Electrophoretic analysis of five enzyme systems. *J. Exp. Zool.* 178:313–318.

Johnston, R. F. (1969) Taxonomy of house sparrows and their allies in the Mediterranean basin. *Condor* 71:129–139.

Jones, M. J. (1973) Effects of thirty years hybridization on the toads *Bufo americanus* and *Bufo woodhousii fowlerii* at Bloomington, Indiana. *Evolution* 27:435–448.

Kaneshiro, K. Y, and Val, F. C. (1977) Natural hybridization between a sympatric pair of Hawaiian *Drosophila. Am. Nat.* 111:897–902.

Kessler, S. (1966) Selection for and against ethological isolation between *Drosophila pseudoobscura* and *D. persimilis. Evolution* 20:634–645.

Koopman, K. F. (1950) Natural selection for reproductive isolation between *Drosophila pseudoobscura* and *Drosophila persimilis. Evolution* 4:135–148.

Lamprecht, H. (1941) Die Artgrenze zwischen *Phaseolus vulgaris* L. und *P. multiflorus* Lam. *Hereditas* 27:51–175.

Lamprecht, H. (1944) Die genisch-plasmatische Grundlage der Artbarriere. *Argic. Hortic. Genet.* 2:75–142.

Lee, D. W. (1975) Population variation and introgression in *Typha. Taxon* 24:633–641.

Levin, D. A. (1967) Variation in *Phlox divaricata. Evolution* 21:92–108.

Levin, D. A. (1970) Reinforcement of reproductive isolation: plants vs. animals. *Am. Nat.* 104:571–581.

Levin, D. A. (1973) The age structure of a hybrid swarm in *Liatris* (Compositae). *Evolution* 27:532–535.

Levin, D. A., and Levy, M. (1971) Secondary intergradation in genome incompatibility in *Phlox pilosa* (Polemoniaceae). *Brittonia* 23:246–265.

Levin, D. A., and Schaal, B. A. (1970) Corolla color as an inhibitor of inter-specific hybridization in *Phlox Am. Nat.* **104**:273–283.

Levin, D. A., and Smith, D. M. (1966) Hybridization and evolution in the *Phlox pilosa* complex. *Am. Nat.* **100**:289–302.

Lewis, H. (1961) Experimental sympatric populations of *Clarkia. Am. Nat.* **95**:155–168.

Lewis, H., and Epling, C. (1959) *Delphinium,* a diploid species of hybrid origin. *Evolution* **13**:511–525.

Lewis, H., and Lewis, M. E. (1955) The genus *Clarkia. Univ. Calif. Publ. Bot.* **20**:241–392.

Lewontin, R. C., and Birch, L. C. (1966) Hybridization as a source of variation for adaptation to new environments. *Evolution* **20**:315–336.

Lindley, J. (1830) *Introduction to the Natural System of Botany.*

Littlejohn, M. J., and Watson, G. F. (1976) Mating-call structure in a hybrid population of the *Geocrinia laevis* complex (*Anura: Leptodactylidae*) over a seven-year period. *Evolution* **30**:848–850.

Lotsy, J. P. (1916) *Evolution by Means of Hybridization.* The Hague: M. Nijhoff.

Lowe, C. H., and Wright, J. W. (1966) Evolution of parthenogenetic species of *Cnemidophorus* (whiptail lizards) in western North America. *J. Ariz. Acad. Sci.* **4**:81–87.

Mangelsdorf, P. C. (1958) The mutagenic effect of hybridizing maize and teosinte. *Cold Spring Harbor Symp. Quant. Biol.* **23**:409–421.

Mayr, E. (1940) Speciation phenomena in birds. *Am. Nat.* **74**:249–278.

Mayr, E. (1942) *Systematics and the Origin of Species.* New York: Columbia University Press.

Mayr, E. (1963) *Animal Species and Evolution.* Cambridge, Mass.: Belknap Press, Harvard University Press.

Meyerhof, P. G., and Haley, L. E. (1975) Ontogeny of LDH isozymes in chicken-quail hybrid embryos. *Biochem. Genet.* **13**:7–18.

Mirov, N. T. (1956) Composition of turpentine of lodgepole and jack pine hybrids. *Can. J. Bot.* **34**:443–457.

Muller, C. (1952) Ecological control of hybridization in *Quercus*: a factor in the mechanism of evolution. *Evolution* **6**:147–161.

Muller, H. J. (1940) Bearings of the *Drosophila* work on systematics. In Huxley, J. (ed.), *The New Systematics.* London: Oxford University Press.

Muller, H. J. (1942) Isolating mechanisms, evolution and temperature. *Biol. Symp.* **6**:71–125.

Nagle, J. J., and Mettler, L. E. (1969) Relative fitness of introgressed and parental populations of *Drosophila mojavensis* and *D. arizonensis. Evolution* **23**:519–524.

Narayan, R. K. J., and Rees, H. (1977) Nuclear DNA divergence among *Lathyrus* species. *Chromosoma* **63**:101–107.

Neaves, W. B. (1969) Adenosine deaminase phenotypes among sexual parthenogenetic lizards in the genus *Cnemidophorus* (Teiidae). *J. Exp. Zool.* **171**:175–184.

Nei, M. (1975) *Molecular Population Genetics and Evolution.* New York: American Elsevier Publishing Company, Inc.

Ogilvie, R. T., and von Rudloff, E. (1968) Chemosystematic studies in the genus *Picea* (Pinaceae). IV. The introgression of white and Engelmann spruce as found along the Bow River. *Can. J. Bot.* **46**:901–908.

Ornduff, R. (1966) Biosystematic survey of the goldfield genus *Lastnenia*. *Univ. Calif. Publ. Bot.* **40**:1–92.

Ornduff, R. (1967) Hybridization and regional variation in Pacific Northwestern Impatiens (*Balsminaceae*). *Brittonia* **19**:122–128.

Ornduff, R. (1969) The systematics of populations of plants. In Blair, F. (ed.), *Systematic Biology*, Washington, D. C.: American Association for the Advancement of Science.

Ornduff, R.; Bohm, B.; and Saleh, N. A. M. (1973) Flavonoids of artificial interspecific hybrids of *Lasthenia*. *Biochem. Syst.* **1**:147–151.

Ozaki, H., and Whitely, A. H. (1970) L-Malate dehydrogenase in the development of the sea urchin *Strongylocentrotus purpuratus*. *Develop. Biol.* **21**:196–215.

Pipkin, S. B. (1968) Introgression between closely related species of *Drosophila* in Panama. *Evolution* **22**:140–156.

Prager, E. M., and Wilson, A. C. (1975) Slow evolutionary loss of the potential for interspecific hybridization in birds: a manifestation of slow regulatory evolution. *Proc. Natl. Acad. Sci., USA.* **72**:200–204.

Pryor, L. D., and Bryant, L. H. (1958) Inheritance of oil characters in *Eucalyptus*. *Proc. Linn. Soc. N. S. W.* **83**:55–64.

Rand, A. L. (1958) Notes on African bulbils. Family *Pyconotidae*: class *Aves*. Fieldiana. *Zool.* **34**:65–70.

Raven, P. H. (1963) *Circaea* in the British Isles. *Watsonia* **5**:262–272.

Raven, P. H. (1976) Systematics and plant population biology. *Syst. Bot.* **1**:284–316.

Raven, P. H. (1978). Hybridization and the nature of species in higher plants. *Can. Bot. Assoc. Bull*: in press.

Raven, P. H., and Raven, T. E. (1976) The genus *Epilobium (Onagraceae)* in Australasia: A systematic and evolutionary study. *N. Z. Dep. Sci. Ind. Res. Bull.* **216**.

Reeves, R. G., and Bockholt, A. J. (1964) Modification and improvement of a maize inbred by crossing it with *Tripsacum*. *Crop Sci.* **4**:7–10.

Remington, C. L. (1954) The genetics of *Colias (Lepidoptera)*. *Adv. Genet.* **6**:403–450.

Remington, C. L. (1968) Suture-zones of hybrid interaction between recently joined biotas. *Evol. Biol.* **2**:321–428.

Rick, C. M. (1963) Differential zygotic lethality in a tomato species hybrid. *Genetics* **48**:1497–1507.

Rick, C. M. (1969) Controlled introgression of chromosomes of *Solanum pennellii* into *Lycopersicon esculentum*: segregation and recombination. *Genetics* **62**:753–768.

Sailer, R. I. (1953) Significance of hybridization among stink bugs of the genus *Euschistus*. *Yearb. Am. Phil. Soc.* 146–149.

Sailer, R. I. (1954). Interspecific hybridization among insects with a report on crossbreeding experiments with stinkbugs. *J. Econ. Entomol.* **47**:377–388.

Scandalios, J. G. (1969) Genetic control of multiple forms of isozymes in plants: a review. *Biochem. Genet.* **36**:37–79.

Schmalhausen, I. I. (1949) *Factors of Evolution*. Philadelphia: The Blakiston Company.

Schultz, R. J. (1977) Evolution and ecology of unisexual fishes. *Evol. Biol.* **10**:277–331.

Sears, J. W. (1947) Studies in the genetics of *Drosophila*. VII. Relationships within the *Quinaria* species group of *Drosophila*. *Univ. Tex. Publ. No.* 4720:137–156.

Short, L. L. (1965) Hybridization in the flickers (*Colaptes*) of North America. *Bull. Am. Mus. Nat. Hist.* **129**:307–428.

Short, L. L. (1969) Taxonomic aspects of avian hybridization. *Auk* **86**:84–105.

Sibley, C. G. (1950) Species formation in the red-eyed towhees of Mexico. *Univ. Calif. Publ. Zool.* **50**:109–194.

Sibley, C. G. (1954) Hybridization in the red-eyed towhees of Mexico. *Evolution* **8**:252–290.

Sibley, C. G., and West, D. A. (1958) Hybridization in the red-eyed towhees of Mexico: the eastern plateau populations. *Condor* **60**:84–104.

Sims, R. W. (1959) The *Ceyx erithacus* and *rufidorsus* species problem. *J. Linn. Soc. London., Zool.* **44**:212–221.

Smith, D. M., and Guard, A. T. (1958) Hybridization between *Helianthus divaricatus* and *H. microcephalus*. *Brittonia* **10**:137–145.

Smith, H. H. and Daly, K. (1959) Discrete populations derived by interspecific hybridization and selection in *Nicotiana*. *Evolution* **13**:476–487.

Smith, S. G. (1966) Natural hybridization in the coccinellid genus *Chilocorus*. *Chromosoma* **18**:380–406.

Sokal, R. R., and Crovello, T. J. (1970) The biological species concept: a critical evaluation. *Am. Nat.* **104**:127–153.

Sokal, R. R., and Taylor, C. E. (1976) Selection at two levels in hybrid populations of *Musca domestica*. *Evolution* **30**:509–522.

Stace, C. A. (ed.), (1975a) *Hybridization and the Flora of the British Isles.* New York: Acadamic Press, Inc.

Stace, C. A. (1975b) Wild hybrids in the British flora. In Walters, S. M. (ed.), *European Floristic and Taxonomic Studies.*

Stebbins, G. L. (1950) *Variation and Evolution in Plants.* New York: Columbia University Press.

Stebbins, G. L. (1957) The hybrid origin of microspecies in the *Elymus glaucus* complex. *Cytologia* suppl. vol.: 336–340.

Stebbins, G. L. (1959) The role of hybridization in evolution. *Proc. Am. Phil. Soc.* **103**:231–251.

Stebbins, G. L. (1977) *Processes of Organic Evolution*, 3rd ed. Englewood Cliffs, N. J.: Prentice-Hall, Inc.

Stebbins, G. L., and Daly, K. (1961) Changes in the variation pattern of a hybrid population of *Helianthus* over an eight-year period. *Evolution* **15**:60–71.

Stebbins, G. L., and Ferlan, L. (1956) Population variability, hybridization and introgression in some species of *Ophrys*. *Evolution* **10**:32–46.

Stephens, S. G. (1950) The internal mechanism of speciation in *Gossypium*. *Bot. Rev.* **16**:115–149.

Stutz, H. C., and Thomas, L. K. (1964) Hybridization and introgression in *Cowania* and *Purshia*. *Evolution* **18**:183–195.

Tretzel, E. (1955) Intragenerische Isolation und interspecifische Konkurrenz bei Spinnen. *Z. Morphol. Onkol. Tiere* **44**:43–62.

Tunner, H. G. (1973) Demonstration of the hybrid origin of the common green frog *Rana esculenta* L. *Naturwiss.* **60**:481–482.

Vaarama, A. (1954) Inheritance of morphological characters and fertility in the progeny of hybrids. *Rubus idaeus* × *R. articus Caryologia* suppl. vol.: 846–850.

Van Haverbeke, D. F. (1968) A population analysis of *Juniperus* in the Missouri River Basin. *Univ. Nebr. Stud. N. A.* **38**:1–82.

Wagner, W. H. (1970) Biosystematics and evolutionary noise. *Taxon* **19**: 146–151.

Wall, J. R. (1968) Leucine aminopeptidase polymorphism in *Phaseolus* and differential elimination of the donor parent genotype in interspecific backcrosses. *Biochem. Genet.* **2**:109–118.

Wallace, A. R. (1889) *Darwinism: An Exposition of the Theory of Natural Selection.* London: Macmillian & Co., Ltd.

Wallace, D. H.; Ozbun, J. L.; and Munger, H. M. (1972) Physiological genetics of crop yield. *Adv. Agron* **24**:97–146.

Wasserman, M., and Koepfer, H. R. (1977) Character displacement for sexual isolation between *Drosophila mojavensis* and *Drosophila arizonensis. Evolution* **31**:812–823.

Whalen, M. D. (1977) A systematic and evolutionary investigation of *Solanum* section *Androceras.* Ph.D. dissertation, University of Texas, Austin.

White, M. J. D. (1954) *Animal Cytology and Evolution,* 2nd ed. Cambridge, England: Cambridge University Press.

White, M. J. D. (1978) *Modes of Speciation.* San Francisco: W. H. Freeman and Company.

Whitt, G. S.; Philipp, P. D. L.; and Childers, W. F. (1977) Aberrant gene expression during the development of hybrid sunfishes (*Perciformes, Teleostei*). *Differentiation* **9**:97–109.

Williams, C. A.; Harborne, J. B.; and Smith, P. (1974) The taxonomic significance of leaf flavonoids in *Saccharum* and related species. *Phytochem.* **13**:1141–1149.

Wilson, A. C.; Maxson, L. R.; and Sarich, V. M. (1974) Two types of molecular evolution. Evidence from studies of interspecific hybridization. *Proc. Natl. Acad. Sci. USA* **1**:2843–2847.

Winge, Ö. (1938) Inheritance of species characters in *Tragopogon.* A cytogenetic investigation. *Compt. Rend. Lab. Carlsberg, ser Physiol.* **22**:155–194.

Wittliff, J. L. (1964) Venom constituents of *Bufo fowleri, Bufo valliceps,* and their natural hybrids analyzed by electrophoresis and chromatography. In Leone, C. A. (ed.), *Comparative Biochemistry and Serology.* New York: The Ronald Press Company.

Woodson, R. E. (1947a) Notes on the "historical" factor in plant geography. *Contribution Gray Herb.* **165**:12–25.

Woodson, R. E. (1947b) Some dynamics of leaf variation in *Ascelpias tuberosa. Ann. Miss. Bot. Gard.* **34**:353–432.

Woodson, R. E. (1962) Butterfly revisited. *Evolution* **16**:168–185.

Yang, S. Y. and Selander, R. K. (1968) Hybridization in the grackle *Quiscalus quiscalus* in Louisiana. *Syst. Zool.* **17**:107–143.

# AUTHOR CITATION INDEX

Adams, R. P., 303, 307
Ahuja, M. R., 303
Alava, R. O., 29
Alexander, R. D., 289, 303
Allan, H. H., 29, 45, 46
Allard, R. W., 251
Allee, W. C., 274
Alston, R. E., 194, 303
Anderson, B. W., 213
Anderson, E., 18, 29, 45, 99, 165, 178, 196, 213, 251, 303
Anderson, P. K., 236, 237
Anderson, R. A., 130, 214
Andrewartha, H. G., 274
Angus, R., 131
Asher, J. H., 129
Askew, R. R., 81
Axelrod, D. I., 29
Axtell, R. W., 129
Ayala, F. J., 303

Babbel, G. R., 251
Babcock, E. B., 29, 99
Bailey, I. W., 99, 100
Bailey, R. M., 157, 158, 274
Baker, H. G., 45, 303
Barkley, F. A., 18
Barrett, D. H. P., 303
Bartholomew, B., 251
Bauer, E., 303
Bayer, M. H., 303
Beadle, G. W., 18
Beckman, G., 129
Beckman, L., 129
Benson, L., 116
Berger, L., 303
Berry, R. J., 236
Bezy, R. L., 130
Bicudo, H. E. M. C., 303
Biddle, F. G., 237
Bigelow, R. S., 213, 236
Birch, L. C., 274, 308

Bjorkman, O., 307
Black, J. D., 157
Blackith, R. E., 251
Blair, A. P., 303
Blair, W. F., 274, 297, 304
Blake, S. F., 179
Bloom, W. L., 251
Bockholt, A. J., 309
Bodmer, W. F., 129
Bogorad, L., 304
Bohm, B., 309
Bourne, W. R. P., 274
Bradshaw, A. D., 237
Breedlove, D. E., 304
Brehm, B. G., 194
Brooks, J. L., 29
Brown, D. E. S., 158
Brown, W. L., Jr., 274, 297, 303, 304
Brubaker, L. L., 236
Brunner, G., 236
Bryant, L. H., 309

Cain, A. J., 274
Cain, S. A., 18
Calvert, A. E., 18
Camp, W. H., 18
Carlquist, S., 304
Carson, H. L., 45, 196, 236
Chapin, J. P., 304
Chapman, V. M., 236, 237, 238, 307
Charles, D. R., 304
Chase, V. C., 304
Childers, W. F., 311
Chipman, R. K., 237
Cimino, M. C., 129
Clarke, B., 236, 237
Clegg, M. T., 251
Cleland, R. E., 99
Cockayne, L., 29, 46
Cockerell, T. D. A., 179
Cole, C. J., 130
Colenso, W., 46

Condon, H. T., 274
Cook, L. M., 237
Cooper, R. C., 46
Condit, C., 29
Couper, R. A., 46
Craddock, E., 196
Crenshaw, J. W., 304
Critchfield, W. B., 304
Crosby, J. L., 213, 237, 304
Crossley, S. A., 304
Crovello, T. J., 310
Crow, J. F., 129
Cuellar, O., 129

Daly, K., 72, 310
Dansereau, P., 18, 29
Danske Meteorologiske Institut, 237
Darevsky, I. S., 131
Darlington, C. D., 72, 165
Darrow, G. M., 18
Darwin, C., 297, 304
Davidson, E. H., 304
Davidson, R. L., 304
Davies, E., 237
Davis, J. H., 46
Day, A., 72
Deam, C. C., 179
DeBach, P., 81, 130
DeFries, J. C., 237
Delisle, A. L., 165, 179
Desmarais, Y., 29
Dessaur, H. C., 213, 304, 305
Dickenson, J. C., 274
Diels, L., 46
Dixon, K. L., 214
Dobzhansky, T., 61, 165, 179, 196, 214, 237, 274, 283, 289, 297, 304, 305
Doerschug, E. B., 305
Dort, W., Jr., 251
Dronamraju, K. R., 297
Duellman, W. E., 129

East, E. M., 18
Eaton, L. C., 251
Eddy, S., 157
Ehrendorfer, F., 46, 305
Ehrman, L., 289, 297, 305
Elens, A. A., 289
Elias, M. K., 29
Ellerton, S., 100
Emerson, A. E., 274
Epling, C., 18, 19, 29, 104, 305, 308
Epstein, C. J., 305
Erbe, L., 297
Eshel, I., 129
Etkin, L. D., 305

Faegri, K., 297
Feldman, M. W., 129
Felsenstein, J., 129
Ferlan, L., 310
Fincham, J. R. S., 237
Finlayson, A. C., 46
Fisher, R. A., 129, 237, 305
Flake, R. H., 305
Flavell, R. B., 303
Flemming, C. A., 46
Focke, W. O., 18, 305
Foot, K., 81
Ford, E. B., 46, 81, 237, 297
Fox, W., 213, 304
Fritts, T. H., 214

Gartner, C., 305
Gartside, D. F., 214
Geisler, F., 179
Gentry, J. B., 130
Gerald, P. S., 130, 214
Gerassimova, H., 72, 305
Gerhold, H. D., 305
Giles, N. H., Jr., 18
Gillard, E. T., 305
Gillett, G. W., 306
Goldblatt, S. M., 131
Goldschmidt, R., 61
Goodwin, R. H., 18, 165, 179, 304
Gordon, H. D., 46
Gosline, W. A., 157
Gottlieb, L. D., 306
Grant, A., 72
Grant, K., 297, 306
Grant, P. R., 306
Grant, V., 29, 72, 99, 104, 116, 196, 214, 289, 297, 306
Guard, A. T., 310
Gustafsson, A., 100, 116

Haffer, J., 214
Hagen, D. W., 214, 237, 306
Haldane, J. B. S., 237
Haley, L. E., 308
Hall, M. T., 29
Hall, W. P., 214, 237, 306
Hamrick, J. L., 251
Harborne, J. B., 311
Harlan, J. R., 306
Harland, S. C., 61, 306
Harrison, G. J., 61
Heine, E. M., 46
Heiser, C. B., 29, 179, 251, 306
Henderson, N. S., 237
Herbert, W., 306
Herriott, E. M., 46

Heslop–Harrison, J., 306
Heslop–Harrison, Y., 307
Hiesey, W. M., 18, 29, 99, 179, 304, 307
Hill, J. L., 236
Hinton, W. F., 307
Hitzeroth, H., 307
Holloway, J. T., 46
Honjo, T., 307
Hornback, E., 18
Hovanitz, W., 307
Howden, H. F., 274
Hubbard, J. P., 214
Hubbs, C. L., 157, 158, 274
Hubbs, L. C., 158
Hubricht, L., 18, 179, 251, 303
Hunt, R. S., 307
Hunt, W. G., 214, 237
Huntington, C. E., 214
Huskins, C. L., 179
Hutchinon, J. B., 61
Hutton, J. J., 237
Huxley, J. S., 46, 237

Irving, R. S., 307
Iyengar, N. K., 61

Jackson, J. F., 214
Jacob, J., 129
Jain, S. K., 237
Jeffrey, E. C., 307
Johnson, D. S., 116
Johnson, K. E., 307
Johnson, R. E., 158
Johnson, W. E., 130, 196, 238
Johnston, R. F., 307
Jones, J. K., Jr., 251
Jones, M. J., 307
Jordan, D. S., 158
Jukes, T. H., 237

Kahler, A. L., 251
Kallman, K., 129
Kaneshiro, K. Y., 307
Kawamura, T., 274
Kay, F. R., 130, 214
Kearney, T. H., 61, 116
Keck, D. D., 18, 29, 99, 100, 104, 179, 304
Kelly, J. P., 297
Kessler, S., 307
Kimura, M., 129, 237
King, J. C., 283
King, J. L., 130, 237
Klose, J., 307
Knight, G. R., 297
Knight, R. L., 61
Koelz, W., 158

Koepfer, H. R., 311
Koller, P. C., 283, 297
Koopman, K. F., 274, 283, 297, 307
Kramer, P. J., 46
Kuechler, A. W., 214
Kuronuma, K., 158

Lack, D., 274
Lagler, K. F., 157
Lamprecht, H., 72, 307
Langdon, L. M., 46
Lankinen, P., 130
Larisey, M. M., 194
Larsen, E. L., 18
Lawrence, G. H. M., 100
Lee, D. W., 307
Levene, H., 289
Levin, D. A., 130, 297, 307, 308
Levy, M., 130, 307
Lewis, D., 46
Lewis, H., 81, 251, 308
Lewis, M. E., 251, 308
Lewontin, R. C., 308
Lindegren, C. C., 18
Lindegren, G., 18
Lindley, J., 308
Linton, R. S., 237
Littlejohn, M. J., 214, 215, 297, 308
Loftus-Hills, J. J., 214, 215
Lokki, J., 130
Lorković, Z., 297
Lotsy, J. P., 308
Lowe, C. H., 130, 131, 214, 215, 308

McCarley, H., 297
McClearn, G. E., 237
Macgregor, H. C., 130
McIndoe, G., 46
McKinney, C. O., 130, 214
McLintock, A. H., 46
McMinn, H. E., 100
McPhail, J. D., 237
McQueen, D. R., 46
MacSwain, J. W., 251
Malogolowkin-Cohen, C., 289
Mangelsdorf, P. C., 29, 308
Manton, I., 29
Marie-Victorin, Fr., 18
Marsden-Jones, E. M., 18, 179
Martin, J. E., 237
Maslin, T. P., 130
Mason, H. L., 18, 29
Mather, K., 46, 61
Matzke, E. G., 19
Maumus, L. T., 305
Maxon, L. R., 311

Maynard Smith, J., 130
Mayr, E., 18, 81, 104, 214, 237, 274, 297, 308
Meacham, W. R., 214
Mecham, J. S., 297
Meise, W., 237
Merrell, D. J., 289
Mettler, L. E., 308
Meyer, H., 158
Meyerhof, P. G., 308
Miksche, J. P., 305
Millener, L. H., 46
Miller, A. H., 274
Miller, R. R., 158
Mirov, N. T., 308
Money, L., 100
Montanucci, R. R., 214
Moore, J. A., 274
Moore, T. E., 303
Moore, W. S., 214
Mosquin, T., 251
Muller, C. H., 214, 308
Muller, H. J., 130, 283, 308
Munger, H. M., 311
Müntzing, A., 72, 165
Murphy, H. M., 236
Murray, J., 237

Nace, G. W., 129
Nagle, J. J., 308
Nair, P., 196
Narayan, R. K. J., 308
Nast, C., 99
Neaves, W. B., 130, 214, 308
Nei, M., 130, 308
Nicholson, A. J., 274
Nickerson, N. H., 29
Nobs, M. A., 29, 100, 307
Nur, U., 130

Ogilvie, R. T., 308
Ohno, S., 307
Ohta, T., 130, 237
Oliver, W. R. B., 46
Ornduff, R., 309
Ownbey, R. P., 18
Ozaki, H., 309
Ozbun, J. L., 311

Parker, E. D., 214
Patterson, J. T., 81, 196
Pavlovsky, O., 289, 297, 305
Peebles, R. H., 116
Perry, M., 283
Petras, M. L., 237
Phillip, P. D. L., 311
Pilj, L. van der, 297
Pipkin, S. B., 196, 309

Plank, G. H., 305
Poole, A. L., 46
Popp, R. A., 237
Popper, K. R., 214
Pough, F. H., 213, 304
Powell, J. R., 305
Prager, E. M., 309
Probber, J., 305
Pryor, L. D., 309
Putt, E. D., 179

Rand, A. L., 309
Rao, S. V., 81
Rattenbury, J. A., 46
Raven, P. H., 251, 304, 309
Raven, T. E., 309
Reeder, R. H., 307
Rees, H., 308
Reeves, R. G., 309
Rehder, A., 18
Reimer, J. D., 237
Remington, C. L., 214, 237, 251, 309
Rendel, J. M., 283
Reyment, R. A., 251
Richardson, R. H., 303
Rick, C. M., 309
Riley, H. P., 18, 165, 179
Rising, J. D., 214
Roberts, H. F., 19
Robertson, A., 297
Roderick, T. H., 236, 237, 238
Rössler, Y., 130
Roychoudhury, A. K., 130
Ruddle, F. H., 236, 237, 238
Rudloff, E. von, 305, 307, 308
Russell, E. S., 305

Sailer, R. I., 81, 309
Saleh, N. A. M., 309
Sarich, V. M., 311
Sauer, C. O., 29
Saura, A., 130, 131
Sax, K., 18, 165
Schaal, B. A., 308
Schafer, B., 18
Schmalhausen, I. I., 309
Schultz, L. P., 158
Schultz, R. J., 131, 214, 309
Schwarz, E., 238
Schwarz, H. K., 238
Scudday, J., 130
Seal, H., 251
Sears, J. W., 81, 310
Selander, R. K., 130, 214, 215, 237, 238, 306, 311
Sene, F., 196
Serventy, D. L., 274

Short, L. L., 215, 238, 310
Shows, T. B., 237, 238
Sibley, C. G., 215, 274, 310
Sibley, F. C., 215
Silow, R. A., 61
Simmons, A. S., 289
Simmons, J., 194
Simpson, G. G., 29
Sims, R. W., 310
Smith, C. E., Jr., 29
Smith, D. M., 297, 308, 310
Smith, H. H., 72, 310
Smith, M. H., 130
Smith, P., 311
Smith, S. G., 81, 130, 310
Snyder, L. A., 29
Sokal, R. R., 251, 310
Spassky, B., 289, 297
Stace, C. A., 310
Stanley, S. M., 130
Stebbins, G. L., 19, 29, 45, 61, 72, 99, 100, 116,
     238, 297, 310
Stephens, S. G., 61, 310
Stern, J., 305
Stone, W. S., 81, 196
Straw, R. M., 104
Strobell, E. C., 81
Sturtevant, A. H., 165
Stutz, H. C., 310
Suomalainen, E., 130, 131
Swamy, B. G. L., 100

Taylor, B. A., 238
Taylor, C. E., 310
Thaeler, C. S., 215
Thibault, R. E., 215
Thomas, L. K., 310
Thomson, G. M., 46
Thorp, R. W., 251
Trautman, M. B., 158
Travers, W. T. L., 46
Tretzel, E., 310
Tschirch, A., 46
Tucker, J. M., 30, 100
Tunner, H. G., 310
Turner, B. L., 194, 297, 303, 305
Turner, C. L., 158
Turrill, W. B., 18, 179

Upcott, M., 165
Urbatsch, L., 305
Ursin, E., 215, 238
Uzzell, T. M., Jr., 130, 131

Vaarama, A., 72, 310
Val, F. C., 307
Valencia, J. I., 30

Valencia, R. M., 30
Van Haverbeke, D. F., 311
Vaurie, C., 275
Viosca, P., 19, 100
Vrijenhoek, R. C., 131

Waddington, C. H., 283, 297
Wagner, W. H., Jr., 251, 311
Walker, B. W., 158
Wall, J. R., 311
Wallace, A. R., 311
Wallace, B., 215, 283, 298
Wallace, D. H., 311
Ward, G. H., 100
Ware, J. O., 61
Wasserman, M., 311
Waterbolk, H. T., 238
Watson, E. E., 179
Watson, G. F., 214, 215, 308
Watt, W. B., 251
Wattiaux, J. M., 289
Weigl, P. G., 237
West, D. A., 215, 310
Weston, J. A., 305
Wet, J. M. J. de, 306
Wetmore, R. H., 179
Whalen, M. D., 311
Wharton, L. T., 81
Wheeler, L. L., 238
Wherry, E. T., 298
Whitaker, T. W., 165
White, M. J. D., 131, 311
Whitely, A. H., 309
Whitt, G. S., 311
Whitten, W. K., 305
Wiegand, K. M., 19, 165
Willet, R. W., 46
Williams, C. A., 311
Williams, G. C., 131
Williamson, L. O., 157
Wilson, A. C., 309, 311
Wilson, E. O., 215, 238, 274, 297, 298, 304
Winge, Ö., 72, 311
Wittliff, J. L., 311
Wolf, U., 307
Woodson, R. E., Jr., 18, 19, 30, 165, 311
Woolf, B., 283
Wright, J. W., 130, 131, 214, 215, 308
Wright, S., 46

Yang, S. Y., 130, 215, 238, 311

Zimmerman, E. C., 30
Zimmermann, K., 238
Zirkle, C., 19
Zweifel, R. G., 129, 131, 215

# SUBJECT INDEX

Adaptive radiation, 22–27, 37, 45
Agamospermy, 82, 105
Albumin, 121, 138
Allozymes, 5, 122, 138, 196, 199, 223–226
*Asclepias*, 200

Biological species concept, 5, 74
Birds, 6, 10, 85, 136, 204, 208, 259, 262–265
Blood proteins, 137
Breeding system, 37, 85
*Bufo*, 137, 204, 207

Calypterdium, 257
Canonical variate analysis, 241
Catostomidae, 145
*Ceanothus*, 27, 97
Cenospecies, 177
Character displacement
  defined, 254, 259
  examples, 209, 259, 273, 293–296
*Chilocorus*, 74, 238
Chromatographic profiles, 184–193
Chromatography, 137, 184–186
*Circaea*, 85
*Cistus*, 24
Climatic factors, 23–27, 40–42, 207, 208, 229, 249
*Cnemidophorus*, 86, 117–129, 207
Coadapted gene complexes, 201, 202, 210, 234–235, 248
Competition
  among animals, 269–271
  for pollinators, 258
Compilospecies, 135
Computer simulation, 206, 232
*Coprosma*, 35
Corolla color, 37, 101–104, 190, 242, 290–296
*Crepis*, 25, 50, 70, 96, 105
*Crina*, 266
Cross-incompatibility, 78, 152, 256, 258, 293
Cytoplasmic factors, 284

*Dacus*, 136
Differentiation-hybridization cycles, 41
Dispersal, long distance, 44, 156
Dosage effect, 121
*Drosophila*, 50, 74, 138, 195–196, 255, 275–289
Dynamic-equilibrium hypothesis, 202

Ecological displacement, 269
Ecological isolation, 9, 15, 16, 140, 155, 160, 167, 206–212, 239, 295
Ecotones, 9, 205, 207–209
Electrophoretic techniques, 3, 121, 221
*Elymus*, 28, 50, 70
Ephemeral-zone hypothesis, 202
Esterase, 121, 221
Ethological isolation, 94, 255–257, 287–288, 293–296
*Euschistus*, 74
Evolution under domestication, 21
Evolutionary
  novelty, 15, 21, 102–103
  potential, 20, 59, 128
  rate, 9, 20, 23, 37, 155
Experimental populations, 49, 58, 276–281, 286–287

Flower constancy, 101, 291
*Formica*, 268
Founder effect, 230
Frogs, 6, 85, 132, 207, 266
*Fuchsia*, 256, 257

*Gambusia*, 142, 152
Gamete wastage, 293–296
Gametic elimination, 48
Gene flow
  asymmetry, 199, 233–236
  cohesion, 201, 250
Gene frequencies, 121–124, 223–235
Genetic identity, 123
Genetic tumors, 3

Geological period, 23–26, 31
*Geospiza*, 262
*Gilia*, 49, 50, 62–70, 87–99
Glutamate oxalate transaminase, 121
α-glycerophosphate dehydrogenase, 121
*Gossypium*, 25, 51–60

Habitat
  disturbance, 9, 15, 16, 128, 135, 211
  novel, 13, 15, 16, 26, 211
  preference, 87–88, 156, 160
Hardy-Weinberg equilibrium, 196
*Helianthus*, 135, 136, 166–178
Herbarium specimens, 160–164, 166, 175
Heteroblasty, 36
Heterosis, 129
Heterozygosity
  genic, 37, 122, 127, 229
  inversion, 170, 195–196
  translocation, 127, 170, 246–250
Histogram, 171, 262
Hitchhiking, 5
Homogamy, 104, 255, 281
Homoselection, 37
Hybrid
  advanced generation, 64–69, 175
  backcross, 54–58, 76–78, 136
  breakdown, 71, 175
  competitive ability, 15, 16 ₊
  ecological requirements, 15, 136, 160, 299
  fitness, 4, 73, 209, 248, 299
  frequency, 145
  index, 135, 210, 227, 271
  meiotic behavior, 64–69, 169, 247
  stabilized derivative, 8, 24, 70, 101–104,
    118
  sterility, 64–70, 112, 143, 170–175
  superiority hypothesis, 202
  swarm, 16, 136, 143, 170–174, 188–193
  weakness, 59, 64–69, 140, 209, 267
  zone
    age, 208, 209, 229, 232
    location, 201–212, 229–232, 239
    width, 205, 208–210, 232, 239, 240
Hybridization
  cyclic, 40–44
  evolutionary role, 20–28, 36–43, 96–98,
    134
  of the habitat, 15, 22
  influence of man, 16, 22, 135
  insular, 10, 39–42, 134
  polyspecies, 180
Hybridogenetic fish, 86, 127

Impregnation, artificial, 155
Incipient species, 85, 284

Introgression
  across chromosome barrier, 246–248
  critique, 4, 239, 300
  process, 12, 113, 159, 246–250
  product, 13, 91, 138, 157, 162–164, 177–
    178, 223–235, 246–250
Introgressive replacement, 249
*Iris*, 16
Isocitrate dehydrogenase, 221
Isolating mechanisms, origin, 2, 211, 269,
  275

*Juniperus*, 200

Lactate dehydrogenase, 121
*Lasius*, 260
*Lepomis*, 139
Leucine aminopeptidase, 49
Leucyl-alanine peptidase, 121
Linkage, 48, 126–128
*Listoria*, 204, 209
Lizards, 86, 118, 121–128, 204, 207
*Lycopersicon*, 49

Malate dehydrogenase, 121, 221
Malic enzyme, 121, 221
Mating call, 137, 266
*Melicope*, 32
*Microhyla*, 265
Microspecies, 27, 109–115
Migration, 125, 233
Minority disadvantage, 8
*Mollienisia*, 141, 152
*Monarcha*, 265
Morphology-viability linkage, 49
Mutation, 118, 128
*Myzartha*, 262–263

Natural selection
  against reproductive isolation, 39, 300
  for reproductive isolation, 57–59, 157,
    202, 205, 209, 211, 233–236, 249, 287–
    289, 293–296
*Nicotiana*, 3, 48, 50, 70
*Notropis*, 148

*Oenothera*, 96, 127
*Opuntia*, 85, 105–116

Parthenogentic organisms, 76, 117, 128, 207
*Passer*, 85
Phenolic compounds, 132, 186–193
*Phlox*, 84, 199, 290–296
6-phosphogluconate dehydrogenase, 121
Phosphoglucomutase, 121
Phosphoglucose isomerase, 121

*Pipilo*, 135, 208, 221–222
Pleiotropy, 48
*Platycercus*, 264
Pleistocene glaciation, 17, 23, 128, 209, 257
*Poeciliopsis*, 86, 127, 203
Pollen flow
   conspecific, 38, 250, 299
   heterospecific, 250, 291–295
Pollen load, heterospecific, 292
Pollinators
   bees, 87, 101–104, 250, 257
   butterflies, 257, 291–296
   hummingbirds, 101–104, 257
   wasps, 102–103
Polygenes, 5, 59, 284
Polymorphic loci, 121–124, 223–236
Polyploidy, 20, 27, 71, 84, 161, 301
*Potentilla*, 48, 84
Preadaptation, 103
Premeiotic endoduplication, 118, 128

*Quercus*, 24
*Quiscalus*, 207

Recombination, 27, 124, 296
Reproductive capacity, 199
Reproductive isolation, 3, 38, 70, 79, 89, 195, 246–250, 254–258, 275–296
Reticulate evolution, 96, 98
Rhytidoponera, 267
*Rubus*, 49, 105

*Salvia*, 22
Scatter diagram, 92
*Sceloporus*, 199, 208–209
Seasonal isolation, 109, 167, 291
Secondary intergradation, 198–202. *See also* Hybrid zone
Seed dispersal, 38, 94, 199, 268
Skewed segregation ratios, 49, 55–58
Selection
   artificial, 50, 63, 70, 255–283
   gametic, 48
   zygotic, 48, 55

Selective fertilization, 52
Self-fertilization, 63–69, 93, 257
Sex ratios, 76, 148
Sexual isolation, 275–282, 284, 287–288
*Sitanion*, 27, 80
*Sitta*, 256, 259–260
*Solanum*, 49, 257
Somatic cell hybrids, 3
Speciation
   allopatric, 79, 196
   recombinational, 69, 103
   saltational, 79
   sympatric, 103–104
Species complex
   agamic, 96
   clonal, 96, 105–116
   heterogamic, 96
   homogamic, 91–96
   polyploid, 91–96
Superoxide dismutase, 121
Suture zones, 198

Taxonomy, 87, 98, 113, 134, 139, 159, 166, 217, 272–273
Terpenes, 137, 200
Time-lapse analysis, 136, 137, 200
*Tradescantia*, 12, 14, 15, 20, 159–165
*Tripsacum*, 49

Unisexuality, 86, 127, 203, 207

Variation
   clonal populations, 105–115, 121–125
   hybrid zone, 85, 142, 235, 246, 263–268
Vegetative polymorphy, 34, 178
Vivipary, 141, 150

Wallace effect, 3, 296
Weed, 28, 178
"Weed" habitat, 207

*Xenopus*, 3

Zea, 21, 49

# About the Editor

DONALD A. LEVIN is Professor of Botany at the University of Texas at Austin, where he has been on the faculty since 1972. He also taught at Yale University and the University of Illinois at Chicago Circle and was a visiting Professor at the State University of New York at Stony Brook. He received the B.S. and Ph.D from the University of Illinois, Urbana. He has published empirical, synthetic, and review papers on various aspects of ecological and evolutionary genetics and chemical ecology of plants. Dr. Levin has served on the editorial boards of *Evolution* and *The American Naturalist* and as an officer in the Society for the Study of Evolution. He was awarded a Guggenheim Fellowship in 1975.